An Introduction to Metric Spaces

An Introduction to Metric Spaces

Dhananjay Gopal
Aniruddha Deshmukh
Abhay S. Ranadive
Shubham Yadav

CRC Press
Taylor & Francis Group
Boca Raton London New York

CRC Press is an imprint of the
Taylor & Francis Group, an **informa** business

A CHAPMAN & HALL BOOK

First edition published 2021
by CRC Press
6000 Broken Sound Parkway NW, Suite 300, Boca Raton, FL 33487-2742

and by CRC Press
2 Park Square, Milton Park, Abingdon, Oxon, OX14 4RN

CRC Press is an imprint of Taylor & Francis Group, LLC

ISBN: 978-0-367-49348-6 (hbk)
ISBN: 978-0-367-49349-3 (pbk)
ISBN: 978-1-003-04587-8 (ebk)

Typeset in CMR10
by Lumina Datamatics Limited

https://www.routledge.com/An-Introduction-to-Metric-Spaces/
Gopal-Deshmukh-Ranadive-Yadav/p/book/9780367493486

Contents

Preface

The term metric is derived from metre (English) or metriqué (French), which means measure. Maurice Fréchet (1878–1973) is usually credited with the first major effort to develop an abstract theory of spaces in his 1906 doctoral thesis. An E-class was later named a "metric space" (une classE) by the German mathematician Felix Hausdorff (1868–1942) in 1914.

The development of the Euclidean geometry has preceded mainly general theory of metric spaces and topological theory of metric spaces. Also the nature of particular metric plays an important role in the investigation of cases such as non-Euclidean geometry, differential geometry, mechanics, and physics.

These days, the study of metric spaces is considered fascinating and highly useful because of its increasing role not only in mathematics but also in applied sciences. It has been increasingly realized that this branch of mathematics is a convenient and very powerful way of examining the behavior of various mathematical models, and many major theorems were given a simpler proof in a more general setting.

Looking at the current scenario and in the interest of students of applied sciences to become acquainted with the basic ideas of metric spaces in their early studies, a semester course on "Metric Spaces" has been prescribed in almost all universities throughout India and abroad either at the undergraduate or graduate level.

There are, indeed, a number of books available on "topology" or "functional analysis" which contain metric spaces only as a chapter, usually too concise to be intelligible to the students, particularly when the course is given at the undergraduate level.

This book is aimed to serve as a textbook for an introductory course in metric spaces for undergraduate or master students. It has been our goal to present the basics of metric spaces in a natural (intuitive) way that encourages geometric thinking and makes the students participate actively in the learning of the subject. Very often, we have tried to give the strategy (sketch) for the proofs, rather than directly giving the "textbook proof", and we motivate the reader to complete the proof on their own. Bits of pertinent history are scattered throughout the text, including brief biographies of some of the central players in the development of metric spaces. The textbook is composed of seven chapters which contain the main materials on metric spaces; namely, introductory concepts, completeness, compactness, connectedness, continuous functions, and metric fixed point theorems with applications.

Some of the noteworthy features of this book are:

- Pictures are included to encourage the reader to think geometrically.

- Focus on systematic strategy to generate ideas for the proofs of theorems.

- A wealth of remarks and observations along with a variety of exercises.

- Historical notes and brief biographies appearing throughout the text.

The chapters in this book are arranged in a logically linear fashion. Throughout the text, we have also tried to keep in mind not to introduce a concept without covering its prerequisites. Therefore, a first time reader would find it easy, reading the text linearly and solving the exercises one-by-one to improve the understanding of the subject. In fact, we strongly recommend the reader to solve all the exercises with geometric intuitions and making appropriate diagrams, wherever needed. Also, the readers should look at hints/solutions only after they have given sufficient time and thought on the problem.

Chapter 1 deals with basic set theory and the theory of functions and relations. In this chapter, the reader is introduced to the notion of sets, relations and functions, and some major outcomes of these notions. The important parts of this chapter include equivalence relations, partial order relations, images and pre-images of sets under functions and countability of sets. Although most of the material presented in this chapter would have been studied by the reader in an analysis course (if the reader has taken one), this would help brush up certain concepts of importance which we will be using further in the text. Also, many books on set theory and basic analysis deal with cardinal numbers and their arithmetic. Since it was not of much use in the theory of metric spaces, we have not included cardinal numbers in this chapter. Rather, we deal with countability and uncountability of sets through bijective functions, which is natural, intuitive, and useful in developing theory of metric spaces.

Chapter 2 introduces the reader to the notion of (arbitrary) metric space. The chapter starts with the set of real numbers, where the reader knows how to measure "distances" and gradually abstracts properties of this distance to define an abstract metric space. A lot of examples are provided and a variety of exercises are given in the text itself to clear the concepts easily. Throughout the text, three sets remain of interest to us: One, the set of real numbers and its Cartesian products (called the n-space), the set of sequences of certain type (see text for more information), and the set of continuous functions. Major remarks about these sets considered as metric spaces are given in this chapter. Immediately after defining metric spaces, we move toward the consequences of the definition. We look at open balls, open sets, closed sets, bounded sets, types of points in a set of a metric space, and other related concepts. At the end of this chapter, we ask a major question concerning different metrics on the same set and try to answer it.

Chapter 3 is where we start developing the theory of metric spaces. First, we go through a recap of sequences and subsequences and then define their convergence in (arbitrary) metric spaces. Indeed, this definition is also motivated from our experience in the set of real numbers. The major portion of this chapter deals with making a metric space "complete", in terms of sequences we have in it. Finally, a theorem due to René–Louis Baire, famously known as the "Baire category theorem" is discussed along with its applications to complete metric spaces.

Chapter 4 deals with compactness of metric space and the consequences of compactness. We start with the basic intuitive definition of compactness, see some examples, and then proceed to define this concept in different ways. In the process of doing so, we come across other concepts such as total boundedness, sequential compactness, and Lebesgue number. Finally, we form a characterization of compact metric spaces in terms of these new-found concepts.

Chapter 5, the shortest chapter reader will encounter in this text, is about connectedness of a metric space. As usual, we start with the intuitive definition of connectedness and then develop relevant theory. Since most of the "remarkable" theorems on connectedness require the knowledge of continuous functions, we postpone these until Chapter 6, and throughout the chapter encourage the reader to think only geometrically and prove facts using the intuitive definition only.

Chapter 6 is again a major chapter in the text. It deals with continuous functions, indeed abstracted from our experience in the set of real numbers. This chapter starts with the usual definition of continuity, and then we form some equivalent definitions which help us proving or disproving the continuity of any function. This is the point where the keen reader is ready to start learning general topology. After looking at sufficient examples and exercises of continuous functions, we "upgrade" this idea to uniform continuity and Lipschitz continuity (although, we only scratch the surface of Lipschitz continuity without going into depth). We then look at the consequences of applying continuous functions on various types of metric spaces (compact, connected, complete, bounded, etc.) dealt with earlier in the text. Finally, as a close of this chapter, the reader is introduced to the homeomorphisms, the very core of general topology and all the theory that develops further.

Chapter 7 is all about introducing the reader to fixed point theory. After covering the preliminaries, a basic fixed point result due to Banach, called the "Banach contraction principle" is stated (and proved). The remaining chapter deals with the applications of this fixed point result in various fields of mathematics such as (but not limited to) numerical analysis, differential equations, integral equations, and the solution of linear algebraic equations.

Without claiming originality of the results, we do claim simplicity and lucidity of presentation coupled with comprehensiveness of the materials. We have been influenced by many books on the same subject and have listed in the bibliography those books which have been of particular assistance to us in preparing this book.

We greatly admire and are deeply indebted to our teachers, colleagues, and students who helped us directly or indirectly in preparing this book. Our deepest gratitude and thanks are also due to our family members who extended moral support throughout the execution of this project.

A special thanks to Mr. Thekku Veettil Abhitjih for making the diagrams and illustrations from our paper-made illustrations.

The authors are very thankful to Aastha Sharma, Shikha Garg, and the staff from CRC press for their unfailing support, cooperation, and patience in publishing this book.

We would like suggestions, edits, and/or corrections of the text at

gopaldhananjay@yahoo.in,   aniruddha480@gmail.com,   asranadive04@ yahoo.co.in, or yshubh305@gmail.com.

https://www.routledge.com/An-Introduction-to-Metric-Spaces/ Gopal-Deshmukh-Ranadive-Yadav/p/book/9780367493486

A Note to the Reader

Two matters require comment: the problems and the proofs.

Working on problems is a crucial part of learning mathematics. No one can learn the subject merely by pouring over the definitions, theorems, and examples that are worked out in the text. One must be a part of it for oneself. To provide that opportunity is the purpose of the exercises.

They vary in difficulty, with the easier ones usually given first. Some are routine verifications designed to test whether you have understood the definitions and examples of the preceding section. Other are less routine. You may, for instance, be asked to prove a theorem. The main purpose of such an exercise is to encourage you to work carefully through the proof in question.

It is a basic principle in the study of mathematics, and one too seldom emphasized, that a proof is not really understood until the stage is reached at which one can grasp it as a whole and see it as a single idea. In achieving this end, much more is necessary than merely following the individual steps in the reasoning. This is only the beginning. A proof should be chewed, swallowed, and digested, and this process of assimilation should not be abandoned until it yields a full comprehension of the overall pattern of thought.

For those readers who are pursuing their undergraduate studies, we would like to mention that there are two "types" of proofs that we usually write: One is that which shows the ideas, steps, and methods which we shall implement. We call this a "rough" proof or a sketch. The other type of proof gives the exact logical sequence of arguments. Such type of proof is given in most of the textbooks, and therefore we call it a "textbook proof". However, one disadvantage of writing textbook proofs directly is that it emphasizes way too much on arguments that it loses the track of intuition. This causes confusion among first time readers, and most students then try to remember the proof. One must understand that this is not a correct way to do mathematics.

Keeping this in mind, in this text, we always try to give the sketch of the proof rather than the textbook proof and leave the writing of textbook proof up to the reader. We suggest the students as well as teachers to use geometric intuitions to sketch the proof first, and then start writing the proof.

Authors

Dr. Dhananjay Gopal has a doctorate in Mathematics from Guru Ghasidas University, Bilaspur, India, and is currently Assistant Professor of Applied Mathematics in S V National Institute of Technology, Surat, Gujarat, India. He is author and co-author of several papers in journals, proceedings, and a monograph titled *Background and Recent Developments of Metric Fixed Point Theory*. He is devoted to general research on the theory of Nonlinear Analysis and Fuzzy Metric Fixed Point Theory.

Mr. Aniruddha Deshmukh is currently a student of (Integrated) MSc Mathematics and is associated to the Applied Mathematics and Humanities Department, S V National Institute of Technology, Surat, Gujarat, India. He has been an active student in the department and has initiated many activities for the benefit of the students, especially as a member of the science community (student chapter), known by the name of SCOSH. During his course, he has also attended various internships and workshop such as the Mathematics Training and Talent Search (MTTS) Programme for two consecutive years (2017–2018) and has also done a project on the qualitative questions on Differential Equations at Indian Institute of Technology (IIT), Gandhinagar, Gujarat, India in 2019. He has also qualified CSIR-NET JRF. Furthermore, his research interest focuses on Linear Algebra and Analysis and their applicability in solving some real-world problems.

Abhay S. Ranadive is a Professor at the Department of Pure & Applied Mathematics Ghasidas Vishwavidyalaya (A Central University), Bilaspur, Chattisgarh, India. He has been teaching at the university for the last 30 years. He is author and co-author of several papers in journals and proceedings. He is devoted to general research on the theory of fuzzy sets and fuzzy logic, modules, and metric fixed point.

Mr. Shubham Yadav is currently a student of (Integrated) M.Sc. Mathematics and is associated to the Applied Mathematics and Humanities Department, S V National Institute of Technology, Surat, Gujarat, India. As a member of SCOSH the student prominent science community in the institute, he has attended and organized various workshops and seminars. He also attended Madhava Mathematics Camp 2017. He did an internship on the calculus of fuzzy numbers at NIT, Trichy and one on operator theory at IIT, Hyderabad. He has also qualified for JRF. His main research interests are functional analysis and fuzzy sets with a knack for learning abstract mathematical concepts.

Chapter 1

Set Theory

In this chapter, we shall see some basic concepts of set theory which will be used further. The aim of this chapter is to make the reader familiar with concepts of sets, relations, and functions, which probably have been covered in the high school level. We shall, in the text, try to give as much intuition as possible with appropriate diagrams and illustrations wherever required. However, the reader is advised to keep making illustrations on their own to have a better understanding of the concepts. We shall start with revising about sets and slowly develop the theory on functions.

1.1   Sets

A *collection of elements* or *objects* is often called a *set*. Although almost no book gives what is called a formal definition of sets, this statement is often taken as a definition. The reason for no formal definition is that if we keep this as the formal definition, we would need to define what is meant by a "collection" and in fact, we will also need to define what is meant by "object" or "element". But, we need to stop somewhere! So henceforth, we shall treat the above statement as our definition of a set.

One should keep in mind that if at all we want to collect objects, we need something from which we can do so. Such a thing is again a set and is often called the *universal set* or sometimes, even referred to as the "*universe of discourse*". Such a universal set is often used to form other sets, given the context. Also, whenever we consider a universal set, we will keep ourselves restricted to the elements of that set itself. In other words, we shall act like we do not know anything "outside" our universe of discourse.

As a matter of notations and conventions that is followed by almost every mathematician and author, we shall use **CAPITAL** letters to denote sets and **small** letters to denote the elements of a set. The elements of any set are enclosed between (curly) braces, i.e., "{" and "}". Also, if an object x belongs to a set X, we shall denote it as $x \in X$. Similarly, if an element does not belong to the set, we write $x \notin X$.

Although we have defined sets and also have seen some intuitions that arise, we have not yet seen how to write the sets. Let us, through examples, see how to write sets.

Example 1.1 (Writing sets). Suppose that our universe of discourse (universal set) is the set of real numbers, $\mathbb{R}$. We would like to collect elements (in this case, numbers) of the form which are positive, integers, and (strictly) less than 6. We name this collection (set) A. The easiest way to write A is $\{1, 2, 3, 4, 5\}$. We often write this as $A = \{1, 2, 3, 4, 5\}$.

Now, moving one step ahead, we now wish to make a collection B of elements of $\mathbb{R}$ which are: positive, integers, and divisible by 2. If we are to list all the elements, our lifetime would prove to be short since there will be no end finding such numbers which are divisible by 2. Hence, we may write a few elements of the set and then use $\cdots$ to denote that the list goes on forever. So, we have $B = \{2, 4, 6, \cdots\}$.

Such a method of writing sets is called the *listing* or *roaster* form of sets. Although this looks simple and convenient, it is not so when we start dealing with different types of sets especially when elements are not really such which can be listed. Let us take an example to look at this.

Example 1.2. Consider the collection C, which we shall make from our universe $\mathbb{R}$, of real numbers of the form greater than or equal to 0 and less than or equal to 1. Let us see if we can list the elements as above. Clearly, 0 is a real number which follows our criterion for being in the collection and so does 1. Is there any other real number in between 0 and 1 which also satisfies the criterion? Yes! One such number is $\frac{1}{2}$. In fact, there are infinitely many such numbers and to list a few elements and then use $\cdots$ in a fashion that explains the criterion well is difficult! So, we shall use the criterion itself to write our set. We write:

$$C = \{x \in \mathbb{R} | 0 \leq x \leq 1\}.$$

This reads as, "collection of all those elements x of $\mathbb{R}$ such that 0 is less than or equal to x and x is less than or equal to 1". Such a way of writing is called the *set-builder* form of writing sets.

The set-builder form is convenient especially when we deal with abstract sets where we do not really know what elements we have in our set, but rather the properties which they follow. We shall come to abstraction later, once we are really familiar with concepts we know by experience.

Considering the two ways of writing sets dealt with in Examples 1.1 and 1.2, we shall make the following sets out of the set of real numbers $\mathbb{R}$.

$$\mathbb{N} = \{1, 2, 3, \cdots\},$$

$$\mathbb{Z} = \{\cdots, -3, -2, -1, 0, 1, 2, 3, \cdots\},$$

$$\mathbb{Q} = \left\{\frac{p}{q} \in \mathbb{R} \middle| p \in \mathbb{Z} \text{ and } q \in \mathbb{N}\right\},$$

$$\mathbb{Q}^+ = \{x \in \mathbb{Q} | x > 0\},$$

$$\mathbb{Q}^- = \{x \in \mathbb{Q} | x < 0\},$$

$$\mathbb{Q}^* = \{x \in \mathbb{Q} | x \neq 0\},$$

$$\mathbb{R}^+ = \{x \in \mathbb{R} | x > 0\},$$

$$\mathbb{R}^- = \{x \in \mathbb{R} | x < 0\},$$

$$\mathbb{R}^* = \{x \in \mathbb{R} | x \neq 0\}.$$

In the complete text that follows, we shall use these notations for the sets defined above. Here, the set $\mathbb{N}$ is called the set of *natural numbers*, $\mathbb{Z}$ is called the set of *integers*, and $\mathbb{Q}$ is called the set of *rational numbers*. A set which we did not yet write and give a notation to is the set of *irrational numbers*. This will be dealt with later.

We have tactfully avoided defining natural numbers, integers, rational numbers, etc. We have seen until now that we can form sets which contain numbers. A natural question arises: Are there any sets which contain elements that are not only numbers? Well, as all of us might have seen in our high school, sets can contain any type of elements: numbers, alphabets, words, or in fact, a set of books or papers is also a set! At this stage though, a better question can be asked: Can the elements of a set be themselves sets? Let us try to figure it out through an example.

Example 1.3 (Family of sets). Consider the set of real numbers, $\mathbb{R}$. We wish to collect all those sets constructed from the elements of $\mathbb{R}$ which contain 0. Now, we are collecting sets instead of individual elements from $\mathbb{R}$. Can we have one such set? Yes, $\mathbb{R}$ itself. Can we have another? Again, the answer is yes! $\{0\}$ is another such set. Clearly, listing all such sets would be impractical. So, we will use the set-builder form to write our collection which we shall name $\mathscr{F}$. We then have

$$\mathscr{F} = \{S | S \text{ is constructed from the elements of } \mathbb{R} \text{ and } 0 \in S\}.$$

As seen in Example 1.3, the elements of sets can themselves be sets. Whenever such a thing happens, i.e., we have a collection of sets, we shall use *script letters* (such as the $\mathscr{F}$ we used in Example 1.3) to write them. Before moving ahead, let us try to get collection of sets, where the sets will be constructed from $\mathbb{N}$.

Example 1.4 (Indexing sets). Let us consider, as our universe, the set of natural numbers $\mathbb{N}$ and for each natural number $n \in \mathbb{N}$, let us try to collect sets (constructed from $\mathbb{N}$) which have all elements from 1 through n. This means to say, we collect sets S_n for every n. Here, if we try to give different symbols (letters) to each of these sets, we will get short of symbols! Hence, we try to "index" these sets. That is, we write $S_n = \{1, 2, \cdots, n\}$, where it is understood that as n changes, the elements of the set S_n also change.

Therefore, $S_1 = \{1\}$, $S_2 = \{1, 2\}$, $S_3 = \{1, 2, 3\}$, and so on. Hence, we write our family of sets as

$$\mathscr{F} = \{S_n | n \in \mathbb{N}\}.$$

Here, we say that $\mathscr{F}$ is a family of sets which is *indexed* by $\mathbb{N}$; the set of natural numbers $\mathbb{N}$ is called the *index set* and n is called the *index*.

Unlike Example 1.4, sets are not always indexed by natural numbers. We can also index sets by other sets such as integers, rational numbers, real numbers, or even by a set which is not necessarily a set of numbers. Most of the time, we shall consider an arbitrary index set, which we denote by Λ (throughout the text), whose elements are not exactly known to us. We shall use capital Greek letters to denote arbitrary index sets and the (corresponding) small Greek letters to denote the elements of the index set. Therefore, in general, an indexed family of sets will be written as

$$\mathscr{F} = \{A_\lambda | \lambda \in \Lambda\}.$$

Exercise 1.1. Give an example of a family of sets which is indexed by the set of real numbers, $\mathbb{R}$.

1.1.1   The empty set

Before moving ahead, let us try to see a special type of collection. Suppose our universe is the set of all humans living on earth. Suppose a person like us wishes to collect all those humans who have 8 hands, 4 legs, and 2 tails. Is there any such living human on earth? The answer is no! So, our collection has no elements at all. A set with no elements is called the *empty set* and is denoted by $\emptyset$.

A person of good experience in logic can ask a question at this point: Everywhere it has been written "*the*" empty set. Is the use of "*the*" justified? In other words, is the empty set unique? We shall address this question later, after we have seen enough on operations and equality of sets.

1.1.2   Operations on sets

Once we have sets, we can now start playing with them. The very first thing we can do at this point is to compare two sets. First, we shall address the question: When can we say that two sets are equal? At the beginning, we defined our sets to be collections. First we notice that while collecting, we do not give importance to the order in which they are collected. As a result, the sets $\{1, 2\}$ and $\{2, 1\}$ are the same. What do we observe? Given any two sets X and Y, when can we say that these are equal? An answer based on complete intuition and observation is: Whenever every element of X is an element of Y and every element of Y is an element of X. The formal (mathematical) definition of equality will be given a bit later.

The next thing we can do is to observe the sets we defined in the above section. If we look carefully, every natural number is also an integer (positive). Are these two sets equal? Intuitively, the answer to this question is: No! 0 is one such element in $\mathbb{Z}$ which is not a natural number. However, the set of integers has all the elements of the set of natural numbers. In such a case, we call the set of natural numbers a *subset* of the set of integers. We are now ready for the formal definitions of subset and equality.

Definition 1.1 (Subset). *A set X is a subset of a set Y if $\forall x \in X, x \in Y$. This is denoted by $X \subseteq Y$.*

Note.

1. If the set Y (from Definition 1.1) has at least one element which is not in X, then X is called a *proper subset* of Y. This is denoted by $X \subset Y$ throughout this book.

2. If X is a subset of Y, then Y is called the *superset* of X.

Definition 1.2 (Equality). *Two sets X and Y are said to be equal if $\forall x \in X, x \in Y$ and $\forall y \in Y, y \in X$. This is equivalent to $X \subseteq Y$ and $Y \subseteq X$. Equality is denoted by $X = Y$.*

Exercise 1.2. Compare for subsethood and/or equality the sets which we made from the set of real numbers above.

Now, we try to make more new sets from the sets which we already have. Given any two sets X and Y, one way to make a new set is to collect all the elements from X and all the elements from Y into a single collection, say Z. Thus, any element of Z is either from X or from Y (or even both, if they have it in common). A set formed in this manner is called the *union* of X and Y. Another way to make a new set is by collecting those elements which are in both X and Y and put them in a single collection, say U. Such a collection is called the *intersection* of X and Y. We now move to the formal definition of union and intersection.

Definition 1.3 (Union). *Given two sets X and Y, their union Z is the collection of elements which are in X or in Y. This is denoted by $X \cup Y$. Formally, we may write*

$$X \cup Y = \{x | x \in X \ or \ x \in Y\}.$$

Definition 1.4 (Intersection). *Given two sets X and Y, their intersection Z is the collection of elements which are both in X as well as Y. This is denoted by $X \cap Y$. Formally, we may write*

$$X \cap Y = \{x | x \in X \ and \ x \in Y\}.$$

Exercise 1.3. From the sets constructed from $\mathbb{R}$, for every pair, give the union and the intersection. What do you observe?

The definitions of union and intersection are only made for two sets. But, we would like to make a general definition for arbitrary collection of sets whose union and intersection we need to find. Simply extending the definitions (made from our intuition), we get the following definitions for unions and intersections of arbitrary family of sets.

Definition 1.5 (Arbitrary union). *Given an arbitrary indexed family of sets $\mathscr{F} = \{A_\lambda | \lambda \in \Lambda\}$, the union of this family is the collection of elements which are in at least one of the sets of the family. We write it as*

$$\bigcup_{\lambda \in \Lambda} A_\lambda = \{x | \exists \lambda_0 \in \Lambda \ such \ that \ x \in A_{\lambda_0}\}.$$

Similarly, arbitrary intersection can be defined as follows.

Definition 1.6 (Arbitrary intersection). *Given an arbitrary indexed family of sets $\mathscr{F} = \{A_\lambda | \lambda \in \Lambda\}$, the intersection of this family is the collection of elements which are in all the sets of the family. We write it as*

$$\bigcap_{\lambda \in \Lambda} A_\lambda = \{x | \forall \lambda \in \Lambda, x \in A_\lambda\}$$

As an observation from Exercise 1.3, we may see that intersection of some sets can be the empty set, i.e., there can be sets X and Y such that $X \cap Y = \emptyset$. Such sets are called *disjoint*. In particular, the reader must have observed that $\mathbb{Q}^+$ and $\mathbb{Q}^-$ are disjoint. If we take the union of such sets (whose intersection is empty), the union is called *disjoint union*. As an immediate observation, we can conclude that $\mathbb{Q}^*$ is the disjoint union of $\mathbb{Q}^+$ and $\mathbb{Q}^-$ (Exercise 1.3).

Similarly, if $\mathscr{F} = \{A_\lambda | \lambda \in \Lambda\}$ is an arbitrary indexed family, then $\mathscr{F}$ is said to be a disjoint family if $\bigcap_{\lambda \in \Lambda} A_\lambda = \emptyset$. Here, we can have another concept, often called, *pairwise disjointness*. The family $\mathscr{F}$ is said to be *pairwise disjoint* if $\forall \lambda_1, \lambda_2 \in \Lambda$ with $\lambda_1 \neq \lambda_2$, we have $A_{\lambda_1} \cap A_{\lambda_2} = \emptyset$.

Exercise 1.4. Prove that every pairwise disjoint family of sets is disjoint, but the converse is not necessarily true.

Exercise 1.5. As an extension to Exercise 1.4, can we have a disjoint family where there is no disjoint pair?

Another way to get new sets from old ones is to collect all those elements which are not in the given set. We call this collection a *complement* of the given set.

Definition 1.7 (Complement). *Given a set A, its complement is the collection of elements which are not in A. We write it as*

$$A^c = \{x | x \notin A\}.$$

Here, we must note that we do not know anything "outside" our universe of discourse. Hence, to define a complement, we need a universal set. We call it, for the time being, X. Since we do not know what is outside X, clearly, $X^c = \emptyset$ and also, $\emptyset^c = X$, since none of the elements of X are in $\emptyset$. Thus, a better way to write complements is

$$A^c = \{x \in X | x \notin A\}.$$

Apart from taking complements, a way to get new sets from two given sets A and B is by collecting those elements which are only in one of the sets and not in other. We call this *relative complement* or *set difference*.

Definition 1.8 (Set difference). *Given two sets A and B, the set difference $A \setminus B$ is the collection of all elements of A which are not in B. We write it as*

$$A \setminus B = \{x | x \in A \text{ and } x \notin B\}.$$

With these definitions made, it is clearly seen that $A^c = X \setminus A$ and for any set A, $A \setminus \emptyset = A$.

Exercise 1.6. For given sets A and B, prove that $A \setminus B = A \setminus (A \cap B) = A \cap B^c$.

We now look at some properties of the definitions we have made above in terms of a theorem. Since the proof is easy and many of us might have gone through it at the high school level, it is left as an exercise for the reader.

Theorem 1.1. *Let A, B, and C be given sets and X be the universal set. Then, the following properties hold:*

1. $(A^c)^c = A$.   *[Involution]*

2. $A \cup B = B \cup A$ *and* $A \cap B = B \cap A$.   *[Commutativity]*

3. $A\cup(B \cup C) = (A \cup B)\cup C$ *and* $A\cap(B \cap C) = (A \cap B)\cap C$. *[Associativity]*

4. $A \cap (B \cup C) = (A \cap B) \cup (A \cap C)$ *and* $A \cup (B \cap C) = (A \cup B) \cap (A \cup C)$. *[Distributivity]*

5. $A \cup A = A$ *and* $A \cap A = A$.   *[Idempotence]*

6. $A \cup (A \cap B) = A$ *and* $A \cap (A \cup B) = A$.   *[Absorption]*

7. $A \cup X = X$ *and* $A \cap \emptyset = \emptyset$.   *[Absorption by X and $\emptyset$]*

8. $A \cup \emptyset = A$ *and* $A \cap X = A$.   *[Identity]*

9. $A \cap A^c = \emptyset$.   *[Law of contradiction]*

10. $A \cup A^c = X$.   *[Law of excluded middle]*

11. $(A \cup B)^c = A^c \cap B^c$ *and* $(A \cap B)^c = A^c \cup B^c$.   *[De Morgan's laws]*

Exercise 1.7. Prove Theorem 1.1.

We now make a few observations on the union and intersection of a family of sets. First, it is clearly seen that the definition of union and intersection of two sets is just a special case, where our family of sets contains only two sets. An important observation that can be made here is that whenever we take union, our resultant set cannot get *smaller*, in a sense that the resultant set has all the elements of at least one of the sets in the family. Similarly, when we take intersection, our resultant set cannot get *bigger* than the sets under consideration, in a sense that it cannot have elements from outside of the sets of the family.

We would like to mention here that when we say "arbitrary index set", it can also be empty! So, we must ask a question: What family of sets can be indexed by the empty set? The answer is the empty family! A family which does not contain any sets at all is called the empty family. So, our next question is: What about the intersection and union of an empty family?

Intuitively and from the observations made above, we are looking for the largest possible set in our universe which could be the result of an intersection and the smallest possible set in our universe which could be the result of a union. Clearly, the largest possible set is the universe itself and the smallest possible set is the empty set, $\emptyset$. We make this intuition as a theorem.

Theorem 1.2. *The union of empty family of sets is empty and the intersection of empty family of sets is the universal set.*

Proof. We wish to prove that $\bigcup_{\lambda \in \emptyset} A_\lambda = \emptyset$ and $\bigcap_{\lambda \in \emptyset} A_\lambda = X$.

Let us see what will happen if this is not true, i.e., $\bigcup_{\lambda \in \emptyset} A_\lambda \neq \emptyset$ and $\bigcap_{\lambda \in \emptyset} A_\lambda \neq X$. In the first case, $\exists x \in \bigcup_{\lambda \in \emptyset} A_\lambda$, hence this x must be in one of A_λ. This means $\exists \lambda \in \emptyset$, such that $x \in A_\lambda$, which is not possible. Hence, $\bigcup_{\lambda \in \emptyset} A_\lambda = \emptyset$.

Similarly, if $\bigcap_{\lambda \in \emptyset} A_\lambda \neq X$, then $\exists x \in X$ such that $x \notin \bigcap_{\lambda \in \emptyset} A_\lambda$. This would mean that x is not in at least one of the A_λ, i.e., $\exists \lambda \in \emptyset$ such that $x \notin A_\lambda$, which is again not possible. Hence, $\bigcap_{\lambda \in \emptyset} A_\lambda = X$. $\qquad\square$

1.1.3 Uniqueness of the empty set

In Section 1.1.1, we asked a question: Is the use of "the" in saying "the empty set" justified? The question means, is there only one empty set? We have seen in that section that when we try to collect humans with features not of humans, we get the empty set. Another way to make an empty set is by collecting all the real numbers whose square is negative. Are these two empty sets different? At first sight, it does seem that these two empty sets are different, since they are constructed in different context. But we must notice that set theoretically, two sets are equal if and only if they are subsets of each other. Suppose that someone wants to prove that there are indeed two different empty sets, say $\emptyset_1$ and $\emptyset_2$, with $\emptyset_1 \neq \emptyset_2$. Then, they must produce an element from $\emptyset_1$ which is not in $\emptyset_2$ (or vice versa). Since this is not possible, the two empty sets considered must be equal. Such statements, which happen to be true due to lack of premise, are called *vacuously true* statements. Hence the empty set is unique, i.e., there is only one empty set. This justifies the use of the word "the" as well as a unique symbol $\emptyset$.

1.1.4 Power sets

Given a set X, we now try to collect all the subsets of X. We call this collection the *power set* and denote it as

$$\mathcal{P}(X) = \{S | S \subseteq X\}$$

A natural question arises: Is $\mathcal{P}(X)$ non-empty for any set X? It is easy to see that it is, since $X \in \mathcal{P}(X)$. Also, $\emptyset \in \mathcal{P}(X)$.

Exercise 1.8. What is $\mathcal{P}(\emptyset)$?

1.1.5 Cartesian products

Given any two sets X and Y, we define the *Cartesian product* as

$$X \times Y = \{(x, y) | x \in X \text{ and } y \in Y\}.$$

An element of $X \times Y$ is called an ordered pair or a 2-tuple. Each entry in a tuple is often called a *coordinate*.[1] Notice that while writing the Cartesian

---

[1] Indeed, the intuition of *coordinates* comes from our experience in $\mathbb{R}^2$ and $\mathbb{R}^3$, which is the Cartesian product of $\mathbb{R}$ with itself.

product, it is important to keep in mind the order of the sets. If X and Y are two different sets, the two Cartesian products $X \times Y$ and $Y \times X$ will have different elements. Indeed, the elements of one will be elements of the other with the order of writing reversed. Therefore, we observe that while writing tuples, the order of writing (or listing) is important. Similarly, we can define an n-tuple as $(x_1, x_2, \cdots, x_n)$, where the order of writing is important. Now, we would not like to restrict ourselves to Cartesian product of two sets (or for that matter, finitely many sets). Therefore, we now define an *arbitrary Cartesian product* for an arbitrarily indexed family of sets. If $\mathscr{F}$ is an indexed family, indexed by the index set Λ, then the Cartesian product for the family is given by $\prod_{\lambda \in \Lambda} A_\lambda$. Although we cannot explicitly write the elements (as tuples) in this product, axiom of choice[2] guarantees us that there will be at least one element in the product, i.e., a "tuple" where each entry (or coordinate) will be from individual sets. Rigorously (and through the axiom of choice), the arbitrary Cartesian product can be defined as

$$\prod_{\lambda \in \Lambda} A_\lambda = \left\{ f : \Lambda \to \bigcup_{\lambda \in \Lambda} A_\lambda \,\middle|\, f(\lambda) \in A_\lambda \right\}$$

Note. Notice that the arbitrary Cartesian product is defined only for a family which does not contain an empty set. In case the family contains the empty set, the Cartesian product remains empty.

Some of the known sets produced from Cartesian products include the plane $\mathbb{R}^2 = \mathbb{R} \times \mathbb{R}$ and the space $\mathbb{R}^3 = \mathbb{R} \times \mathbb{R} \times \mathbb{R}$. Similarly, we can define $\mathbb{R}^n$ for any natural number n.

Exercise 1.9. Are the sets $\mathbb{R}^3$, $\mathbb{R} \times (\mathbb{R} \times \mathbb{R})$ and $(\mathbb{R} \times \mathbb{R}) \times \mathbb{R}$ equal? Justify.

Exercise 1.10. What can you say about the empty product, i.e., $\prod_{\lambda \in \emptyset} A_\lambda$?

1.2   Relations

When we have two sets, say X and Y, we would like to see if the elements of X are somehow related to the elements of Y. For example, if X is the set of all living males on earth and Y is the set of all living females on earth, we would like to see which males are brothers of which females. Clearly, one man can have more than one sister, and hence, one element of X can be related to

---

[2]Axiom of choice states that if X is a non-empty collection of non-empty sets, then there is a function $f : X \to \bigcup_{A \in X} A$ such that for every set $A \in X$, $f(A) \in A$. Such a function is called a *choice function*.

one or more than one elements of Y. Suppose we collect all the ordered pairs of the form (males, females) which are related (as brother-sister); we get a subset of $X \times Y$. We make this intuition a definition of *relation*.

Definition 1.9 (Relation). *A relation between two sets X and Y is a subset R of $X \times Y$.*

Note.

1. Given any two non-empty sets X and Y, we always have at least two relations:

   (a) The empty relation $\emptyset \subset X \times Y$, which suggests that no element of X is related to any element of Y.

   (b) The universal relation $X \times Y$ itself, which suggests that every element of X is related to every other element of Y.

2. If we take a subset $R \subseteq X \times X$, we call it a relation *on X*.

If $(x, y) \in R$, then we say that x is related to y through the relation R, and to make the notations short, we often write it as xRy. There are a few relations we know by experience and have already been using.

Example 1.5 (Known relations). We know a relation R on $\mathbb{R}$ defined by xRy if and only if $x \leq y$. Another relation we know is on $P(X)$, for any set X, given as ARB if and only if $A \subseteq B$. These two relations are special types of relations, which we will deal with later.

Another special type of relation known to us is the equality relation, i.e., on any set X, define a relation R as xRy if and only if $x = y$. This is also called an *identity relation* or *diagonal relation*.

If we take $\mathbb{N}$ as our set, we can define a relation $R \subseteq \mathbb{N} \times \mathbb{N}$ as mRn if and only if $m|n$, i.e., n is a multiple of m. As another example, for a fixed $n \in \mathbb{N}$, we can define a relation[3] R as $m_1 R m_2$ if and only if $n|(m_1 - m_2)$.

As the next step, we would like to get new relations from old. In particular, we would like to get a relation on $X \times Y$ if two relations R and S are given on X and Y respectively. We must note here that we need a relation **on $X \times Y$**, i.e., if we name our relation T, then $T \subseteq (X \times Y) \times (X \times Y)$. Also, we have been provided with $R \subseteq X \times X$ and $S \subseteq Y \times Y$. What is the most natural method to use these two relations for our purpose? We can define T as $(x_1, y_1) T (x_2, y_2)$ if and only if $x_1 R x_2$ and $y_1 S y_2$. However, this is not the only method! Before getting another way, let us look at an example.

---

[3]The readers who have some experience in elementary number theory might have guessed that this relation is the *congruence relation*.

Example 1.6 (Dictionary order). Suppose we are searching a dictionary for words with two letters . We want to locate the words: an, be, and am. First we see the words "an" and "be". We know that "a" comes prior to "b" in the English alphabet. So, we deduce that "an" must come first. Then, we look at the words "an" and "am". Since "m" comes prior to "n", we deduce that "am" comes first. Thus, we complete our search by saying that "am" comes first, followed by "an" and then "be".

What did we do here? We were given relation on the set of letters, namely the usual order, which tells us which letter comes prior. Then, given words with two letters, we first compared the first letters. If the first letter of the first word came prior to the first letter of the other word, we deduced that the first word came prior to the second. If the first letters were the same for both the words, we checked the second letters. If the second letters are the same, then the words are the same! If the second letters are not same, then one of the words would have the second letter which came prior to the second letter of the other word and then we placed the words accordingly.

Such a system of ordering or relating words (or elements), is called *dictionary order* or the *lexicographic order*.

We wish to do the same for arbitrary relations on arbitrary sets. If we are given two relations R and S on the sets X and Y respectively, we define the *lexicographic relation*, L on $X \times Y$ as follows

$$(x_1, x_2) \, L \, (y_1, y_2) \text{ if and only if } ((x_1 \neq y_1 \Rightarrow x_1 R y_1) \text{ and } (x_1 = y_1 \Rightarrow x_2 S y_2)).$$

Exercise 1.11. Write the lexicographic relation of a finite Cartesian product $X_1 \times X_2 \times X_3 \times \cdots \times X_n$ for any $n \in \mathbb{N}$, given relations R_i on X_i for $i = 1, 2, \cdots, n$.

1.2.1 Types of relations

Relations which follow certain special properties have been given special names. A few of them are listed below.

Definition 1.10 (Reflexive relation). *A relation R on X is said to be reflexive if $\forall x \in X$, xRx, i.e., every element of X is related to itself.*

Definition 1.11 (Symmetric relation). *A relation R on X is said to be symmetric if $xRy \Rightarrow yRx$.*

Definition 1.12 (Anti-symmetric relation). *A relation R on X is said to be anti-symmetric if $(xRy \text{ and } yRx) \Rightarrow x = y$.*

Definition 1.13 (Transitive relation). *A relation R on X is said to be transitive if $(xRy \text{ and } yRz) \Rightarrow xRz$.*

Remark. Sometimes, the definition of anti-symmetric relations seems counter intuitive. As the name suggests, an anti-symmetric relation should be just the opposite of a symmetric relation. So, while the symmetric relation forces y to be related to x, whenever x is related to Y, an anti-symmetric relation should never allow y to be related to x whenever x is related to y. However, we must note that there might be some element which is related to itself, i.e., xRx. If we go by our intuition, then x will not be allowed to be related to itself! We do not want this. Therefore, we have made a statement where we can allow an element to be related to itself. However, in our definition, we do take care that if x and y are distinct elements and x is related to y, then y cannot be related to x, thereby satisfying our intuition.

Exercise 1.12. From the known relations in Example 1.5, find out which relations are reflexive, symmetric, anti-symmetric, and transitive.

1.2.2 Equivalence relations

Definition 1.14 (Equivalence relation). *A relation R on X is said to be an equivalence relation if R is reflexive, symmetric, and transitive.*

Clearly, the equality relation is an equivalence relation on any set. Also, the relation R on $\mathbb{N}$ (for a fixed $n \in \mathbb{N}$) defined by $m_1 R m_2$ if and only if $n|(m_1 - m_2)$ is an equivalence relation. We call it the *congruence relation modulo n* and write as $m_1 \equiv m_2 \pmod{n}$, which is read as "m_1 is congruent to m_2 modulo n".

Once we have relations, we would like to see what are the elements that are related to a particular element $x \in X$. Although such a collection can be defined for any relation, we shall define it only for equivalence relations for its use later. We will call this collection *equivalence class* of x.

Definition 1.15 (Equivalence class). *If R is an equivalence relation defined on X, then the equivalence class of $x \in X$, denoted by $[x]$, is defined as the collection of all those elements which are related to x through the relation R.*

$$[x] = \{y \in X | xRy\}.$$

The first question that we should ask is: Is $[x]$ non-empty for any x and any equivalence relation R? The answer is yes! Since R is an equivalence relation, in particular, it is reflexive. Hence, xRx always holds for any x. Therefore, we always have at least one element in each equivalence class. Let us look at an example to see how can we construct equivalence classes.

Example 1.7. Consider the congruence (modulo) relation. We shall extend it to the set of integers, $\mathbb{Z}$. We define the relation for $n = 2$ and the reader is advised to do it for any $n \in \mathbb{N}$. Let the relation R be defined as $m_1 R m_2$ if and only if $2|(m_1 - m_2)$. First, we will check if this relation is an equivalence relation.

Clearly, for any $m \in \mathbb{Z}$, $2|(m - m)$. Hence, R is **reflexive**. Similarly, if $m_1 R m_2$ holds, we have $2|(m_1 - m_2)$. So, $(m_1 - m_2) = 2k$ for some integer k. Therefore, $(m_2 - m_1) = -2k = 2k'$, where $k' = -k \in \mathbb{Z}$. Hence, $2|(m_2 - m_1)$, i.e., $m_2 R m_1$ also holds. Thus, R is **symmetric**. Finally, if $m_1 R m_2$ and $m_2 R m_3$ holds, then $2|(m_1 - m_2)$ and $2|(m_2 - m_3)$ would give us that $2|(m_1 - m_2) + (m_2 - m_3) = (m_1 - m_3)$. Hence, $m_1 R m_3$ must also hold. Thus, R is also **transitive**.

Therefore, we conclude that R is an equivalence relation. We will keep the same notation for the congruence relation, i.e., $m_1 \equiv m_2 \pmod{2}$. Now, let us look at the equivalence class of 0. By definition, $[0] = \{m \in \mathbb{Z}|\ 2|(m - 0)\}$, which is same as saying that $[0] = \{m \in \mathbb{Z}|\ 2|m\}$. Thus, the equivalence class of 0 under this relation is the set of all even integers. Similarly, let us look at the equivalence class of 1. By definition, we have $[1] = \{m \in \mathbb{N}|\ 2|(m - 1)\}$. If $2|(m - 1)$, then $m = 2k + 1$ for some integer k. Thus, the equivalence class of 1 is the set of all odd integers.

Now, let us take any other even integer and denote it by $2k$. Then, $[2k] = \{m \in \mathbb{Z}|\ 2|(m - 2k)\}$. But this would mean that $m = 2k' + 2k$, i.e., m is even. Thus, the equivalence class of any even integer is the set of all even integers (same as $[0]$). Similarly, we can see that the equivalence class of any odd integer is the set of all odd integers (same as $[1]$).

Thus, under this relation, there are only two equivalence classes, namely $[0]$ and $[1]$. The equivalence class of any other integer is one of these two as shown above.

Exercise 1.13. Generalize the equivalence relation of congruence modulo n for any $n \in \mathbb{N}$ on the set of integers, $\mathbb{Z}$. Also, find all the equivalence classes.

Exercise 1.14. For the relations which are equivalence relations from Example 1.5, give all the equivalence classes.

Let us make certain observations from the example and exercises above. First, the distinct equivalence classes of an equivalence relation are all pairwise disjoint. Second, if an element y is in the equivalence class of x, then their equivalence classes are the same. Third, all the equivalence classes form a disjoint union of the set X on which the relation is defined. We make these observations as a theorem.

Theorem 1.3. *If R is an equivalence relation on X, then*

1. *For any two elements $x, y \in X$, either $[x] = [y]$ or $[x] \cap [y] = \emptyset$.*

2. *For an element $y \in X$, $y \in [x]$ if and only if $[x] = [y]$.*

3. *X is the disjoint union of equivalence classes.*

Proof.

1. Both the conditions cannot hold simultaneously, since $[x] = [y]$ and $[x] \cap [y] = \emptyset$ would imply that $[x] = [y] = \emptyset$. But, we know that $x \in [x]$ and $y \in [y]$. Hence, only one of the two conditions can hold at a time.

Given two equivalence classes, $[x]$ and $[y]$, they can either be disjoint or not. If they are disjoint, then we are done! So, we consider the case where $[x] \cap [y] \neq \emptyset$ and then prove that the second condition must hold.

Let $z \in [x] \cap [y]$, i.e., $z \in [x]$ and $z \in [y]$. Also, for any element $u \in [x]$, we have zRx and uRx. Now, by symmetry and transitivity, we have zRu. But, zRy is also true since $z \in [y]$. Therefore, again by symmetry and transitivity uRy holds true, i.e., $u \in [y]$. Hence, $[x] \subseteq [y]$.

Similarly, we can show $[y] \subseteq [x]$. By the definition of equality, we get $[x] = [y]$.

2. First we will prove that if $y \in [x]$, then $[x] = [y]$.

Since $y \in [x]$ and $y \in [y]$, we have, $[x] \cap [y] \neq \emptyset$. Hence, from above $[x] = [y]$.

Conversely, if $[x] = [y]$, then we have $y \in [y] = [x]$.

3. Clear from 1 and 2.

$\square$

1.2.3 Partition of sets

We observe that the set X can be expressed as a disjoint union of equivalence classes on X due to a given equivalence relation. We consider this property of equivalence classes and make a definition.

Definition 1.16 (Partition). *A family $\mathscr{F} = \{A_\lambda \subseteq X \mid \lambda \in \Lambda\}$ of subsets of a set X is said to form a partition of X if $\mathscr{F}$ is pairwise disjoint and $\bigcup_{\lambda \in \Lambda} A_\lambda = X$.*

Clearly, from the discussion above, every equivalence relation gives a partition of X. A natural question arises: Given a partition, can we form an equivalence relation?

Let $\mathscr{F} = \{A_\lambda \subseteq X \mid \lambda \in \Lambda\}$ be a partition of X. We define a relation R on X as xRy if and only if $\exists \lambda_0 \in \Lambda$ such that $x \in A_{\lambda_0}$ and $y \in A_{\lambda_0}$. As $\mathscr{F}$ is a partition of X, any $x \in X$ is in some A_λ, so xRx by definition. Therefore, R is **reflexive**. Also, if xRy holds, then x and y are in some A_λ for some $\lambda \in \Lambda$, but then y and x are in same A_λ. This means yRx, i.e., R is **symmetric**. Finally, if xRy and yRz holds, then all three elements x, y, and z are in the same set A_{λ_0} of the partition. Therefore, xRz also holds, and R is **transitive**. Hence R is an equivalence relation. What is the partition of X corresponding to this equivalence relation? It is the same as $\mathscr{F}$. We summarize this discussion in the form of a theorem.

Theorem 1.4. *For any set X,*

1. *Every equivalence relation on X induces a partition of X.*

2. *Corresponding to every partition of X, there is an equivalence relation whose equivalence classes form the given partition.*

1.2.4 Partial order relations

We have seen a relation on $\mathbb{R}$ given by $\leq$. We call it the (usual) order on $\mathbb{R}$. By experience, we know that given two elements, this relation tells us which one comes first. We would now like to generalize this notion to arbitrary sets. Before generalizing, we see that in Example 1.5 and the exercise that followed, we have proved that $\leq$ is reflexive, anti-symmetric, and transitive. Taking into consideration these properties of the order relation, we define it in an abstract setting and call it a *partial order relation*. The reason for naming it partial order will be dealt with later.

Definition 1.17 (Partial order relation). *A relation R on a set X is said to be a partial order relation if it is reflexive, anti-symmetric, and transitive.*

As a matter of notations and conventions accepted by the mathematical community, we shall write a partial order relation as $\leq$ or $\preccurlyeq$ (when the context is confusing). The ordered pair $(X, \leq)$, i.e., a set X equipped with a partial order relation, is called a *partially ordered set* or as a shorthand notation, a *POSET*.

We have seen from Exercise 1.12 that the natural order $\leq$ on $\mathbb{R}$ forms a partial order. In fact, the definition of partial order takes its motivation from here! Also, the relation of subsethood, $\subseteq$ on the power set $\mathcal{P}(X)$ of a set X forms a partial order. Is there a difference in these two? In the usual order on $\mathbb{R}$, given any two elements, $x, y \in \mathbb{R}$, either $x \leq y$ or $y \leq x$ always holds. Thus, we see that in $(\mathbb{R}, \leq)$, any two elements are comparable.

But this is not true in the case of $\subseteq$. To see this, consider the power set of the set $X = \{1, 2, 3\}$, i.e., $\mathcal{P}(X) = \{\emptyset, \{1\}, \{2\}, \{3\}, \{1, 2\}, \{1, 3\}, \{2, 3\}, X\}$. From the definition of $\subseteq$, we have $\{1\} \not\subseteq \{2\}$ and $\{2\} \not\subseteq \{1\}$. This shows us that in $(\mathcal{P}(X), \subseteq)$, two elements need not always be comparable. Hence, in general, we call the relation a partial order relation.

A partial order relation $\leq$ defined on a set X, for which any two elements of X are comparable, i.e., given $x, y \in X$ either $x \leq y$ or $y \leq x$, is called *total order* and $(X, \leq)$ is called a totally ordered set. Before moving further, let us see another example of partial order.

Example 1.8. Consider the set $\mathbb{N}$ of natural numbers. Define a relation R on $\mathbb{N}$ as mRn if and only if $m|n$. Clearly, the relation is **reflexive** since $\forall m \in \mathbb{N}$, we always have $m|m$. Now, if $m|n$ and $n|m$ is true, we have $m \leq n$ and $n \leq m$, i.e., $m = n$. Therefore, R is **anti-symmetric**. Also, if $m|n$ and $n|l$, then $m|l$, i.e., R is **transitive**.[4] Hence, R is a partial order relation.

Clearly R is not a total order relation since neither $2 \nmid 3$ nor $3 \nmid 2$.

---

[4]The readers who are facing some difficulty in understanding how we reached the three conclusions of reflexive, transitive, and anti-symmetry are advised to *attack* the definition of divisibility. We know that $m|n$ if and only if there is an integer k such that $n = mk$. Write all the divisibility relations mentioned in terms of this definition to notice that all conclusions are indeed true.

From this example, it is evident that there are many ways to define a partial order relation on a set. We have exhibited (explicitly) two different partial orders on $\mathbb{N}$, one being the natural order ($\leq$) and the other being the divisibility relation ($|$).

Often Hasse diagrams are used to represent partially ordered sets. In such diagrams, we "join" elements which are related to each other through the partial order. Hasse diagrams of the partially ordered sets discussed above are shown in the Figure 1.1a–c.

As an immediate consequence of partial order, we make the following definitions for the elements of a POSET.

Definition 1.18 (Least element). *An element $x \in X$ is called the least element of X if $\forall y \in X$, we have $x \leq y$.*

Definition 1.19 (Greatest element). *An element $x \in X$ is called the greatest element of X if $\forall y \in X$, we have $y \leq x$.*

Definition 1.20 (Maximal element). *An element $m \in X$ is called a maximal element of X if $\forall x \in X$, we have $m \leq x \Rightarrow m = x$.*

Definition 1.21 (Minimal element). *An element $m \in X$ is called a minimal element of X if $\forall x \in X$, we have $x \leq m \Rightarrow m = x$.*

Note.

1. The definition of maximal element suggests us that there is no element greater than this element. Some elements may still be incomparable to this element. Similarly, the definition of minimal element suggests that there is no element smaller than this element.

2. An alternative way to define maximal element is, "An element $m \in X$ is called a maximal element if $\forall y \in X$, either y is not comparable to m or $y \leq m$". Similarly, we can define minimal element in this fashion.

3. The definition of least (greatest) element suggests that this element is smaller (greater) than all other elements of the POSET.

Theorem 1.5. *The least and the greatest elements of a POSET, if they exist, are unique.*

Proof. Let $(X, \leq)$ be a POSET and if possible, let $l_1, l_2 \in X$ be least elements and $g_1, g_2 \in X$ be greatest elements. Then, considering l_1 as a least element, we have $l_1 \leq l_2$. Similarly considering l_2 as a least element, $l_2 \leq l_1$. Since $\leq$ is anti-symmetric, we have $l_1 = l_2$.

In the same fashion, considering g_1 as a greatest element, we have $g_2 \leq g_1$ and considering g_2 as a greatest element, we get $g_1 \leq g_2$. Therefore, $g_1 = g_2$. Hence, the least element and greatest element of a set, if they exist, are unique. □

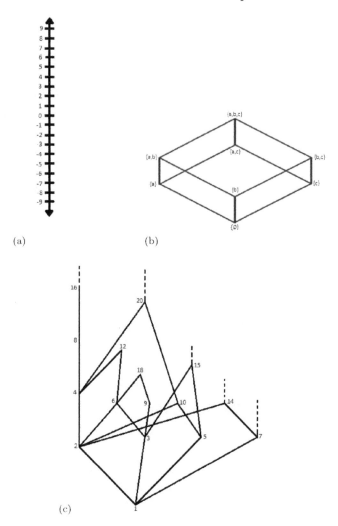

FIGURE 1.1: Hasse diagrams for different POSETs. (a) Hasse diagram for $\mathbb{R}$ with usual ordering. (b) Hasse diagram for $P(X)$ with $X = \{a, b, c\}$ and ordered by subsethood. (c) Hasse diagram for $\mathbb{N}$ with ordering given by divisibility.

Exercise 1.15. Give an example of a POSET where the least and greatest elements do not exist.

Exercise 1.16. Give an example of a POSET where there are more than one maximal elements and more than one minimal elements.

If we look at the set of natural numbers, $\mathbb{N}$, we have a least element, namely 1 but no greatest element. This is because if there was a greatest element, say

$g \in \mathbb{N}$, then $g + 1$, which is a natural number, would be less than or equal to g, a contradiction! In fact, any non-empty subset of $\mathbb{N}$ always has a least element but may or may not have a greatest element. Taking this property, we define a new property of partially ordered sets.

Definition 1.22 (Well ordered sets). *For a given POSET $(X, \leq)$, if every subset of X has a least element, then X is said to be well-ordered.*

Exercise 1.17. As an immediate consequence of the definition of well-ordered sets, we can see that every well-ordered set has a least element. Is the converse true, i.e., is every POSET with a least element well-ordered?

Since we have been generalizing our experiences with $\mathbb{R}$ and its usual order on arbitrary partially ordered sets, the next natural thing to ask for is a generalization of upper bound, lower bound, supremum, and infimum. Since these terminologies are defined for sets (subsets of $\mathbb{R}$), we shall do the same for subsets of a POSET. For the definitions, we shall consider $(X, \leq)$ as our POSET and $A \subseteq X$.

Definition 1.23 (Upper bound). *An element $u \in X$ is said to be an upper bound of A if $\forall x \in A$, we have $x \leq u$.*

Definition 1.24 (Lower bound). *An element $l \in X$ is said to be a lower bound of A if $\forall x \in A$, we have $l \leq x$.*

We note here that there can be more than one upper (or lower) bound of a set, just as in the case of $\mathbb{R}$ with usual ordering.

Definition 1.25 (Supremum). *An element $l \in X$ is said to be the supremum or least upper bound of A if it is the least element of the set of upper bounds of A. It is often denoted by $\sup A$ or $lub\,(A)$.*

Definition 1.26 (Infimum). *An element $g \in X$ is said to be the infimum or greatest lower bound of A if it is the greatest element of the set of lower bounds of A. It is often denoted by $\inf A$ or $glb\,(A)$.*

It is clear from the definitions that a set may or may not have bounds, supremum or infimum.

If we go back to $\mathbb{R}$, we make another note that every non-empty subset of $\mathbb{R}$ which is bounded above (has an upper bound in $\mathbb{R}$) has a supremum. We call this the order completeness property (or the LUB property). In an abstract setting, a POSET $(X, \leq)$ is said to have the order completeness property axiom if and only if every non-empty subset of X which has an upper bound in X, has a supremum.

Another observation we can make out of the definitions is that since supremum and infimum are the least and greatest elements of some sets, they are unique (Theorem 1.5).

Finally, we shall generalize a last concept that arises due to the ordering on $\mathbb{R}$, namely *intervals*. We first see that intervals in $\mathbb{R}$ are sets that satisfy certain properties. In particular, as a part of an elementary analysis course, one may have observed that there are only the following types of intervals: $[a,b]$, $[a,b)$, $(a,b]$, (a,b), (a,∞), $[a,\infty)$, $(-\infty,b)$, $(-\infty,b]$, and $\mathbb{R}$, where $a \leq b$ (or more preferably, $a < b$). Here, the first four intervals are bounded intervals in $\mathbb{R}$ and the next five intervals are unbounded intervals in $\mathbb{R}$. On a similar note, we shall define intervals in our POSET $(X, \leq)$ for two elements $a, b \in X$ with $a \leq b$ as follows:

$$[a,b] = \{x \in X | a \leq x \text{ and } x \leq b\},$$

$$(a,b] = \{x \in X | a \leq x \text{ and } x \neq a \text{ and } x \leq b\},$$

$$[a,b) = \{x \in X | a \leq x \text{ and } x \leq b \text{ and } x \neq b\},$$

$$(a,b) = \{x \in X | a \leq x \text{ and } x \leq b \text{ and } x \neq a \text{ and } x \neq b\}.$$

Here, we must note that to define the unbounded intervals, we cannot use the symbols $-\infty$ and ∞ since they are specifically used in the context of number systems. However, in a similar manner, unbounded intervals can be defined by just giving one of the bounds and not mentioning anything about the other. Let us look at an example to see how can we actually construct intervals in any partially ordered set.

Example 1.9 (Intervals in $\mathbb{R}^2$). Consider the set $\mathbb{R}^2$ with the partial order defined by the lexicographic ordering, i.e., $(x_1, y_1) \leq (x_2, y_2)$ if and only if $x_1 \leq x_2$ and if $(x_1 = x_2)$ then $(y_1 \leq y_2)$. Now, we shall construct the interval $[(0,0),(1,1)]$. Since this is a bounded interval of the first type (as mentioned above), we need to find all those $(x,y) \in \mathbb{R}^2$ such that $(0,0) \leq (x,y)$ and $(x,y) \leq (1,1)$. First, we ask ourselves a question: Is this interval non-empty? For if this is empty, we cannot really construct and exhibit it! We observe that we have at least elements $(0,0)$ and $(1,1)$ in our interval and hence it is non-empty.

Now, we look for other elements, if there are any in the interval. What type of elements (x,y) satisfy the first condition $(0,0) \leq (x,y)$? By the definition of ordering, we see that for $y \geq 0$, $(0,y)$ satisfies the criterion. Hence, whole of non-negative Y-axis satisfies the criterion. Other elements include (x,y) for $x > 0$. Therefore, the totality of elements (x,y) that satisfy $(0,0) \leq (x,y)$ is the right half plane (open) including the non-negative Y-axis.

Similarly, if we look at the elements that satisfy the second criterion, we see that any element $(1,y)$ for $y \leq 1$ satisfies the criterion. Other elements include (x,y) for $x < 1$. Thus, the totality of elements which satisfy the second criterion is the part of plane left of the line $x = 1$ (open) including the line $x = 1$ for $y \leq 1$.

Hence, the interval $[(0,0),(1,1)]$ is the intersection of these two parts of the plane. Geometrically it is exhibited as in Figure 1.2a–c.

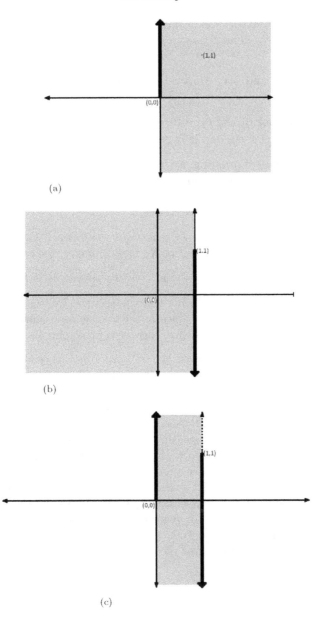

FIGURE 1.2: Constructing the interval $[(0,0),(1,1)]$ with the lexicographic ordering for $\mathbb{R}^2$. (a) Elements in $\mathbb{R}^2$ satisfying the first criterion. (b) Elements in $\mathbb{R}^2$ satisfying the second criterion. (c) The required interval.

Exercise 1.18. Construct the interval $((0,0),(1,1))$ for the POSET $(\mathbb{R}^2, \leq)$, where the ordering is defined as $(x_1, y_1) \leq (x_2, y_2)$ if and only if

$$x_1^2 + y_1^2 \leq x_2^2 + y_2^2 \text{ and } \left(x_1^2 + y_1^2 = x_2^2 + y_2^2 \Rightarrow \tan^{-1}\left(\frac{y_1}{x_1}\right) \leq \tan^{-1}\left(\frac{y_2}{x_2}\right)\right)$$

and for any element of the type $(0, y)$, we have $\tan^{-1}\left(\frac{y}{x}\right) = \frac{\pi}{2}$ for $y > 0$, $\tan^{-1}\left(\frac{y}{x}\right) = -\frac{\pi}{2}$ for $y < 0$ and it is not defined for $y = 0$.

As a consequence of ordering, we have two theorems (principles) which use the axiom of choice in their proof. We shall not prove these statements since it is beyond the scope of the text. However, the reader is encouraged to look for the proofs which are readily available.

Theorem 1.6 (Zorn's lemma). *In a partially ordered set X, if every totally ordered subset has an upper bound in X, then X has a maximal element.*

Theorem 1.7 (Well-ordering theorem). *Every set can be well-ordered.*

In fact, the axiom of choice, Zorn's lemma, and the well-ordering principle can be proved to be equivalent. Since in set theory, we assume the axiom of choice to be true, by the equivalence, these two principles also hold!

Remark. If one carefully observes, the well-ordering theorem states that any set can be well-ordered. In particular, $\mathbb{R}$ can be well-ordered. Is it true with the usual order? No! $\mathbb{R}$ has no least element in the usual order. Sometimes, this can be confusing for first-time readers. Although the well-ordering theorem guarantees a well-ordering for any set, the proof of the statement does not tell us how to construct one! Thus, we can say that there is indeed a well-ordering of $\mathbb{R}$ but we do not know, as of now, how to construct it. Similarly, the proof of Zorn's lemma is an existential proof and not a constructive proof, i.e., it guarantees us the existence of a maximal element (whenever the hypothesis holds) but does not tell us how to construct it. As a consequence, we can prove in linear algebra that every vector space has a basis (using Zorn's lemma), but we do not know, as of now, how to construct a basis for vector spaces such as $\mathscr{C}[a, b]$ or even $\mathbb{R}$ when considered as a vector space over $\mathbb{Q}$.

1.3 Functions

Earlier, we compared two given sets by means of subsethood. However, this is possible only when the elements in the given sets are of the same type (otherwise they cannot be subsets). Now, we would like to compare any two sets in terms of the number of elements they contain. In this section, we shall develop some prerequisites that will help us in comparing given sets.

Definition 1.27 (Function). *A function from a set X to a set Y, often denoted by $f : X \to Y$, is a rule that assigns to each element $x \in X$ a unique element $y \in Y$.*

Frequently, a function is referred to as a "special type of relation" between two sets X and Y, where the ordered pair (x, y) occurs exactly once for each $x \in X$, i.e., in this type of relation an element $x \in X$ can be related to only one element from Y and every element in X must be related to some element of Y. However, we shall not use this as our definition.

If for a particular $x \in X$, a function f assigns $y \in Y$ to it, we denote it as $f(x) = y$. It is read as, "The value of f at x is y". Here, y is called the *image* of x and x is called the *pre-image* of y. Also, the set X is called the *domain* of f, denoted by $\mathscr{D}(f)$ and the set Y is called the *codomain* of f. The set of all the images of elements of X is called the *range* of f, denoted by $\mathscr{R}(f) = \{f(x) \in Y \mid x \in X\}$. We shall also use the notation $f(X)$ to denote the range of f.

We note here that although the definition of function demands every element of X to be associated to a single (unique) element of Y, it does not specify as to how many elements of X can be associated to a single element of Y, neither does it specify how many elements in Y are assigned to elements of X. This gives rise to two new definitions.

Definition 1.28 (Injection). *A function $f : X \to Y$, is said to be injective, or one-to-one, if $f(x_1) = f(x_2) \Rightarrow x_1 = x_2$.*

Definition 1.29 (Surjection). *A function $f : X \to Y$, is said to be surjective, or onto, if for any $y \in Y$, $\exists x \in X$, such that $f(x) = y$.*

Remark.

1. The contrapositive of the definition of injection is $x_1 \neq x_2 \Rightarrow f(x_1) \neq f(x_2)$. This suggests that for injective functions, distinct elements of X are assigned distinct elements of Y. Sometimes, this is also taken as a definition of injective functions.

2. A function which is both injective and surjective is called a **bijection** or **bijective function**.

Exercise 1.19. Is $f : \mathbb{R} \to \mathbb{R}$ defined as $\forall x \in \mathbb{R}$, $f(x) = \frac{1}{x}$ a function? Justify.

Exercise 1.20. Give an example of a function $f : \mathbb{R} \to \mathbb{R}$ which is

1. Injection but not surjection.

2. Surjection but not injection.

3. Neither surjection nor injection.

4. Bijection.

Exercise 1.21. If possible, exhibit a function with a non-empty domain and empty co-domain. If it is not possible, give a justification.

Before moving ahead, let us answer one question: When are two functions equal? From a natural intuition of equality, we can define that two functions $f : X \rightarrow Y$ and $g : X \rightarrow Y$ are equal if and only if $\forall x \in X$, we have $f(x) = g(x)$. This means that not only the domain of these functions should be the same, but also at each point, these two functions must take the same values.

1.3.1   Composition of functions

If we are given two functions $f : X \rightarrow Y$ and $g : Y \rightarrow Z$, we wish to construct another function $h : X \rightarrow Z$ which makes use of the given ones. In other words, we know how the elements of X are associated to element(s) of Y and similarly, we know how elements of Y are associated to element(s) of Z. What is more natural than to assign each element x of X, that element z of Z which is assigned, through the function g, to the image y in Y of x under the function f? We call such an assignment *composition* of the maps. Thus, for given f, g, the composition is defined as $\forall x \in X$, $(g \circ f)(x) = g(f(x))$, where $g \circ f$ is called the *composite map*. A diagrammatic representation of composite maps is shown in Figure 1.3.

A natural question to ask here is that what properties of f and g are preserved through their composition? We answer this question in the theorem that follows.

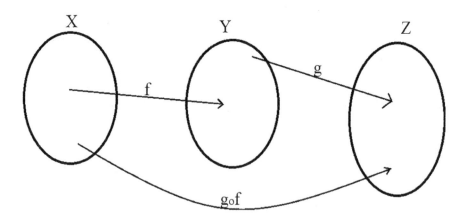

FIGURE 1.3: Diagrammatic representation of composition of functions.

Theorem 1.8. *Let $f : X \to Y$ and $g : Y \to Z$ be functions.*

1. *If f and g are injective, then so is $g \circ f$.*

2. *If f and g are surjective, then so is $g \circ f$.*

3. *If f and g are bijective, then so is $g \circ f$.*

4. *If $g \circ f$ is injective, then f is injective.*

5. *If $g \circ f$ is surjective, then g is surjective.*

Proof. The general strategy to prove these properties is by "utilizing" the definitions we have made.

1. Since f and g are injective, we have $f(x_1) = f(x_2) \Rightarrow x_1 = x_2$ and $g(y_1) = g(y_2) \Rightarrow y_1 = y_2$. Therefore, $(g \circ f)(x_1) = (g \circ f)(x_2) \Rightarrow g(f(x_1)) = g(f(x_2)) \Rightarrow f(x_1) = f(x_2) \Rightarrow x_1 = x_2$. Hence, $g \circ f$ is also injective.

2. Since f and g are surjective, we have for any $y \in Y$, $\exists x \in X$ such that $f(x) = y$ and for any $z \in Z$, $\exists y \in Y$ such that $g(y) = z$. From these two statements, we have for any $z \in Z$, $\exists x \in X$ such that $g(f(x)) = z$. Hence, $g \circ f$ is surjective.

3. Since f and g are bijective, they are both injective and surjective. From above two parts (1) and (2), we get that $g \circ f$ is both injective and surjective and hence, bijective.

4. Since $g \circ f$ is injective, we have $(g \circ f)(x_1) = (g \circ f)(x_2) \Rightarrow x_1 = x_2$. Now, we have the following: $f(x_1) = f(x_2) \Rightarrow g(f(x_1)) = g(f(x_2)) \Rightarrow (g \circ f)(x_1) = (g \circ f)(x_2) \Rightarrow x_1 = x_2$. Hence, f is injective.

5. Since $g \circ f$ is surjective, for any $z \in Z$, $\exists x \in X$ such that $(g \circ f)(x) = z$. Therefore, for any $z \in Z$, $\exists y (= f(x)) \in Y$ such that $g(y) = z$. Hence, g is surjective.

$\square$

Exercise 1.22. Give examples of functions f and g such that

1. $g \circ f$ is injective but g is not.

2. $g \circ f$ is surjective but f is not.

3. $g \circ f$ is bijective but neither is f nor is g.

It is natural to ask now, what properties does composition satisfy? Is it associative? Commutative? Is there any function whose composition with any other function gives back the other function? Clearly, the answer to the last two questions is no! In general, notice that for arbitrary sets X, Y, Z, both $f \circ g$ and $g \circ f$ may not be defined. However, we shall answer the first question.

Theorem 1.9. *If $f : W \to X$, $g : X \to Y$ and $h : Y \to Z$ are functions, then $h \circ (g \circ f) = (h \circ g) \circ f$. In other words, composition of functions is associative.*

Proof. Let $w \in W$. Then,

$$(h \circ (g \circ f))(w) = h((g \circ f)(w))$$
$$= h(g(f(w)))$$
$$= (h \circ g)(f(w))$$
$$= ((h \circ g) \circ f)(w)$$

Since $w \in W$ was arbitrary, we have $\forall w \in W$, $((h \circ g) \circ f)(w) = (h \circ (g \circ f))(w)$. Hence, $h \circ (g \circ f) = (h \circ g) \circ f$. $\qquad\square$

Exercise 1.23. Let f and g be functions such that both $f \circ g$ and $g \circ f$ are well-defined. Is it true, in general, that $f \circ g = g \circ f$? In other words, whenever the composition from both sides is well-defined, is it commutative?

1.3.2 Inverse of a function

In the previous sections, we have seen the general idea of a function and even some special classes, namely injective, surjective, and bijective functions. Of these three, bijective functions have a very special property that we will be exploring in this section. For bijective functions, every element of the domain set has a unique image and every element of the codomain has a unique pre-image. Thus, if $f : X \to Y$ is a bijective function, then for each $y \in Y$, $\exists! x \in X$ such that $f(x) = y$.[5] We now define another mapping $g : Y \to X$ given by for each $y \in Y$, $g(y) = x$ where $f(x) = y$. Is this a well-defined function? By well-defined, we mean to ask does every element of Y have a unique image in X, or is there some element which have two or more images (or no image) under g? If an element $y \in Y$ has two images x_1 and x_2, then by the definition of g, $f(x_1) = y = f(x_2)$ and by the fact that f is bijective, and hence in particular, injective, $x_1 = x_2$. Thus, every element of Y has at most one image in X. Is there any element in Y which has no image under g? If an element $y \in Y$ has no image under g, then by the definition of g, it would imply that y is not an image of any element of X under f, i.e., f is not a surjection. But, this contradicts the fact that f was considered to be a bijection. Hence, under g every element in Y has a unique image in X, i.e., the function g is well-defined.

Thus, we have seen that if we have a bijection from X to Y, then we can define another function from Y to X, which practically takes back the images under f to their pre-images. Such a function is called the *inverse* of f.

We now make a formal definition for *inverse* of a function.

---

[5]The symbol $\exists!$ is read as "there exists a unique".

Definition 1.30 (Inverse). *Given a bijective function $f : X \to Y$, a function $g : Y \to X$ is said to be the inverse of f if $\forall y \in Y$, $g(y) = x$, where $f(x) = y$. We denote the inverse of f by f^{-1} and write $x = f^{-1}(y)$, whenever $f(x) = y$.*

Exercise 1.24. Give two examples of a function $f : X \to Y$ for which the mapping $g : Y \to X$ defined as above is not a well-defined function. Of the two examples given, one example should be for the failure of well-defined g due to an element having two (or more) distinct images under g and the other example should be for the failure of well-defined g due to an element having no image.

It is natural to ask about the properties of inverse of a function. Is it injective? Surjective? We answer these questions in the following theorem.

Theorem 1.10. *If $f : X \to Y$ is a bijection, then $f^{-1} : Y \to X$ is also a bijection.*

Proof. First, we prove that f^{-1} is injective. If $f^{-1}(y_1) = f^{-1}(y_2) = x$, then since f is well-defined, we have $f(x) = y_1 = y_2$. Hence, f^{-1} is injective.

Now, since f is well-defined, $\forall x \in X$, $\exists! y \in Y$ such that $f(x) = y$. Hence, by the definition of inverse $\forall x \in X$, $\exists y \in Y$ such that $f^{-1}(y) = x$, i.e., f^{-1} is surjective.

Therefore, f^{-1} is bijective. $\qquad\qquad\square$

If we carefully observe, in the text, we have been using the word *"the"* for inverse of a function. Is it justified? We answer this in the following theorem.

Theorem 1.11. *Inverse of a function is unique.*

Proof. Let $f : X \to Y$ be a bijection and $g : Y \to X$ and $h : Y \to X$ be inverses of f. Let $y \in Y$. If $g(y) = x_1$ and $h(y) = x_2$, then by the definition of inverse $f(x_1) = y = f(x_2)$. Since f is a bijection, in particular, it is injective. Hence, $x_1 = x_2$, i.e., $g(y) = h(y)$. Since $y \in Y$ was arbitrarily chosen, we have $g = h$, i.e., the inverse of a function is unique. $\qquad\square$

Thus, we have justified the use of *"the inverse"*. Another natural question to ask is about the result when function is composed with its inverse. Before answering it, we define a special kind of function.

Definition 1.31 (Identity function). *The identity function is defined as, id : $X \to X$, where $\forall x \in X$, $id(x) = x$.*

Theorem 1.12. *If $f : X \to Y$ is a bijection, then $f \circ f^{-1} = id_Y$ and $f^{-1} \circ f = id_X$, where id_X and id_Y denote identity functions on the sets X and Y respectively.*

Proof. By the definition of f^{-1}, $\forall y \in Y$, $f^{-1}(y) = x$, where $f(x) = y$.

Hence, $\forall y \in Y$, $(f \circ f^{-1})(y) = f(f^{-1}(y)) = f(x) = y$. Therefore, $f \circ f^{-1} = id_Y$.

Using the same argument, $\forall x \in X$, $(f^{-1} \circ f)(x) = f^{-1}(f(x)) = f^{-1}(y) = x$. Therefore, $f^{-1} \circ f = id_X$. $\qquad\qquad\square$

Exercise 1.25. Let $f : X \to Y$ and $g : Y \to X$ be functions. If $f \circ g = id_Y$ and $g \circ f = id_X$, then prove that f and g are bijections and $g = f^{-1}$. Hence conclude that inverse of inverse is the function itself.

Exercise 1.26. Give an example of functions $f : X \to Y$ and $g : Y \to X$ such that $f \circ g = id_Y$ but f is not a bijection. Similarly, give an example where $g \circ f = id_X$ but f is not a bijection. (In both the cases, g cannot be the inverse of f.)

Now that we have formulated quite a lot of theory about bijective functions and their inverses, let us try to construct bijections between sets which are known to us.

Example 1.10 (Bijection between intervals). As a start, let us consider a few cases for the type of intervals we can have. We may have bounded, unbounded, closed, or open (or their combinations) intervals. In this example, we shall take closed and bounded intervals. The reader is encouraged to try forming bijections between intervals not included in this example.

First, let us consider two intervals $[0, 1]$ and $[0, 4]$. How can we form a bijection between them? First we see that one end of the interval is fixed while the *length* of the interval is made 4 times the original one. Thus, a natural way to form a function between these two intervals is $f : [0, 1] \to [0, 4]$ given by $\forall x \in [0, 1], f(x) = 4x$. Under this function, 0 is associated to 0 and 1 is associated to 4. Is this a bijection? Before asking about bijection, one should always ask: Is this function well-defined? In this case, it is! Every element in $[0, 1]$ has a unique image. Is f injective? To check if it is injective, we consider $f(x_1) = f(x_2)$, i.e., $4x_1 = 4x_2$, which implies $x_1 = x_2$ (since $4 \neq 0$). Thus, f is injective. Is f surjective? If $y \in [0, 4]$, then $0 \leq y \leq 4$, i.e., $0 \leq \frac{y}{4} \leq 1$ (since $4 > 0$) and $f\left(\frac{y}{4}\right) = y$. Thus, $\forall y \in [0, 4], \exists x \left(= \frac{y}{4}\right) \in [0, 1]$ such that $f(x) = y$. Hence, f is also surjective and therefore a bijection.

Now we have a way to construct bijection between two intervals when one end of the interval is fixed to 0 and the length is changed. Such a bijection is also called *scaling*. This is shown in Figure 1.4.

Now, suppose we have the intervals $[0, 1]$ and $[2, 3]$. First we observe that the lengths of the two intervals are the same. However, their end points are shifted. One can ask: by how much have they shifted? A simple observation

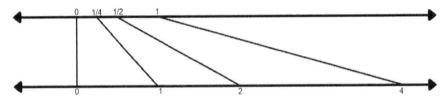

FIGURE 1.4: Scaling between the intervals $[0, 1]$ and $[0, 4]$.

leads to the answer that the complete interval seems to be shifted by 2 units. How can we describe this effect as a function? We construct $f : [0, 1] \rightarrow [2, 3]$ as $\forall x \in [0, 1]$, $f(x) = x + 2$. The reader should check that this is in fact, a bijection!

Therefore, we have a method of constructing bijection between any interval with unit length and the interval $[0, 1]$. Such shifting is called *translation*. This is shown in Figure 1.5.

But, we would like to have a more general method to construct bijection between any two intervals $[a, b]$ and $[c, d]$, where $a < b$ and $c < d$. To have it, first we have to somehow get $[a, b]$ to $[0, 1]$ and then by using the above methods and composition of functions, we can have the desired bijection.

First, we construct a bijection $f : [a, b] \rightarrow [0, 1]$. To do so, we first translate our interval $[a, b]$ so that the left end coincides with 0 by constructing a function $f_1 : [a, b] \rightarrow [0, b - a]$ given by $\forall x \in [a, b]$, $f_1(x) = x - a$. Then, we scale the interval $[0, b - a]$ to $[0, 1]$ by the bijection $f_2 : [0, b - a] \rightarrow [0, 1]$ given by $\forall x \in [0, b - a]$, $f_2(x) = \dfrac{x}{b - a}$. Hence, the required bijection f is $f_2 \circ f_1$.

Now, we wish to have a bijection $g : [0, 1] \rightarrow [c, d]$. To get g, we perform the exact reverse process as above. First we scale $[0, 1]$ to $[0, d - c]$ by means of the bijection $g_1 : [0, 1] \rightarrow [0, d - c]$ given by $\forall x \in [0, 1]$, $g_1(x) = (d - c)x$. Now, we translate the interval $[0, d - c]$ to $[c, d]$ by means of the bijection $g_2 : [0, d - c] \rightarrow [c, d]$ given as $\forall x \in [0, d - c]$, $g_2(x) = x + c$. Hence, the required bijection $g = g_2 \circ g_1$.

Finally, the bijection between $[a, b]$ and $[c, d]$ is given by $(g \circ f)(x) = \left(\dfrac{d-c}{b-a}\right) x + c$, as shown in Figure 1.6.

Exercise 1.27. In Example 1.10 what would happen if $a = b$ or $c = d$?

Exercise 1.28. Give a bijection for open bounded intervals.

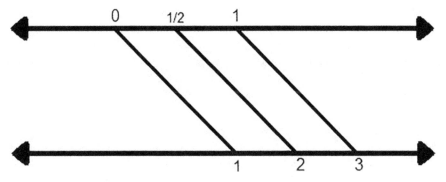

FIGURE 1.5: Translation between the intervals $[0, 1]$ and $[2, 3]$.

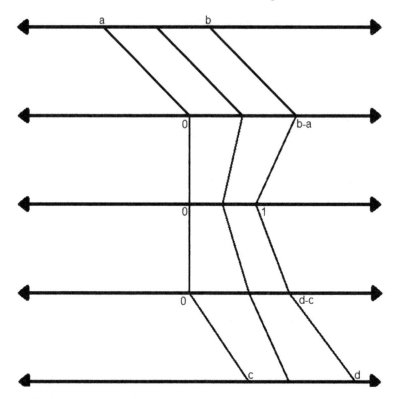

FIGURE 1.6: Bijection between any two intervals $[a, b]$ and $[c, d]$.

From above, we have method(s) to construct bijection between two closed (open) and bounded intervals. Is it possible to have a bijection between one open interval and one closed interval? In particular, is it possible to have a bijection between $[0, 1]$ and $(0, 1)$? We answer this question in the next example.

Before beginning with the construction of aforementioned bijection, we shall see an interesting property of the set of natural numbers, $\mathbb{N}$. Suppose we want to construct a bijection between $\mathbb{N}$ and $\mathbb{N} \cup \{0, -1\}$. Is it possible? A simple thought would tell us that if we map every natural number to a natural number ahead of it by 2 units, we are left with 1 and 2 which can be assigned to 0 and -1 respectively, thereby making a bijection.

Example 1.11. For the interval $(0, 1)$, we see that $\frac{1}{n} \in (0, 1)$ and $\frac{1}{n} \in [0, 1]$ for every $n \in \mathbb{N} \setminus \{1\}$. The only two numbers we cannot accommodate are 0 and 1 from the interval $[0, 1]$. To accommodate those two, we shift $\frac{1}{n}$ by two units. Thus, 0 is associated to $\frac{1}{2}$ and $\forall n \in \mathbb{N}$, $\frac{1}{n}$ is associated to $\frac{1}{n+2}$. All other elements from $[0, 1]$ are mapped to themselves in $(0, 1)$. Thus, the required bijection is $f : [0, 1] \to (0, 1)$ given by

$$f(x) = \begin{cases} \dfrac{1}{2}, & x = 0. \\ \dfrac{1}{n+2}, & x = \dfrac{1}{n} \text{ for } n \in \mathbb{N}. \\ x, & \text{otherwise.} \end{cases}$$

The reader is encouraged to check that f is indeed a bijection and find its inverse.

Exercise 1.29. Exhibit a bijection between the intervals (a, b) and $[c, d]$ where $a < b$ and $c < d$. What will happen in the case when $a = b$ or $c = d$?

As a closing to this article of bijections and inverses, we try to answer a question: Can we have a bijection between a set X and its power set $\mathcal{P}(X)$? Before answering the question, we look at a classic example, known as the barber's paradox.

Example 1.12 (Barber's paradox). Suppose in a village, a barber is a man who shaves those and only those who do not shave themselves. The question is: Who shaves the barber? Here, we have a definition of barber and we cannot deviate from it while analyzing the situation. We observe that the barber has only two choices: either he shaves himself or he doesn't. If he shaves himself, he violates the definition. This is because he is supposed to shave only those who do not shave themselves! However, if he does not shave himself it is concluded that another barber must shave him. But this means that he does not shave himself and being a barber, he should have shaven himself. Hence, in such a definition, a barber cannot be shaven!

It will not be out of context to give an elegant and symbolic proof using the same paradox.

Theorem 1.13 (Cantor's theorem). *We cannot have a bijection between a set X and its power set $\mathcal{P}(X)$.*

Proof. We shall prove this theorem by contradiction. So, let $f : X \to \mathcal{P}(X)$ be a bijection. First, observe that to each element of X, there is associated a unique subset of X. Now, we look at one subset of X given by the definition

$$B = \{x \in X \,|\, x \notin f(x)\}.$$

Such a definition is possible since $f(x)$ is a set. Now, since f is bijection, in particular, it is surjective. Hence, $\exists b \in X$, such that $f(b) = B$. Since $B \subseteq X$, we have two choices: $b \in B$ or $b \notin B$. If $b \in B$, then by the definition of B, $b \notin f(b) = B$, which is a contradiction. Next if $b \notin B = f(b)$, we get by definition of B that $b \in B$, again a contradiction. Since there cannot be an element which is both in a set and not in it, our assumption was wrong, i.e., f cannot be a bijection. □

1.3.3   Images of sets under functions

Until now, we have been looking at what happens to elements, given a function. Now, we move a step ahead to look at what happens to sets (subsets), given a function. Such analysis will be useful in the concepts that follow, especially when we consider special properties of metric spaces later. In this section, we try to cover as many proofs as possible with all the explanation required. However, the reader is advised to complete some of the proofs which will be left as an exercise.

In all the text that follows, we shall assume that a function $f : X \to Y$ is given. Also, before moving ahead, we must answer an important question: How do we define the image of a (sub)set (of X)? It is quite natural to make use of the elements in the set. Thus, the definition of image of a set $A \subseteq X$ is given as follows

$$f(A) = \{f(x) \in Y | x \in A\}.$$

Note here that when we write $f(x)$ in Y, we mean to say "collect all those elements of Y, whose pre-image(s) under f are in A".

Whenever we talk about subsets, there are two special subsets that should always come to our mind: the empty set, $\emptyset$, and the whole universe, X. Clearly, $f(X)$ is the range of the function and we have already defined it. We need to see what is $f(\emptyset)$. What kind of elements are in $f(\emptyset)$? Are there any elements at all?

To answer these questions, let us consider $y \in f(\emptyset)$. But, by the definition of images of sets, this would mean $\exists x \in \emptyset$ such that $f(x) = y$, which is impossible! Hence, $f(\emptyset) = \emptyset$. Also, when can we say that for a set $A \subseteq X$, $f(A) = \emptyset$? If $A \neq \emptyset$, then $\exists x \in A$ for which $f(x) \in f(A)$, by the definition. Hence, if $f(A) = \emptyset$, then $A = \emptyset$.

Therefore, we have proved here $f(A) = \emptyset$ *if and only if* $A = \emptyset$.

Let us see what happens to the relation of subsethood when a function is applied on sets. If $A_1 \subseteq A_2 \subseteq X$, can we conclude $f(A_1) \subseteq f(A_2)$? To conclude so, we need to show that every element in $f(A_1)$ is also in $f(A_2)$. Let $y \in f(A_1)$. Then, $\exists x \in A_1$ such that $f(x) = y$. Since $A_1 \subseteq A_2$, we have $\exists x \in A_2$ such that $f(x) = y$. Therefore, $y \in f(A_2)$, i.e., $f(A_1) \subseteq f(A_2)$.

When can we say that $f(A_1) = f(A_2)$? Clearly, if $A_1 = A_2$, we have $f(A_1) = f(A_2)$. Can we have an example where $A_1 \subset A_2$ and $f(A_1) = f(A_2)$? The next exercise asks the reader to get such an example.

Exercise 1.30. Construct a function $f : X \to Y$ (your choice of X and Y) and two subsets A_1 and A_2 of X such that $A_1 \subset A_2$ and $f(A_1) = f(A_2)$.

Next we look at what happens to union. If we have $A_1, A_2 \subseteq X$, then we expect $f(A_1 \cup A_2) = f(A_1) \cup f(A_2)$ to hold. We try to prove it. If at some point during the proof, we get stuck, we can look for a way of constructing a counterexample to disprove the statement. To prove equality of two sets, we need to show that all the elements in one set are also in the other. First, let $y \in f(A_1 \cup A_2)$. This means that $\exists x \in A_1 \cup A_2$ such that $f(x) = y$, which further means that $\exists x \in A_1$ such that $f(x) = y$ or $\exists x \in A_2$ such that $f(x) = y$. By our definition of images of sets, we reach a conclusion that $y \in f(A_1)$ or $y \in f(A_2)$, i.e., $y \in f(A_1) \cup f(A_2)$. Thus, we have obtained a subsethood relation, $f(A_1 \cup A_2) \subseteq f(A_1) \cup f(A_2)$. Similarly, we take an arbitrary $y \in f(A_1) \cup f(A_2)$, i.e., $y \in f(A_1)$ or $y \in f(A_2)$. This would mean, $\exists x \in A_1$ such that $f(x) = y$ or $\exists x \in A_2$ such that $f(x) = y$, i.e., $\exists x \in A_1 \cup A_2$ such that $f(x) = y$. Hence, we also have $y \in f(A_1 \cup A_2)$, i.e., $f(A_1) \cup f(A_2) \subseteq f(A_1 \cup A_2)$. Therefore, we have proved our statement.

In the same way, we would want $f(A_1 \cap A_2) = f(A_1) \cap f(A_2)$. Let $y \in f(A_1 \cap A_2)$. Then, $\exists x \in A_1 \cap A_2$ such that $f(x) = y$, i.e., $\exists x \in X$ such that $x \in A_1$ and $x \in A_2$ and $f(x) = y$. Thus, $y \in f(A_1)$ and $y \in f(A_2)$, i.e., $y \in f(A_1) \cap f(A_2)$. Therefore, we have obtained a subsethood relation, $f(A_1 \cap A_2) \subseteq f(A_1) \cap f(A_2)$. Now, we would try to prove the reverse subsethood. Let $y \in f(A_1) \cap f(A_2)$, i.e., $y \in f(A_1)$ and $y \in f(A_2)$. Then, $\exists x_1 \in A_1$ such that $f(x_1) = y$ and $\exists x_2 \in A_2$ such that $f(x_2) = y$. Now, we are stuck! Nothing guarantees us that $x_1 = x_2$ and therefore, a common element from $A_1 \cap A_2$ has the image y need not be obtained. Therefore, in general, we cannot have the reverse subsethood. However, this try also gives us an idea about construction of a counterexample.

Example 1.13. Consider a function $f : \mathbb{R} \to \mathbb{R}$ defined as $f(x) = x^2$ for every $x \in \mathbb{R}$. We take two subsets $A_1 = [-2, -1] \subseteq \mathbb{R}$ and $A_2 = [1, 2] \subseteq \mathbb{R}$. Clearly, $A_1 \cap A_2 = \emptyset$ and hence, $f(A_1 \cap A_2) = \emptyset$. Now, $f(A_1) = [1, 4] = f(A_2)$. Thus, $f(A_1) \cap f(A_2) = [1, 4] \neq \emptyset$. Thus, we get $f(A_1 \cap A_2) \subset f(A_1) \cap f(A_2)$. This is shown in Figure 1.7.

In the try that we made above to prove $f(A_1 \cap A_2) = f(A_1) \cap f(A_2)$, we could not prove one subsethood because in general, two different elements (from two different sets) can have the same image. If we do not allow this to happen by defining f in such a way, then is it possible to get the equality?

Exercise 1.31. Prove that if $f : X \to Y$ is a function, then for any two subsets $A_1, A_2 \subseteq X$, $f(A_1 \cap A_2) = f(A_1) \cap f(A_2)$ if and only if f is injective.

Next, we would like to establish a relation between the image of complement and complement of image, i.e., if $A \subseteq X$, then we would like to relate $f(A^c)$ and $(f(A))^c$. First, we try to see if $f(A^c) \subseteq (f(A))^c$ is possible. As a try,

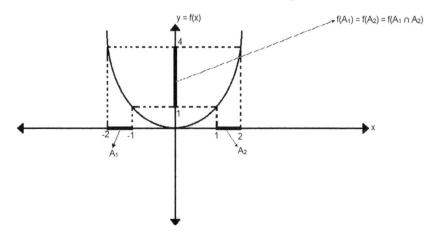

FIGURE 1.7: Image of intersection need not necessarily contain intersection of images of sets.

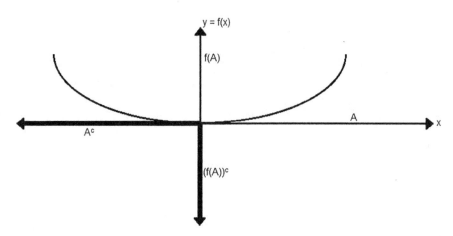

FIGURE 1.8: Image of complement and complement of image need not contain one another.

let $y \in f(A^c)$. Then, $\exists x \in A^c$ such that $f(x) = y$. Does this mean that $y \in (f(A))^c$? If so, then $y \notin f(A)$ would further mean $\forall x \in A$, $f(x) \neq y$. Is it true in general? The next example answers this question negatively.

Example 1.14. For the function defined in Example 1.13, we consider $A = [0, \infty)$. Then, $A^c = (-\infty, 0)$. Thus, $f(A^c) = (0, \infty)$. Also, $f(A) = [0, \infty]$ and hence $(f(A))^c = (-\infty, 0)$. Clearly, $f(A^c) \nsubseteq (f(A))^c$. Here, we can also see that $(f(A))^c \nsubseteq f(A^c)$. This is shown in Figure 1.8.

Now, we try to look at conditions when we can get the required subsethood. We first observe that in our try, we were not guaranteed the non-existence of another element from X whose image is y. This is because, in general, we do not consider f to be injective. What happens when f is injective? Do we get the desired subsethood? Clearly, if f is injective and $\exists x_1 \in A^c$ such that $f(x_1) = y$ and $\exists x_2 \in A$ such that $f(x_2) = y$, we would have $x_1 = x_2 \in A$ and $x_1 = x_2 \in A^c$, which is not possible. Hence, if f is injective, then $f(A^c) \subseteq (f(A))^c$. One can, at this point, ask the dual question: What if for every subset $A \subseteq X$, we have $f(A^c) \subseteq (f(A))^c$? Can we conclude that f is injective?

Let $x_1 \neq x_2$ and $f(x_1) = f(x_2)$ hold. Consider $A = \{x_1\}$. Clearly, $x_2 \in A^c$. Hence, $f(x_1) = f(x_2) \in f(A^c) \subseteq (f(A))^c$, i.e., $f(x_1) \notin f(A)$, which is a contradiction! Hence, $f(x_1) = f(x_2) \Rightarrow x_1 = x_2$, i.e., f is injective. Therefore, we have characterized injective functions in another way as: *For any set $A \subseteq X$, $f(A^c) \subseteq (f(A))^c$ if and only if f is injective.*

Looking at Example 1.14, we can say that $(f(A))^c \not\subseteq f(A^c)$ in general. When can we get this subsethood? Is it true when f is injective? Let us try! Suppose that f is injective and let $y \in (f(A))^c$. This means that $y \notin f(A)$, i.e., $\forall x \in A$, $f(x) \neq y$. Does this mean that $\exists x \in A^c$ such that $f(x) = y$? In general, the answer is no! As seen in Example 1.14, $(f(A))^c \not\subseteq f(A^c)$ because there is no real number whose square is negative. This gives us a hint that if, instead of an injection, we have a surjection, the subsethood might hold. If f is surjective and $\forall x \in A$, $f(x) \neq y$, then $\exists x \in A^c$ such that $f(x) = y$, i.e., $y \in f(A^c)$, thus giving the relation. Again asking the dual question, if the subsethood holds for every set A, then we would want f to be surjective.

To prove a function to be surjective, we just need to prove that the range is the same as the codomain, i.e., $f(X) = Y$. Clearly, $f(X) \subseteq Y$. Now, consider $A = \emptyset$. Then, $A^c = X$. Therefore, from our hypothesis, $Y = \emptyset^c = (f(\emptyset))^c \subseteq f(\emptyset^c) = f(X)$. Hence, $f(X) = Y$, i.e., f is surjective. Therefore, we have also characterized surjections as: *For any set $A \subseteq X$, $(f(A))^c \subseteq f(A^c)$ if and only if f is a surjection.* As an immediate consequence, we get the following corollary: $f(A^c) = (f(A))^c$ if and only if f is a bijection.

We conclude the discussion by summarizing as a theorem.

Theorem 1.14. *Let $f : X \to Y$ be a function and $A, A_1, A_2 \subseteq X$. Then,*

1. $f(A) = \emptyset$ *if and only if* $A = \emptyset$.

2. $f(A_1 \cup A_2) = f(A_1) \cup f(A_2)$.

3. $f(A_1 \cap A_2) \subseteq f(A_1) \cap f(A_2)$ *and* $f(A_1 \cap A_2) = f(A_1) \cap f(A_2)$ *if and only if f is injective.*

4. $f(A^c) \subseteq (f(A))^c$ *if and only if f is injective.*

5. $(f(A))^c \subseteq f(A^c)$ *if and only if f is surjective.*

6. $f(A^c) = (f(A))^c$ *if and only if f is a bijection.*

Exercise 1.32. Let $f : X \to Y$ be a function and $\{A_i \subseteq X | i \in I\}$ be an family of subsets of X indexed by the index set I. Prove that:

1. $f\left(\bigcup_{i \in I} A_i\right) = \bigcup_{i \in I} f(A_i)$.

2. $f\left(\bigcap_{i \in I} A_i\right) \subseteq \bigcap_{i \in I} f(A_i)$.

1.3.4 Inverse images of sets under functions

Earlier, we defined images for elements of the domain and pre-images for elements of the codomain, when provided with a function. In the previous section, we have seen images of sets under functions. Now, we will see pre-images of (sub)sets (of codomain). We call them inverse images and define for a subset $B \subseteq Y$, where $f : X \to Y$ is a function,

$$f^{-1}(B) = \{x \in X | f(x) \in B\}.$$

Note. Writing f^{-1} is an abuse of notation and it is not related to the inverse of the function. While the inverse of a function associates to every element in the codomain, a unique element in the domain, here, by writing $f^{-1}(B)$, we are trying to collect all the pre-images of all the elements of the set B. The definition so made is natural and makes use of the concept of pre-images of elements.

Let us try to establish relations similar to those discussed for images of sets. For the complete discussion, we shall make use of a function $f : X \to Y$ and sets $B, B_1, B_2 \subseteq Y$. First, let us see what happens to the empty set. Just as in case of images of sets, our guess is that $f^{-1}(\emptyset) = \emptyset$. If $\exists x \in f^{-1}(\emptyset)$, then $f(x) \in \emptyset$, which is not possible. Hence, $f^{-1}(\emptyset) = \emptyset$. Asking the dual question, when is $f^{-1}(B) = \emptyset$? The following example answers this question.

Example 1.15. Consider the function $f : \mathbb{R} \to \mathbb{R}$ defined by $\forall x \in \mathbb{R}$, $f(x) = x^2$. If we take $B = (-\infty, 0)$, clearly $f^{-1}(B) = \emptyset$ but $B \neq \emptyset$.

Is it possible to have some functions for which $f^{-1}(B) = \emptyset \Rightarrow B = \emptyset$? Looking at the above example, one can guess that for a function which is not surjective, this implication cannot hold. To prove it, let $y \in Y$ be an element for which there is no pre-image image (and hence, f is not a surjection). For $B = \{y\}$, we have $f^{-1}(B) = \emptyset$ and $B \neq \emptyset$. What happens when f is a surjection? Can we say that $f^{-1}(B) = \emptyset \Rightarrow B = \emptyset$? Let, if possible, $\exists y \in B$. Since f is a surjection, $\exists x \in X$ such that $f(x) = y \in B$, i.e., $x \in f^{-1}(B) = \emptyset$, a contradiction! Thus, the implication $f^{-1}(B) = \emptyset \Rightarrow B = \emptyset$ holds exactly when

f is a surjection and in no other conditions. Therefore, we have character-ized surjections in a way that $f^{-1}(B) = \emptyset \Rightarrow B = \emptyset$ *if and only if f is a surjection*. Now, let us see if subsethood is preserved in the inverse images. If $B_1 \subseteq B_2$, we would like to have $f^{-1}(B_1) \subseteq f^{-1}(B_2)$. We consider an arbitrary element $x \in f^{-1}(B_1)$. By our definition of inverse image of sets, we have $f(x) \in B_1 \subseteq B_2$. Hence, $f(x) \in B_2$, i.e., $x \in f^{-1}(B_2)$. Therefore, we conclude that subsethood is preserved.

Exercise 1.33. Is it possible to have proper subsets $B_1 \subset B_2$ such that $f^{-1}(B_1) = f^{-1}(B_2)$?

Now, we look at what happens to the union and intersection of (sub)sets under inverse images. As always, our first try will be to prove $f^{-1}(B_1 \cup B_2) = f^{-1}(B_1) \cup f^{-1}(B_2)$ and $f^{-1}(B_1 \cap B_2) = f^{-1}(B_1) \cap f^{-1}(B_2)$. If we get stuck somewhere in the proof, we get a way to find a counter example to disprove our claim!

For the first equality, let $x \in f^{-1}(B_1 \cup B_2)$. Then, $f(x) \in B_1 \cup B_2$, i.e., $f(x) \in B_1$ or $f(x) \in B_2$. Therefore, $x \in f^{-1}(B_1)$ or $x \in f^{-1}(B_2)$, i.e., $x \in f^{-1}(B_1) \cup f^{-1}(B_2)$. Hence, $f^{-1}(B_1 \cup B_2) \subseteq f^{-1}(B_1) \cup f^{-1}(B_2)$. Similarly, let $x \in f^{-1}(B_1) \cup f^{-1}(B_2)$, i.e., $x \in f^{-1}(B_1)$ or $x \in f^{-1}(B_2)$. Therefore, $f(x) \in B_1$ or $f(x) \in B_2$, i.e., $f(x) \in B_1 \cup B_2$. Hence, $x \in f^{-1}(B_1 \cup B_2)$ and the desired equality is achieved.

On similar lines, the other equality can be proved and hence is left as an exercise for the reader.

Exercise 1.34. Prove that $f^{-1}(B_1 \cap B_2) = f^{-1}(B_1) \cap f^{-1}(B_2)$.

We may now ask: What happens to complements of sets under inverse images? Let $x \in f^{-1}(B^c)$. Then, $f(x) \in B^c$, i.e., $f(x) \notin B$. Hence, $x \notin f^{-1}(B)$, i.e., $x \in (f^{-1}(B))^c$. Thus, we have proved, without imposing any conditions on f, that $f^{-1}(B^c) \subseteq (f^{-1}(B))^c$. Similarly, if $x \in (f^{-1}(B))^c$, i.e., $x \notin f^{-1}(B)$, then $f(x) \notin B$, i.e., $x \notin f^{-1}(B)$. Therefore, we have proved $f^{-1}(B^c) = (f^{-1}(B))^c$ without imposing any conditions on f. Unlike the images of sets, this equality holds true for any f.

Note. Since all the set operations, i.e., union, intersection, complements, subsethood, are preserved under the inverse images, we say that these operations are "well-behaved" under the inverse images. On the other hand, only subsethood and union are well-behaved under images of sets.

Now, we try to establish relations between images and inverse images of sets. If $A \subseteq X$ and $B \subseteq Y$, what can we say about $f^{-1}(f(A))$ and $f(f^{-1}(B))$? Let us try to see if $f^{-1}(f(A)) = A$. Let $x \in f^{-1}(f(A))$. Then, $f(x) \in f(A)$. Does this mean that $x \in A$? In general, the answer is no as can be seen in the following example.

Example 1.16. Define a function $f : \mathbb{R} \to \mathbb{R}$ as $f(x) = x^2$ for every $x \in \mathbb{R}$. Let $A = [-2, -1]$. Then, $f(A) = [1, 4]$ and $f^{-1}(f(A)) = [-2, -1] \cup [1, 2]$. Thus, we get $A \subset f^{-1}(f(A))$.

In the example, the problem occurs when we can choose an element in $f(A)$ for which we get two distinct pre-images, like 2 (in the example) for which two distinct pre-images are $\sqrt{2}$ and $-\sqrt{2}$ one of which is not in A. However, in the same example, we see that $A \subset f^{-1}(f(A))$. Is it true in general? If $x \in A$, then clearly, $f(x) \in f(A)$ and hence $x \in f^{-1}(f(A))$. Therefore, for any function f, $A \subseteq f^{-1}(f(A))$.

When can the equality be guaranteed? From the example itself, it seems that the equality did not hold because our function was not injective. Our guess is that if we consider an injective function, then the equality should hold. Let $x \in f^{-1}(f(A))$. Then, $f(x) \in f(A)$. By the definition of $f(A)$, $\exists a \in A$ such that $f(a) = f(x)$. Since f is injective, $x = a \in A$. Hence, the equality is obtained. Also, if for every set $f^{-1}(f(A)) = A$ holds, then we would like our function to be injective. Let $f(x_1) = f(x_2)$. We take $A = \{x_1\}$. Since $f^{-1}(f(A)) = A = \{x_1\}$, we have $f^{-1}(\{f(x_1)\}) = f^{-1}(\{f(x_2)\}) = \{x_1\}$, i.e., $x_2 \in \{x_1\}$ and hence $x_1 = x_2$. Therefore, we have now characterized injection in another way: $f^{-1}(f(A)) = A$ *if and only of f is injective*.

Similarly, let us try to establish $f(f^{-1}(B)) = B$. If $y \in f(f^{-1}(B))$, then $\exists x \in f^{-1}(B)$ such that $f(x) = y$. By the definition of $f^{-1}(B)$, we have $f(x) = y \in B$. Hence, for any function f, $f(f^{-1}(B)) \subseteq B$. We now try to establish the reverse subsethood. Let $y \in B$. To say that $y \in f(f^{-1}(B))$, we would require an $x \in f^{-1}(B)$ such that $f(x) = y$. Is it always possible?

Example 1.17. In Example 1.16, if we take $B = (-\infty, 0)$, then $f(f^{-1}(B)) = f(\emptyset) = \emptyset$ and hence $f(f^{-1}(B)) \subset B$.

Why did this happen? Since B had no pre-images, we could not get equality. Thus, our guess is that if f is surjective, then the equality should hold. Now, if f is surjective and $y \in B$, then $\exists x \in X$ such that $f(x) = y$. Clearly, $x \in f^{-1}(B)$ and hence $f(x) = y \in f(f^{-1}(B))$, thereby yielding the equality. Now, if $f(f^{-1}(B)) = B$ holds for every subset B of Y, can we conclude that f is surjective? Clearly, $f^{-1}(Y) = X$. Then, $Y = f(f^{-1}(Y)) = f(X)$, i.e., f is indeed surjective. Thus, we have also characterized surjections in another way: $f(f^{-1}(B)) = B$ *if and only if f is surjective*.

We conclude the discussion by summarizing it in the form of a theorem.

Theorem 1.15. *Let $f : X \to Y$ be a function, $A \subseteq X$ and $B, B_1, B_2 \subseteq Y$. Then,*

1. $f^{-1}(\emptyset) = \emptyset$.

2. $f^{-1}(B) = \emptyset \Rightarrow B = \emptyset$ *if and only if f is surjective.*

3. $B_1 \subseteq B_2 \Rightarrow f^{-1}(B_1) \subseteq f^{-1}(B_2)$.

4. $f^{-1}(B_1 \cup B_2) = f^{-1}(B_1) \cup f^{-1}(B_2)$.

5. $f^{-1}(B_1 \cap B_2) = f^{-1}(B_1) \cap f^{-1}(B_2)$.

6. $f^{-1}(B^c) = (f^{-1}(B))^c$.

7. $A \subseteq f^{-1}(f(A))$ and $A = f^{-1}(f(A))$ *if and only if f is injective.*

8. $f(f^{-1}(B)) \subseteq B$ and $f(f^{-1}(B)) = B$ *if and only if f is surjective.*

Exercise 1.35. Let $f : X \to Y$ be a function and $\{B_i \subseteq Y | i \in I\}$ be a family of subsets of Y indexed by the index set I. Prove that:

1. $f^{-1}\left(\bigcup_{i \in I} B_i\right) = \bigcup_{i \in I} f^{-1}(B_i)$.

2. $f^{-1}\left(\bigcap_{i \in I} B_i\right) = \bigcap_{i \in I} f^{-1}(B_i)$.

1.4 Countability of Sets

As promised in the previous section, we will now start comparing two sets by means of functions.

Humans, by nature, have always had a habit of counting whatever we get. Keeping this habit, we shall now try to count the number of elements in a given set. Before proceeding, we shall deal with a few important theorems in set theory which shall help us in developing theory on counting the number of elements.

Theorem 1.16. *Let A be any set, $C \subseteq A$ and $f : A \to A$ be an injective map. If $f(A) \subseteq C$, then A is bijective with C.*

Proof. First, we see that since $C \subseteq A$, we have $f(C) \subseteq f(A)$. Also, $f(A) \subseteq C \subseteq A$. So, f maps A into a smaller version of itself (with maybe some elements shifted from place). We want to harness this property of f to construct a bijection between A and C. So, first we construct a set $X_1 = A \setminus C$ and then define $X_{n+1} = f(X_n)$ for $n \in \mathbb{N}$. This is depicted in the Figure 1.9 by shaded portions of A.

Now, let $X = \bigcup_{n \in \mathbb{N}} X_n$. We define a function $g : A \to A$ as

$$g(a) = \begin{cases} f(a), & a \in X \\ a, & a \notin X \end{cases}$$

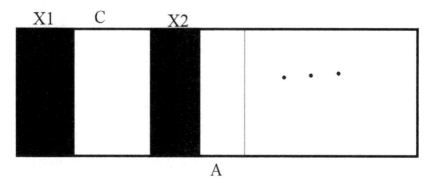

FIGURE 1.9: Pictorial representation of situation.

We claim that this is the required bijection between A and C. First, we prove that g is injective. We note that g maps X into X and X^c into X^c. This mapping is indeed injective because the individual mappings f, which maps X into X, and the identity map, which maps X^c into X^c, are injective.

Now, for any $a \in A$, either $a \in X$ or $a \notin X$. If $a \in X$, then $g(a) = f(a) \in f(A) \subseteq C$ and if $a \notin X$, then $a \in X^c \subseteq C$. Therefore, in any case, $g(a) \in C$ or, in other words $g(A) \subseteq C$.

Now, let $c \in C$ be any arbitrary element. Clearly, $c \notin X_1$, since $X_1 = A \backslash C$. Again, we have two choices here: $c \in X$ or $c \notin X$. Note that when we say $c \in X$, we mean $c \in X_n$ for some $n \geq 2$. For a pictorial representation, see Figure 1.9. If $c \in X$, then $\exists n \in \mathbb{N} \setminus \{1\}$ such that $c \in X_n$. Therefore, $\exists x \in A$ such that $g(x) = f(x) = c$. If $c \notin X$, then $g(c) = c \in C$. Thus, in any case $c \in g(A)$ (since it has a pre-image in A). Therefore, $g(A) = C$.

Therefore, we have exhibited a function $g : A \to C$ which is a bijection, and hence these two sets are bijective. $\square$

Theorem 1.17 (Schröder-Berstein). *Let A and B be two sets. If $f : A \to B$ and $g : B \to A$ are two injective functions, then A and B are bijective.*

Proof. Define a function $g \circ f : A \to A$. Clearly, $(g \circ f)(A) \subseteq g(B) \subseteq A$ and $g \circ f$ is injective. Hence, using Theorem 1.16 there is a bijection $h : A \to g(B)$. Since g is injective, it has in inverse $g^{-1} : g(B) \to B$. Therefore, $g^{-1} \circ h : A \to B$ is the required bijection. $\square$

Theorem 1.18 (Knaster-Tarski). *Let X be any set and $f : \mathcal{P}(X) \to \mathcal{P}(X)$ be a map. Let the map f satisfy the property that for every $A \subseteq B$ we have $f(A) \subseteq f(B)$. Then, $\exists S \subseteq X$ such that $f(S) = S$.*

Proof. Let $\mathscr{C} = \{C \subseteq X | C \subseteq f(C)\}$. Clearly, $\emptyset \in \mathscr{C}$ and hence $\mathscr{C} \neq \emptyset$. Now, let S be the union of all members of $\mathscr{C}$. Then, $\forall C \in \mathscr{C}$, we have $C \subseteq S$, which implies that $f(C) \subseteq f(S)$. Also, $C \subseteq f(C)$ and hence $C \subseteq f(S)$.

Therefore, $S \subseteq f(S)$ (why?). This further gives $f(S) \subseteq f(f(S))$ and therefore $f(S) \in \mathscr{C}$. But then as S is the union of all members of $\mathscr{C}$, we have $f(S) \subseteq S$ and therefore $f(S) = S$.

Here, we call S a *fixed point*[6] of f. $\qquad\square$

We note that in all the above discussion, we are concerned more about bijections rather than arbitrary functions between sets. Why so? Let us look at how we learned counting. In the very beginning, when mathematical tools were not so developed, humans used to count using pebbles. Given a set of objects, we used to assign each object a pebble. Then, we used to say that there are as many objects as we have assigned pebbles. We are trying to do the same with sets. The only difference in our case is that the sets are arbitrary and we do not really know how many elements they contain.

By means of bijection between two sets A and B, we assign to each member of A a unique member of B and no member of B is left out. This, by intuition from the experience of counting, tells us that A has as many elements as B. However, yet we have not devised any method to give the exact number of elements. We shall do so in the proceeding sections.

1.4.1   Finite sets

As discussed above, we started counting for finitely many objects. The resultant pebbles which were assigned were finite and we gave it a number (notation), say n. We shall apply the same strategy for sets. As a notation, we define for any $n \in \mathbb{N}$, $I_n = \{1, 2, \cdots, n\}$.

Definition 1.32 (Finite sets). *A set A is said to be finite if either $A = \emptyset$ or $\exists n \in \mathbb{N}$ and $\exists f : A \to I_n$ such that f is a bijection.*

A set which is not finite is called an *infinite set*. We shall deal with infinite sets a bit later. First, we shall develop some theory on finite sets.

We also note that for each $n \in \mathbb{N}$, the set I_n is finite. Also, for finite sets, by definition there is a bijection to I_n or in other words, every finite set contains as many elements as I_n for some $n \in \mathbb{N}$. Thus, for our interest, it is the same if we study arbitrary finite sets or the sets I_n. Keeping this in mind, we shall now develop some intuitive results on I_n rather than arbitrary finite sets.

Let us first consider two sets I_m and I_n. By the definition, if $m = n$, we have $I_m = I_n$ and hence this does not interest us. Therefore, we shall consider $m \neq n$. In particular, we shall consider $m < n$ without loss of generality. Clearly, there are less number of elements in I_m than those in I_n. What will happen if we start assigning to each element in I_m a unique element in I_n? We will exhaust all the elements of I_m but there will be some elements left in I_n which will not be assigned. What happens when we start assigning to each

---

[6]Indeed, for any function $f : X \to X$, an element $x \in X$ is called a fixed point if $f(x) = x$. We shall study fixed points in some depth in this text later.

element in I_n a unique element in I_m? If we try to assign distinct elements in I_m to distinct elements in I_n, we will again exhaust all elements of I_m before we exhaust all elements of I_n and then we will have no choice but to repeat the assignment for remaining elements. We now present this intuition formally as theorems.

Theorem 1.19. *If $m < n$ then there is no injective map from I_n to I_m.*

Proof. We prove this statement by principle of strong mathematical induction on m. Our statement $P(m)$ will be "$m < n \Rightarrow f : I_n \to I_m$ cannot be a injection".

For $m = 1$, we have $I_m = \{1\}$ and for $n > 1$, any function $f : I_n \to I_1$ gives $f(1) = f(n) = 1$. Therefore, f is not injective. Hence, $P(1)$ is true.

Now, we make our induction hypothesis as $\forall k < m$, let $P(k)$ be true. We now wish to prove $P(m)$ to be true.

Let, if possible, $f : I_n \to I_m$ be an injective function. If $f(n) = m$, we define $g : I_{n-1} \to I_{m-1}$ as $g(i) = f(i)$. Clearly, g is injective. But this contradicts the inductive hypothesis. Therefore, $f(n) \neq m$. Then, since f is injective, there is atmost one $k < n$ such that $f(k) = m$. We define a function $g : I_{n-1} \to I_{m-1}$ as $g(i) = \begin{cases} f(i) & i \neq k \\ f(n) & i = k \end{cases}$. Again, this function is also injective, thereby contradicting the inductive hypothesis. Hence, f cannot be an injective function. □

Theorem 1.20. *If $m < n$ then there is no surjective map from I_m to I_n.*

Proof. Let, if possible, $f : I_m \to I_n$ be a surjection. Define $g : I_n \to I_m$ by $g(i) = \min(f^{-1}(i))$.[7] Since minimum is unique and f is surjection, g is well-defined. Now, we look at what happens when $g(i) = g(j)$. This means that $\min(f^{-1}(i)) = \min(f^{-1}(j))$. Clearly, $f^{-1}(i) \cap f^{-1}(j) \neq \emptyset$. Therefore, $\exists k \in f^{-1}(i) \cap f^{-1}(j)$, or $f(k) = i = j$. Thus, g is injective. However, this is a contradiction to Theorem 1.19. Hence, $f : I_m \to I_n$ cannot be a surjection. □

As an immediate consequence of the above two theorems, we get the following theorem.

Theorem 1.21. *If $f : I_m \to I_n$ is a bijection, then $m = n$.*

Exercise 1.36. Prove Theorem 1.21.

We are now in a position to define what is the number of elements in a finite set.

---

[7] Here, $f^{-1}(i)$ denotes the set of all pre-images of i and not the inverse of f. The readers should not confuse themselves with this abuse of notations.

Definition 1.33 (Cardinality). *A set A is said to have n elements if there is a bijection $f : A \to I_n$. We denote it as $|A| = n$, called the cardinality of A.*

As a matter of convention, we take $|\emptyset| = 0$. Also, we note that $|I_n| = n$.

We now make certain observations about arbitrary finite sets and their subsets. What can we say about the number of elements in a set A if we have an injective map from A to I_n? This means that for every element of A, we have assigned a unique element of I_n and in fact, no two elements of A are assigned the same element of I_n. Intuitively, it seems that the number of elements in A can be at most n. We make this intuition a theorem.

Theorem 1.22. *If $f : A \to I_n$ is an injection, then $|A| \le n$.*

Proof. Let $r_1 = \min\{f(a) \,|a \in A\}$, $r_2 = \min(\{f(a) \,|a \in A\} \setminus \{r_1\})$, and for $k \ge 2$, we have $r_k = \min(\{f(a) \,|a \in A\} \setminus \{r_1, r_2, \cdots, r_{k-1}\})$. Clearly, $i < j \Rightarrow r_i < r_j$ and in fact, $r_1 < r_2 < \cdots < r_k < \cdots$. Since I_n is finite, this process must stop somewhere, say at k. We claim that $k \le n$. By the construction of r_k, we have $r_1 \ge 1$, $r_2 > r_1$ and hence $r_2 \ge 2$. Proceeding in this manner, we can conclude that $r_k \ge k$. If $k > n$, then $r_k > n$ and $r_k \notin I_n$, which is a contradiction. Hence, the process stops at $k \le n$.

We now wish to show that $|A| = k$ so that $|A| \le n$. First we note that for every $a \in A$, $f(a) \in \{r_1, r_2, \cdots, r_k\}$. Thus, we define a function $g : A \to I_k$ such that $g(a) = i$, where $f(a) = r_i$. Since f is injective, so is g and for every r_i, we have a unique a and consequently for every $i \in I_k$, we have a unique $a \in A$ for which $g(a) = i$. Thus, g is a bijection and we have our result. $\square$

Exercise 1.37. Prove that if $f : I_n \to A$ is surjection, then $|A| \le n$.

What can we say about subset(s) of a finite set? Can it have more elements than its superset? Intuitively, the answer is no!

Theorem 1.23. *If A is a finite set and $B \subseteq A$, then $|B| \le |A|$.*

Proof. Since A is finite, $\exists n \in \mathbb{N}$ such that there is a bijection $f : A \to I_n$ and $|A| = n$. Define the inclusion map $g : B \to A$ as $g(x) = x$. Clearly, g is injective. Therefore, the composite map $f \circ g : B \to I_n$ is injective. Therefore by Theorem 1.22, $|B| \le n = |A|$. $\square$

What can we say about proper subsets? Intuitively, if A is a finite set and B is a proper subset of A, then there is at least one element in A which is not in B. Also, from previous theorem, we know that $|B| \le |A|$. With at least one element missed out, there is no possibility of equality. Thus, we expect $|B| < |A|$.

Theorem 1.24. *If A is a finite set and $B \subset A$, then $|B| < |A|$.*

Proof. Let us first take the case where a proper subset C misses exactly one element of A, i.e., $C = A \setminus \{a\}$ for some $a \in A$. Since A is finite, there is a bijection $f : A \to I_n$ for some $n \in \mathbb{N}$. Let $f(a) = i$. We define another map $g : C \to I_{n-1}$ as $g(x) = \begin{cases} f(x) & \text{when } f(x) < i \\ f(x) - 1 & \text{when } f(x) > i \end{cases}$. First, we check g is injective. If we take distinct element $x_1 \neq x_2$ in C, we have the following cases:

Case I: $g(x_1) = f(x_1)$ and $g(x_2) = f(x_2)$. Since f is injective, $g(x_1) = f(x_1) \neq f(x_2) = g(x_2)$.
Case II: $g(x_1) = f(x_1)$ and $g(x_2) = f(x_2) - 1$. This means $f(x_1) < i$ and $f(x_2) \geq i$. Hence, $g(x_1) = f(x_1) \neq f(x_2) = g(x_2)$.
Case III: $g(x_1) = f(x_1) - 1$ and $g(x_2) = f(x_2) - 1$. Again, since f is injective, $g(x_1) \neq g(x_2)$.

Therefore, in all the cases, distinct elements of C are mapped to distinct elements of I_{n-1}. Therefore, g is injective. Let $j \in I_{n-1}$. If $j < i$, then $\exists x \in A$ such that $f(x) = j$. Also, $x \neq a$. Therefore, $\exists x \in C$ such that $g(x) = j$. If $j \geq i$, then $\exists x \in A$ such that $f(x) = j + 1 > i$. Again, $x \neq a$. Therefore, $\exists x \in C$ such that $f(x) = j + 1$. By the definition of g, $\exists x \in C$ such that $g(x) = f(x) - 1 = j$. Therefore, $\forall j \in I_{n-1}$, $\exists x \in C$ such that $g(x) = j$, i.e., g is surjective. Hence, $|C| = n - 1 < n = |A|$.

Now, if B is any proper subset of A, then $\exists a \in A \, (a \notin B)$, i.e., $B \subseteq A \backslash \{a\}$. Therefore, by Theorem 1.23, $|B| \leq |A \backslash \{a\}| < |A|$. $\qquad \square$

Exercise 1.38. If A and B are disjoint finite sets, then prove that $|A \cup B| = |A| + |B|$.

1.4.2 Countable sets

Moving a step ahead, we now deal with infinite sets. An important question is: How do we count the number of elements in an infinite set? Note here that by counting, we now mean in terms of bijections. Let us look at sets which we know by experience. First, we know the set of natural numbers $\mathbb{N}$, then the set of integers $\mathbb{Z}$, set of rational numbers $\mathbb{Q}$, and also, the set of real numbers $\mathbb{R}$. Of these sets, we give special attention to $\mathbb{N}$ and $\mathbb{R}$.

We have seen that if we can have a bijection between two sets, then they have the same number of elements. As discussed earlier, the intervals having more than a single element are all bijective and hence have the same number of elements. It can also be shown easily that $\mathbb{R}$ and $(0,1)$ have the same number of elements.

Exercise 1.39. Exhibit a bijection between $(0,1)$ and $\mathbb{R}$ and hence conclude that they have the same number of elements.

Let us see if $\mathbb{R}$ and $\mathbb{N}$ have the same number of elements. If so, then $(0,1)$ and $\mathbb{N}$ will also have the same number of elements.

Example 1.18 ($\mathbb{R}$ and $\mathbb{N}$ are not bijective). Let, if possible, there be a bijection $f : \mathbb{N} \to (0,1)$. Every number in $(0,1)$ has a decimal expansion of the form $0.a_1a_2a_3\ldots$. Let $f(i) = 0.a_{i1}a_{i2}a_{i3}\ldots$. We construct a new number $0.b_1b_2b_3\ldots$ where $b_n = \begin{cases} a_{nn}+1 & \text{if } a_{nn} \leq 8 \\ 5 & \text{if } a_{nn} = 9 \end{cases}$. Note that for $a_{nn} = 9$, we can choose any digit. There is nothing special about 5! Clearly, this number is different from all other $f(i)$ from its construction, since it differs at the nth place after decimal. Hence, this does not have any pre-image under f, which is a contradiction to the fact that f is a bijection. Therefore, $(0,1)$ and $\mathbb{N}$ are not bijective. This also leads us to the conclusion that $\mathbb{R}$ and $\mathbb{N}$ are not bijective.

What can we say about $\mathbb{Z}$ and $\mathbb{N}$?

Exercise 1.40. Prove that $\mathbb{Z}$ and $\mathbb{N}$ are bijective. **Hint**: Use the trick used in proving $[0,1]$ and $(0,1)$ are bijective. However, this time we would have to shift infinitely many elements.

Now, it remains to be seen if $\mathbb{N}$ and $\mathbb{Q}$ are bijective. Before that, we see another example of a non-trivial set which has as many elements as $\mathbb{N}$.

Example 1.19. Consider the set $\mathbb{N} \times \mathbb{N} = \{(m,n) \,|\, m,n \in \mathbb{N}\}$. Consider a function $f : \mathbb{N} \to \mathbb{N} \times \mathbb{N}$ given by $f(n) = (n,1)$ for every $n \in \mathbb{N}$. Clearly, f is injective. Also, consider the function $g : \mathbb{N} \times \mathbb{N} \to \mathbb{N}$ given by $g(m,n) = 2^m 3^n$. The reader should check that g is injective. Therefore, by Theorem 1.17, $\mathbb{N}$ is bijective to $\mathbb{N} \times \mathbb{N}$.

Now, we are ready to check, in fact prove, that $\mathbb{Q}$ is bijective to $\mathbb{N}$. We observe that any rational number can be written in the form $\frac{p}{q}$, where $p \in \mathbb{Z}$ and $q \in \mathbb{N}$. Therefore, there is a natural bijection $f : \mathbb{Q} \to \mathbb{Z} \times \mathbb{N}$ given by $f\left(\frac{p}{q}\right) = (p,q)$. Also, from Exercise 1.40, we have a bijection $g : \mathbb{Z} \times \mathbb{N} \to \mathbb{N} \times \mathbb{N}$ given by $g(m,n) = (g_1(m),n)$, where $g_1 : \mathbb{Z} \to \mathbb{N}$ is a bijection. Finally, from Example 1.19, we have $h : \mathbb{N} \times \mathbb{N} \to \mathbb{N}$, a bijection. Therefore, the composition $h \circ g \circ f : \mathbb{Q} \to \mathbb{N}$ is a bijection and we conclude that $\mathbb{Q}$ and $\mathbb{N}$ have the same number of elements.

From all the above discussion, we know (by our experience) a few sets which have as many elements as the set of natural numbers and a set which does not hold this property. Based on this, we make the following definition.

Definition 1.34 (Countable sets). *A set A is said to be countable if either A is finite or there is a bijection $f : A \to \mathbb{N}$. A set which is not countable is called uncountable.*

Therefore, $\mathbb{N}$, $\mathbb{Z}$, $\mathbb{Q}$ are all countable, while $\mathbb{R}$ is uncountable and so is $(0,1)$.

The number of elements in a set has been called *cardinality*. However, in the case of infinite sets, this is not possible. We cannot have an actual number for cardinality. However, we do have a method to know if a set contains as many elements as natural numbers or not. Therefore, we define the cardinality of $\mathbb{N}$ and denote it by $|\mathbb{N}| = \aleph_0$.[8] Any set which is bijective to $\mathbb{N}$ is said to have cardinality $\aleph_0$.

By Cantor's theorem, we know that there cannot be any bijection between $\mathcal{P}(\mathbb{N})$ and $\mathbb{N}$. However, there is one question to ask: Is $\mathcal{P}(\mathbb{N})$ finite? To see this, observe that the singleton sets $\{n\}$, where $n \in \mathbb{N}$ are all in $\mathcal{P}(\mathbb{N})$. Since there are infinitely many such singleton sets, we can conclude that $\mathcal{P}(\mathbb{N})$ is also infinite. Since $\mathcal{P}(\mathbb{N})$ is neither finite nor countable, it is uncountable. We now look at another special set $2^{\mathbb{N}} = \{f : \mathbb{N} \to \{0,1\}\}$ of all functions from $\mathbb{N}$ to $\{0,1\}$. Consider a function $\phi : 2^{\mathbb{N}} \to \mathcal{P}(\mathbb{N})$ defined as $\phi(f) = \{n \in \mathbb{N} | f(n) = 0\}$. Now, consider $\phi(f) = \phi(g)$. Then, $\forall n \in \mathbb{N}$ such that $f(n) = 0$, we have $g(n) = 0$. Therefore, $\forall n \in \mathbb{N}$, $f(n) = g(n)$, i.e., $f = g$. Hence, ϕ is injective. Similarly, consider a function $\psi : \mathcal{P}(\mathbb{N}) \to 2^{\mathbb{N}}$ defined by $\psi(S) = f$, where $f(n) = \begin{cases} 1, & \text{for } n \in S \\ 0, & \text{otherwise} \end{cases}$. Now, if $\psi(S_1) = \psi(S_2)$, then we can easily verify that $S_1 = S_2$. Therefore, ψ is also injective. By Theorem 1.17, we conclude that $2^{\mathbb{N}}$ and $\mathcal{P}(\mathbb{N})$ are bijective.

Exercise 1.41. Prove that $\mathcal{P}(\mathbb{N})$ and $[0,1)$ are bijective. **Hint:** Use decimal expansion to have two injective functions and then use Schröder-Berstein theorem.

From above discussions, we can conclude that $\mathcal{P}(\mathbb{N}), 2^{\mathbb{N}}, \mathbb{R}$, and any other interval with more than one point all have the same cardinalities.

Now, we develop some theory about infinite sets through theorems.

Theorem 1.25. *For a set A, the following statements are equivalent:*

1. *A is countable.*

2. *There is an injection from A to $\mathbb{N}$.*

3. *There is a surjection from $\mathbb{N}$ to A.*

Proof. $(1) \Rightarrow (2)$ is true since A is countable implies that there is a bijection $f : A \to \mathbb{N}$, which, in particular, is an injection.

We prove $(2) \Rightarrow (3)$. Let $f : A \to \mathbb{N}$ be the given injection. Fix a point $a \in A$. Define a function $g : \mathbb{N} \to A$ as $g(n) = \begin{cases} f^{-1}(n), & \text{if } n \in f(A) \\ a, & \text{if } n \notin f(A) \end{cases}$.

Clearly, g is a surjection.

To prove $(3) \Rightarrow (1)$, let $n_a = \min(f^{-1}(a))$. Then, either the collection $\{n_a | a \in A\}$ is finite or infinite. In either case, using the well-ordering principle, we arrange the elements in the form $n_1 < n_2 < \cdots < n_k$ or $n_1 < n_2 < \cdots$. Therefore, in the first case, we can define a bijection between A and I_k, while in the other case, we can define a bijection between A and $\mathbb{N}$. Therefore, A is countable. $\square$

---

[8]The symbol $\aleph_0$ is called *"aleph-naught"* or *"aleph-null"*.

Exercise 1.42. Give the bijections mentioned in the last part of the above theorem.

Exercise 1.43. Prove the following:

1. If $f : A \to B$ is injective and B is countable, then so is A.

2. If $f : A \to B$ is surjective and A is countable, then so is B.

3. A subset of a countable set is countable.

Theorem 1.26. *Countable union[9] of countable sets is countable.*

Proof. Let $\{A_i | i \in I\}$ be a countable collection of countable sets, i.e., I is countable and $\forall i \in I$, A_i is countable. This means we have $f : I \to \mathbb{N}$ and $g_i : A_i \to \mathbb{N}$ for each $i \in I$ as bijections. Let $A = \bigcup_{i \in I} A_i$. Define a function $g : A \to \mathbb{N} \times \mathbb{N}$ as $g(a) = (n, g_n(a))$, where $n = \min\{f(i) | a \in A_i\}$. Clearly, g is injective since g_n is a bijection. Hence, A is countable. $\square$

Exercise 1.44. From the above discussion conclude that $\mathbb{Q}^c$, the set of irrational numbers is uncountable.

Problem Set

1. Identify the following sets:

   (a) $\left\{ x \in \mathbb{R} \middle| x + \dfrac{1}{x} > 2 \right\}$.

   (b) $\{ x \in \mathbb{R} | \exists y \in \mathbb{R} \text{ such that } y^{2n} = x \}$, where $n \in \mathbb{N}$ is fixed.

   (c) $\left\{ \dfrac{m}{n} \middle| m, n \in \mathbb{Z} \text{ where } m \text{ and } n \text{ have the same sign and } n \text{ is a divisor of } m \right\}$.

   (d) $S \subseteq \mathbb{N}$ such that $1 \in S$ and for every $k \in S$, $k + 1 \in S$.

   (e) $\{ x \in \mathbb{R} | \exists a, b \in \mathbb{Z} \text{ with } a > 0 \text{ and } x^2 + ax + b = 0 \}$.

   (f) $\{ x \in \mathbb{R} | e^x = 0 \}$.

   (g) $\{ x \in \mathbb{R} | |x - a| + |x - b| = c \}$, where $a, b, c \in \mathbb{R}$ with $a < b$.

   (h) $\{ x \in \mathbb{R} | ||x - a| - |x - b|| = c \}$, where $a, b, c \in \mathbb{R}$ with $a < b$ and $c \geq 0$.

---

[9] A countable union indeed means that the family whose union is considered is indexed by a countable set.

(i) $\{(x, y) \in \mathbb{R}^2 | (1 - x)(1 - y) > 1 - x - y\}$. Draw a picture of this set.

(j) $\{(x, y) \in \mathbb{R}^2 | |x| \leq |y|\}$. Draw a picture of this set.

(k) $\bigcap\limits_{n \in \mathbb{N}} \left(0, \dfrac{1}{n}\right)$.

2. Let A be the set of all integers divisible by m and B be the set of all integers divisible by n, where $m, n \in \mathbb{N}$ are fixed. What is $A \cap B$?

3. If $B \subseteq C$, then show that:

(a) $A \cup B \subseteq A \cup C$.

(b) $A \cap B \subseteq A \cap C$.

4. (a) For a set A, what is $A \setminus A$?

(b) If A and B are disjoint, what is $A \setminus B$?

(c) If $A \subseteq B$, what is $A \setminus B$?

5. Let A be the set of all $n \times n$ real symmetric matrices and B be the set of skew-symmetric matrices. Describe $A \setminus B$.

6. For two sets A, B, show that $(A \cup B) \setminus (A \cap B) = (A \setminus B) \cup (B \setminus A)$.

For two sets A and B, we define the *symmetric difference* as

$$A \Delta B = (A \setminus B) \cup (B \setminus A).$$

Clearly, this set contains the elements exactly in one set and not in the other. In some sense, this forms an *exclusive union*.

(a) Prove that symmetric difference is commutative.

(b) Prove that symmetric difference is associative.

7. For two sets A and B, show that the following are equivalent:

(a) $A \subseteq B$.

(b) $A \cup B = B$.

(c) $A \cap B = A$.

(d) $B^c \subseteq A^c$.

8. Let $\mathscr{C}$ be the set of all circles. Express this as an indexed family of sets of $\mathbb{R}^2$, indexed by $\Lambda = \mathbb{R}^2 \times (0, \infty)$. Is this family disjoint?

9. If the set A contains n elements, show that the power set $\mathcal{P}(A)$ contains 2^n elements.

10. Let $C = \{(x, y) \in \mathbb{R}^2 | x^2 + y^2 = 1\}$. Can this set be written as a Cartesian product of two sets?

11. Let $A, C \subseteq X$ and $B, D \subseteq Y$. Prove or disprove the following:

    (a) $A \times B \subseteq C \times D$ if and only if $A \subseteq C$ and $B \subseteq D$.
    (b) $(A \cap C) \times (B \cap D) = (A \times B) \cap (C \times D)$.
    (c) $(A \cup C) \times (B \cup D) = (A \times B) \cup (C \times D)$.

12. For sets A, B, C prove or disprove the following:

    (a) $A \times (B \cup C) = (A \times B) \cup (A \times C)$.
    (b) $A \times (B \cap C) = (A \times B) \cap (A \times C)$.
    (c) $A \times (B \setminus C) = (A \times B) \setminus (A \times C)$.
    (d) $A \times B = A \times C$ implies $B = C$.
    (e) $\mathcal{P}(A \times B) = \mathcal{P}(A) \times \mathcal{P}(B)$.

13. Are the functions $f : \mathbb{R} \to \mathbb{R}$ and $g : \mathbb{R} \to [0, \infty)$ defined as $f(x) = x^2$ and $g(x) = x^2$ equal?

14. For a function $f : X \times Y$, we define the graph of f as

$$G(f) = \{(x, y) \in X \times Y | y = f(x)\}.$$

Are the following sets graphs of functions?

    (a) $\{(x, y) \in \mathbb{R}^2 | x^2 + y^2 = 1\}$.
    (b) $\{(x, y) \in \mathbb{R}^2 | x^2 + y = 0\}$.
    (c) $\{(x, y) \in \mathbb{R}^2 | x = |y|\}$.

15. Check if the following function is injective. $f : \mathbb{N} \to \mathbb{N}$ defined as

$$f(n) = \begin{cases} n + 2, & n \text{ is odd.} \\ 2n, & n \text{ is even.} \end{cases}$$

16. Let $M(2, \mathbb{R})$ denote the set of all 2×2 matrices with real entries. Consider determinant of these matrices as a function $\det : M(2, \mathbb{R}) \to \mathbb{R}$. Is it injective?

17. Which of the following functions are surjective?

    (a) $f : \mathbb{Z} \to \mathbb{N}$ defined by $f(m) = m^2$.
    (b) $f : \mathbb{R} \to \mathbb{R}$ defined by $f(x) = x^2 + x + 1$.

(c) $f : \mathbb{R} \to \mathbb{R}$ defined by $f(x) = e^x$.

(d) $f : (0, \infty) \to \mathbb{R}$ defined by $f(x) = \ln x$.

(e) $f : \mathbb{R} \to \mathbb{R}$ defined by $f(x) = \sin x$.

18. Let $f : \mathbb{N} \to A$ and $g : \mathbb{N} \to B$ be surjective for some sets A and B. Is there an onto map $h : \mathbb{N} \to A \cup B$?

19. Show that the following functions are bijections and hence find the inverse.

   (a) $f : \mathbb{R} \to \mathbb{R}$ defined as $f(x) = x + 1$.

   (b) $f : [0, 1] \to [0, 1]$ defined as $f(x) = \dfrac{1 - x}{1 + x}$.

   (c) $f : [0, \infty) \to [0, 1)$ defined as $f(x) = \dfrac{x^2}{1 + x^2}$.

20. For two bijective functions $f : X \to Y$ and $g : Y \to Z$, prove that $(g \circ f)^{-1} = f^{-1} \circ g^{-1}$.

21. Let $f : X \to Y$ and $g : Y \to Z$ be such that $g \circ f$ is surjective and g is injective. What can we say about injectivity and surjectivity of f?

22. Let $f : \mathbb{R} \to \mathbb{R}$ be defined as $f(x) = \sin x$. What is $f^{-1}([0, 1])$?

23. Let $f : M(n, \mathbb{R}) \to M(n, \mathbb{R})$ be defined as $f(A) = A^n$. What is $f^{-1}(\{\mathbf{0}\})$? Here, $M(n, \mathbb{R})$ denotes the set of all $n \times n$ real matrices and $\mathbf{0}$ denotes the matrix with all entries zero.

24. Consider the set

$$R = \{(x, y) \in \mathbb{R}^2 | xy > 0\} \cup \{(0, 0)\}.$$

   Is R an equivalence relation? If so, what are the equivalence classes?

25. Consider the relation R defined on $\mathbb{R}^2$ as $(x_1, y_1) R (x_2, y_2)$ if and only if $x_1^2 + y_1^2 = x_2^2 + y_2^2$. Is this an equivalence relation? If so, what are the equivalence classes?

26. Can there by an injective map $f : \mathcal{P}(X) \to X$ for any set X?

27. For a finite set X and a function $f : X \to X$ prove that the following are equivalent:

   (a) f is bijective.

   (b) f is injective.

   (c) f is surjective.

   Can we generalize this to infinite sets?

28. Is the set $\mathbb{N}^{\mathbb{N}}$ of all functions from $\mathbb{N}$ to $\mathbb{N}$ countable?

29. If X is uncountable and A is countable, can $X \setminus A$ be countable?

30. For $n \in \mathbb{N}$, define $H_n = \{kn | k \in \mathbb{Z}\}$. Let $X = \{H_n | n \geq 2\}$ and let $\subseteq$ be the partial order relation on X. What are the maximal elements in X?

31. Show that every finite totally ordered set has a maximum and a minimum.

32. Are there any partially ordered sets in which maximal elements are the same as minimal elements?

Biographical Notes

Helmut Hasse (25 August, 1898 to 26 December, 1979) was a German mathematician working in algebraic number theory, known for fundamental contributions to class field theory, the application of p-adic numbers to local class field theory and diophantine geometry (Hasse principle), and to local zeta functions. He studied at the University of Göttingen, and then at the University of Marburg under Kurt Hensel, writing a dissertation in 1921 containing the Hasse–Minkowski theorem, as it is now called, on quadratic forms over number fields. Hasse diagrams are named after him; they are so called because of the effective use Hasse made of them. However, Hasse was not the first to use these diagrams. One example that predates Hasse can be found in Henri Gustav Vogt (1895).

Max August Zorn (June 6, 1906 to March 9, 1993) was a German mathematician. He was an algebraist, group theorist, and numerical analyst. He is best known for Zorn's lemma, a method used in set theory that is applicable to a wide range of mathematical constructs such as vector spaces, ordered sets, and the like. Zorn's lemma was first postulated by Kazimierz Kuratowski in 1922, and then independently by Zorn in 1935.

Bertrand Arthur William Russell, (18 May, 1872 to 2 February, 1970) was a British philosopher, logician, mathematician, historian, writer, essayist, social critic, political activist, and Nobel laureate. The barber's paradox is derived from Russell's paradox, which he attributes to an unnamed person to have suggested it to him as an illustration of paradoxes.

Georg Ferdinand Ludwig Philipp Cantor (March 3, 1845 to January 6, 1918) was a German mathematician. He created set theory, which has become a fundamental theory in mathematics. Cantor established the importance of one-to-one correspondence between the members of two sets, defined infinite and well-ordered sets, and proved that the real numbers are more numerous than the natural numbers. In fact, Cantor's method of proof of this theorem implies the existence of an "infinity of infinities". He defined the cardinal and ordinal numbers and their arithmetic. Cantor's work is of great philosophical interest, a fact he was well aware of.

Friedrich Wilhelm Karl Ernst Schröder (25 November, 1841 to 16 June, 1902) was a German mathematician mainly known for his work on algebraic logic. He is a major figure in the history of mathematical logic (a term he may have invented), by virtue of summarizing and extending the work of George Boole, Augustus De Morgan, Hugh MacColl, and especially Charles Peirce. He is best known for his monumental *Vorlesungen über die Algebra der Logik* (Lectures on the algebra of logic), in three volumes, which prepared the way for the emergence of mathematical logic as a separate discipline in the twentieth century by systematizing the various systems of formal logic of the day.

Felix Bernstein (24 February, 1878 to 3 December, 1956) was a German Jewish mathematician known for proving in 1896 the Schröder–Bernstein theorem, a central result in set theory, and less well known for demonstrating in 1924 the correct blood group inheritance pattern of multiple alleles at one locus through statistical analysis.

Bronisaw Knaster (22 May, 1893 to 3 November, 1980) was a Polish mathematician. He is known for his work in point-set topology and in particular for his discoveries in 1922 of the hereditarily indecomposable continuum or pseudo-arc and of the Knaster continuum, or buckethandle continuum.

Alfred Tarski (January 14, 1901 to October 26, 1983), born Alfred Teitelbaum, was a Polish-American logician and mathematician of Polish-Jewish descent. Educated in Poland at the University of Warsaw, and a member of the Lwów-Warsaw school of logic and the Warsaw school of mathematics, he immigrated to the United States in 1939 where he became a naturalized citizen in 1945. Tarski taught and carried out research in mathematics at the University of California, Berkeley, from 1942 until his death in 1983. A prolific author best known for his work on model theory, metamathematics, and algebraic logic, he also contributed to abstract algebra, topology, geometry, measure theory, mathematical logic, set theory, and analytic philosophy.

Chapter 2

Metric Spaces

In this chapter, we shall first review our experience in the real number system and measuring what we call in natural language as "distances", and then generalize this concept to define metric spaces. Once we achieve this, all other concepts that arise due to notion of distances, such as sequences and their convergence and continuity of functions, will be generalized in the succeeding chapters.

2.1 Review of Real Number System and Absolute Value

In a course of analysis (or, what is often called Real Analysis), the reader might have studied various algebraic properties of real numbers, sequences and series of real numbers, real valued functions, and their continuity. In this section, we specifically focus on an important function from real analysis, called the *absolute value* of a real number. Formally, it is defined as $|\cdot| : \mathbb{R} \to [0, \infty)$, and satisfies the following properties

1. $|x| = 0 \Leftrightarrow x = 0$.

2. $\forall x \in \mathbb{R}, \ |-x| = |x|$.   [Symmetry]

3. $\forall x, y \in \mathbb{R}, \ |xy| = |x| \, |y|$.   [Homogeneity]

4. $\forall x, y \in \mathbb{R}, \ |x + y| \leq |x| + |y|$.   [Triangle inequality]

Physically, it measures the distance of any real number x from the real number 0. Often, it is called the *length* of x. The readers who have taken a course in linear algebra might know that this notion of length has been generalized to the notion of norm in a vector space, which is often denoted by $||\cdot|| : V \to [0, \infty)$, where V is the vector space under consideration.

Although we shall not use the notion of norms as long as possible, we advise the reader to go through the concept since in many applications, distances can be measured by the help of norms easily, in addition to certain other special properties which norms provide to us. In this text, however, we shall assume that the reader is not familiar with norm and construct all the theory accordingly.

Coming back to the concept of absolute value, let us see what insights can we get from the properties discussed above. From the four properties discussed above, the first two properties and the last property are of interest to us. The third property, called the *homogeneity* is of interest when one studies norms explicitly. We write the properties of interest in words for a better understanding.

1. The only real number whose length or distance from 0 is 0 is the number 0 itself.

2. It does not matter on what side of 0 a real number x lies. The length remains same if we measure toward the left or toward the right.

3. This property is called the *triangle inequality* and as the name suggests, it tells that the sum of lengths of two sides of a triangle is greater than or equal to the length of the third side.[1]

As studied in Real Analysis, we make use of this absolute value function to measure distances between points, especially to study convergent sequences, Cauchy sequences, and continuity of functions. The question is: How do we do so? By experience, we know that to measure the distance between two real numbers x and y, it is enough to measure the length of the line segment joining the two points. But, the only way we know to measure the length is measuring from 0. Therefore, we shift one of the points, say x, to 0. Consequently, y gets shifted to $y - x$. Then, the length of the real number $y - x$ is given by $|y - x|$, which is precisely, the distance between the real numbers x and y as shown in Figure 2.1. Since it does not matter on what side of zero we measure the length, we often write the distance between x and y as $|x - y|$. Here, the readers are advised to make an illustration by shifting y to 0, instead of x, and convince themselves that the distance still remains the same.

With this notion, what properties do we observe of the distance between two real numbers x and y?

1. $\forall x, y \in \mathbb{R}$, $|x - y| \geq 0$ and $|x - y| = 0$ if and only if $x = y$.

2. $\forall x, y \in \mathbb{R}$, $|x - y| = |y - x|$.

3. $\forall x, y, z \in \mathbb{R}$, $|x - y| \leq |x - z| + |z - y|$.

---

[1]This will be geometrically more intuitive to the reader if they look at distances in $\mathbb{R}^2$.

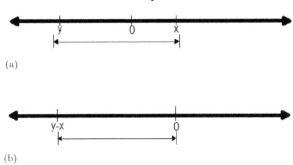

FIGURE 2.1: Measuring the "distance" between two real numbers by means of absolute value function. (a) Line segment joining two points x and y in $\mathbb{R}$. (b) Line segment after shifting x to 0.

Exercise 2.1. Prove the properties of distance between two real numbers discussed above.

With this in mind, we are now ready to generalize the notion of distance to any set. But, before doing so, we go through some important inequalities which will be helpful further.

2.2   Young, Hölder, and Minkowski Inequalities

In this section, we shall prove important inequalities for real numbers, finite sum of real numbers, series of real numbers, and integrals of real valued functions. These inequalities will further help us in making examples for metric spaces in the next section. Throughout the section, we shall make use of two notations for exponents p and q, where $p > 1$ and $\frac{1}{p} + \frac{1}{q} = 1$. These exponents are often mentioned in textbooks as *conjugate exponents*.

Theorem 2.1 (Young's inequality). *Let x, y be non-negative real numbers. Then,*

$$xy \leq \frac{x^p}{p} + \frac{y^q}{q}. \tag{2.1}$$

The equality holds if and only if $x^p = y^q$.

Proof. If $x = 0$ or $y = 0$, then the inequality holds trivially since the left side is 0 and the right side is the sum of non-negative real numbers, which remains non-negative. Therefore, in the proof, we shall assume that $x \neq 0$ and $y \neq 0$.

Let $y > 0$ be fixed. We construct a function $f : (0, \infty) \to \mathbb{R}$ as $f(x) = \frac{x^p}{p} + \frac{y^q}{q} - xy$. Now, consider $f'(x) = x^{p-1} - y$. For $x = y^{\frac{1}{p-1}}$, we have $f'(x) = 0$, i.e., $x = y^{\frac{1}{p-1}}$ is a point of minima or maxima for the function f. Now, using the second derivative test, $f''(x) = (p-1)x^{p-2} > 0$ since $p > 1$ and $x > 0$. Hence $x = y^{\frac{1}{p-1}}$ is a point of minima, i.e., $\forall x > 0$, we have $f(x) \geq f(y^{\frac{1}{p-1}})$. This gives us

$$\frac{x^p}{p} + \frac{y^q}{q} - xy \geq \frac{y^{\frac{p}{p-1}}}{p} + \frac{y^q}{q} - y^{\frac{1}{p-1}}y.$$

$$\therefore \frac{x^p}{p} + \frac{y^q}{q} - xy \geq \frac{y^q}{p} + \frac{y^q}{q} - y^q = 0.$$

$$\therefore xy \leq \frac{x^p}{p} + \frac{y^q}{q}.$$

Now, the equality holds if and only if $x = y^{\frac{1}{p-1}}$, i.e., if and only if $x^p = y^{\frac{p}{p-1}} = y^q$. $\qquad\square$

Theorem 2.2 (Hölder's inequality for finite sum). *Let $x_1, x_2, \cdots, x_n$ and $y_1, y_2, \cdots, y_n \in \mathbb{R}$. Then,*

$$\sum_{i=1}^{n} |x_i| \, |y_i| \leq \left(\sum_{i=1}^{n} |x_i|^p \right)^{\frac{1}{p}} \left(\sum_{i=1}^{n} |y_i|^q \right)^{\frac{1}{q}}. \tag{2.2}$$

The equality holds if and only if $\exists c_1, c_2 \in \mathbb{R}$, with $c_1 \neq 0$ and $c_2 \neq 0$ such that $c_1 |x_k|^p = c_2 |y_k|^q$ for $k = 1, 2, \cdots, n$.

Proof. We consider $\left(\sum_{i=1}^{n} |x_i|^p \right)^{\frac{1}{p}} \neq 0$ and $\left(\sum_{i=1}^{n} |y_i|^p \right)^{\frac{1}{p}} \neq 0$, because if any of them is zero, then the inequality is trivially true (why?). Clearly,

$$\frac{|x_i|}{\left(\sum_{j=1}^{n} |x_j|^p \right)^{\frac{1}{p}}} \geq 0 \text{ and } \frac{|y_i|}{\left(\sum_{j=1}^{n} |y_j|^q \right)^{\frac{1}{q}}} \geq 0 \text{ for } i = 1, 2, \cdots, n.$$ Applying Young's inequality for each i, we get

$$\frac{|x_i| \, |y_i|}{\left(\sum_{j=1}^{n} |x_j|^p \right)^{\frac{1}{p}} \left(\sum_{j=1}^{n} |y_i|^q \right)^{\frac{1}{q}}} \leq \frac{1}{p} \frac{|x_i|^p}{\sum_{j=1}^{n} |x_j|^p} + \frac{1}{q} \frac{|y_i|^q}{\sum_{j=1}^{n} |y^j|^q}.$$

Summing over $i = 1$ through n, we get

$$\sum_{i=1}^{n} \frac{|x_i| \, |y_i|}{\left(\sum_{j=1}^{n} |x_j|^p\right)^{\frac{1}{p}} \left(\sum_{j=1}^{n} |y_j|^q\right)^{\frac{1}{q}}} \leq \sum_{i=1}^{n} \left(\frac{1}{p} \frac{|x_i|^p}{\sum_{j=1}^{n} |x_j|^p} + \frac{1}{q} \frac{|y_i|}{\sum_{j=1}^{n} |y^j|^q}\right).$$

$$\therefore \frac{\sum_{i=1}^{n} |x_i| \, |y_i|}{\left(\sum_{i=1}^{n} |x_i|^p\right)^{\frac{1}{p}} \left(\sum_{i=1}^{n} |y_i|^q\right)^{\frac{1}{q}}} \leq \frac{1}{p} \left(\frac{\sum_{i=1}^{n} |x_i|^p}{\sum_{i=1}^{n} |x_i|^p}\right) + \frac{1}{q} \left(\frac{\sum_{i=1}^{n} |y_i|^q}{\sum_{i=1}^{n} |y_i|^q}\right)$$

$$= \frac{1}{p} + \frac{1}{q} = 1.$$

$$\therefore \sum_{i=1}^{n} |x_i| \, |y_i| \leq \left(\sum_{i=1}^{n} |x_i|^p\right)^{\frac{1}{p}} \left(\sum_{i=1}^{n} |y_i|^q\right)^{\frac{1}{q}}.$$

Since the inequality occurs due to the Young's inequality, the equality in this case shall occur if and only if the equality occurs in the use of Young's inequality. This part of the proof is left as an exercise for the reader. □

Exercise 2.2. Prove the second part of Theorem 2.2 by finding the constants c_1 and c_2.

Theorem 2.3 (Minkowski's inequality for finite sum). *For real numbers* $x_1, x_2, \cdots, x_n, y_1, y_2, \cdots, y_n,$ *we have*

$$\left(\sum_{i=1}^{n} |x_i + y_i|^p\right)^{\frac{1}{p}} \leq \left(\sum_{i=1}^{n} |x_i|^p\right)^{\frac{1}{p}} + \left(\sum_{i=1}^{n} |y_i|^p\right)^{\frac{1}{p}}. \tag{2.3}$$

Proof. If either $\sum_{i=1}^{n} |x_i|^p = 0$ or $\sum_{i=1}^{n} |y_i|^p = 0$, then Equation (2.3) holds trivially (why?). Therefore, we assume that $\sum_{i=1}^{n} |x_i|^p \neq 0$ and $\sum_{i=1}^{n} |y_i|^p \neq 0$ and hence proceed for the proof.

$$\sum_{i=1}^{n} |x_i + y_i|^p = \sum_{i=1}^{n} |x_i + y_i| \, |x_i + y_i|^{p-1}$$

$$\leq \sum_{i=1}^{n} (|x_i| + |y_i|) \, |x_i + y_i|^{p-1} \qquad \text{(Triangle inequality)}$$

$$= \sum_{i=1}^{n} |x_i| \, |x_i + y_i|^{p-1} + \sum_{i=1}^{n} |y_i| \, |x_i + y_i|^{p-1}$$

$$\leq \left(\sum_{i=1}^{n} |x_i|^p\right)^{\frac{1}{p}} \left(\sum_{i=1}^{n} |x_i + y_i|^{q(p-1)}\right)^{\frac{1}{q}} +$$

$$\left(\sum_{i=1}^{n}|y_i|^p\right)^{\frac{1}{p}}\left(\sum_{i=1}^{n}|x_i+y_i|^{q(p-1)}\right)^{\frac{1}{q}}$$

[From Equation (2.2)]

$$=\left(\left(\sum_{i=1}^{n}|x_i|^p\right)^{\frac{1}{p}}+\left(\sum_{i=1}^{n}|y_i|^p\right)^{\frac{1}{p}}\right)\left(\sum_{i=1}^{n}|x_i+y_i|^p\right)^{\frac{1}{q}}.$$

$$\therefore \left(\sum_{i=1}^{n}|x_i+y_i|^p\right)^{1-\frac{1}{q}}\leq\left(\sum_{i=1}^{n}|x_i|^p\right)^{\frac{1}{p}}+\left(\sum_{i=1}^{n}|y_i|^p\right)^{\frac{1}{p}}.$$

$$\therefore \left(\sum_{i=1}^{n}|x_i+y_i|^p\right)^{\frac{1}{p}}\leq\left(\sum_{i=1}^{n}|x_i|^p\right)^{\frac{1}{p}}+\left(\sum_{i=1}^{n}|y_i|^p\right)^{\frac{1}{p}}.$$

$\square$

Before moving to the case of infinite series, let us state a fact about convergent series in which the terms are non-negative. The proof of this fact can be found in a typical analysis book and hence is not covered here.

Theorem 2.4. *If $\sum_{i=1}^{\infty}a_i$ is a series where $\forall n\in\mathbb{N},a_n\geq 0$, then the series is convergent if and only if the sequence of partial sums $\sum_{i=1}^{n}a_i$ is bounded and the value of the series is given by $\sup_{n\in\mathbb{N}}\left\{\sum_{i=1}^{n}a_i\right\}$.*

Now, we prove the Minkowski's inequality for an infinite sum.

Theorem 2.5 (Minkowski's inequality for infinite sum). *Let $(x_i)_{i\in\mathbb{N}}$ and $(y_i)_{i\in\mathbb{N}}$[2] such that $\sum_{i=1}^{\infty}|x_i|^p$ and $\sum_{i=1}^{\infty}|y_i|^p$ converges. Then,*

$$\left(\sum_{i=1}^{\infty}|x_i+y_i|^p\right)^{\frac{1}{p}}\leq\left(\sum_{i=1}^{\infty}|x_i|^p\right)^{\frac{1}{p}}+\left(\sum_{i=1}^{\infty}|y_i|^p\right)^{\frac{1}{p}}. \qquad (2.4)$$

Proof. From Equation (2.3), we know that

$$\left(\sum_{i=1}^{n}|x_i+y_i|^p\right)^{\frac{1}{p}}\leq\left(\sum_{i=1}^{n}|x_i|^p\right)^{\frac{1}{p}}+\left(\sum_{i=1}^{n}|y_i|^p\right)^{\frac{1}{p}}.$$

---

[2]We shall use this notations to denote sequences of real numbers. Notice that when we write $(x_i)_{i\in\mathbb{N}}$, it means the list $(x_1,x_2,\cdots)$ which goes on forever. The explanation for this notation will be provided shortly. Until then, the reader is advised to accept the notation and proceed for the proof.

Since $\sum\limits_{i=1}^{\infty} |x_i|^p$ and $\sum\limits_{i=1}^{\infty} |y_i|^p$ are convergent, we have

$$\sum_{i=1}^{n} |x_i|^p \leq \sum_{i=1}^{\infty} |x_i|^p,$$

$$\sum_{i=1}^{n} |y_i|^p \leq \sum_{i=1}^{\infty} |y_i|^p.$$

Also, since both the sums are non-negative, we have

$$\left(\sum_{i=1}^{n} |x_i|^p\right)^{\frac{1}{p}} \leq \left(\sum_{i=1}^{\infty} |x_i|^p\right)^{\frac{1}{p}},$$

$$\left(\sum_{i=1}^{n} |y_i|^p\right)^{\frac{1}{p}} \leq \left(\sum_{i=1}^{\infty} |y_i|^p\right)^{\frac{1}{p}}.$$

Therefore, we have

$$\left(\sum_{i=1}^{n} |x_i + y_i|^p\right)^{\frac{1}{p}} \leq \left(\sum_{i=1}^{n} |x_i|^p\right)^{\frac{1}{p}} + \left(\sum_{i=1}^{n} |y_i|^p\right)^{\frac{1}{p}} \leq \left(\sum_{i=1}^{\infty} |x_i|^p\right)^{\frac{1}{p}} + \left(\sum_{i=1}^{\infty} |y_i|^p\right)^{\frac{1}{p}}.$$

Therefore, for every $n \in \mathbb{N}$, the partial sum $\sum\limits_{i=1}^{n} |x_i + y_i|^p$ is bounded above by $\left(\left(\sum\limits_{i=1}^{\infty} |x_i|^p\right)^{\frac{1}{p}} + \left(\sum\limits_{i=1}^{\infty} |y_i|^p\right)^{\frac{1}{p}}\right)^p$ and therefore using Theorem 2.4, the series $\sum\limits_{i=1}^{\infty} |x_i + y_i|^p$ is convergent, with the value of the sum given by $\sup\limits_{n\in\mathbb{N}} \left\{\sum\limits_{i=1}^{n} |x_i + y_i|^p\right\}$. Since $\left(\left(\sum\limits_{i=1}^{\infty} |x_i|^p\right)^{\frac{1}{p}} + \left(\sum\limits_{i=1}^{\infty} |y_i|^p\right)^{\frac{1}{p}}\right)^p$ is an upper bound for the set $\left\{\sum\limits_{i=1}^{n} |x_i + y_i|^p \,\middle|\, n \in \mathbb{N}\right\}$, we have

$$\sum_{i=1}^{\infty} |x_i + y_i|^p = \sup_{n\in\mathbb{N}} \left\{\sum_{i=1}^{n} |x_i + y_i|^p\right\} \leq \left(\left(\sum_{i=1}^{\infty} |x_i|^p\right)^{\frac{1}{p}} + \left(\sum_{i=1}^{\infty} |y_i|^p\right)^{\frac{1}{p}}\right)^p.$$

Hence, we get the Minkowski's inequality for an infinite sum,

$$\left(\sum_{i=1}^{\infty} |x_i + y_i|^p\right)^{\frac{1}{p}} \leq \left(\sum_{i=1}^{\infty} |x_i|^p\right)^{\frac{1}{p}} + \left(\sum_{i=1}^{\infty} |y_i|^p\right)^{\frac{1}{p}}.$$

$\square$

Now, we shall consider a special set $\mathscr{C}[a, b]$, the set of all real valued continuous functions on the interval $[a, b]$. It is to be noted here that every element of $\mathscr{C}[a, b]$ is a function $f : [a, b] \to \mathbb{R}$ and being a continuous function on a closed interval, it is Riemann integrable.

Now, we prove the Hölder's and Minkowski's inequalities for these integrable functions on similar lines as above.

Theorem 2.6 (Hölder's inequality for integrals). *Let $f, g \in \mathscr{C}[a, b]$. Then,*

$$\int_a^b |f(x)| |g(x)| \, dx \le \left(\int_a^b |f(x)|^p \, dx \right)^{\frac{1}{p}} \left(\int_a^b |g(x)|^q \, dx \right)^{\frac{1}{q}} . \tag{2.5}$$

Proof. Clearly, if one of the functions f or g is identically zero, then the inequality holds trivially (why?). Therefore, we assume that both these functions are not identically zero so that none of the integrals involved in Equation (2.5) are zero. Now, we have from Equation (2.1)

$$\int_a^b \frac{|f(x)|}{\left(\int_a^b |f(x)|^p \, dx \right)^{\frac{1}{p}}} \frac{|g(x)|}{\left(\int_a^b |g(x)|^q \, dx \right)^{\frac{1}{q}}} \, dx$$

$$\le \int_a^b \left(\frac{1}{p} \frac{|f(x)|^p}{\int_a^b |f(x)|^p \, dx} + \frac{1}{q} \frac{|g(x)|^q}{\int_a^b |g(x)|^q \, dx} \right) \, dx$$

$$= \frac{1}{p} + \frac{1}{q} = 1.$$

$$\therefore \int_a^b |f(x)| |g(x)| \, dx \le \left(\int_a^b |f(x)|^p \, dx \right)^{\frac{1}{p}} \left(\int_a^b |g(x)|^q \, dx \right)^{\frac{1}{q}} .$$

$\square$

Theorem 2.7 (Minkowski's inequality for integrals). *Let $f, g \in \mathscr{C}[a, b]$. Then,*

$$\left(\int_a^b |f(x) + g(x)|^p \, dx \right)^{\frac{1}{p}} \le \left(\int_a^b |f(x)|^p \, dx \right)^{\frac{1}{p}} + \left(\int_a^b |g(x)|^p \, dx \right)^{\frac{1}{p}} . \tag{2.6}$$

Proof. If one of f or g is identically zero, then the inequality holds trivially (why?). Therefore, we assume that both are non-zero (identically) and proceed for the proof.

Since the exponentiation, x^α, is a convex function[3] for non-negative inputs, we have

$$\left| \frac{1}{2} f(x) + \frac{1}{2} g(x) \right|^p \leq \frac{1}{2} |f(x)|^p + \frac{1}{2} |g(x)|^p .$$

$$\therefore |f(x) + g(x)|^p \leq 2^{p-1} (|f(x)|^p + |g(x)|^p).$$

Which means that $|f(x) + g(x)|^p$ is bounded and being continuous, it is also Riemann integrable, i.e., $\int_a^b |f(x) + g(x)|^p \, colorreddx$ exists.

$$\int_a^b |f(x) + g(x)|^p \, dx = \int_a^b |f(x) + g(x)| \, |f(x) + g(x)|^{p-1} \, dx$$

$$\leq \int_a^b (|f(x)| + |g(x)|) \, |f(x) + g(x)|^{p-1} \, dx$$

(Triangle inequality)

$$= \int_a^b |f(x)| \, |f(x) + g(x)|^{p-1} \, dx +$$

$$\int_a^b |g(x)| \, |f(x) + g(x)|^{p-1} \, dx$$

$$\leq \left(\int_a^b |f(x)|^p \, dx \right)^{\frac{1}{p}} \left(\int_a^b |f(x) + g(x)|^{q(p-1)} \, dx \right)^{\frac{1}{q}}$$

$$+ \left(\int_a^b |g(x)|^p \, dx \right)^{\frac{1}{p}} \left(\int_a^b |f(x) + g(x)|^{q(p-1)} \, dx \right)^{\frac{1}{q}}$$

[From Equation (2.5)]

$$= \left(\left(\int_a^b |f(x)|^p \, dx \right)^{\frac{1}{p}} + \left(\int_a^b |g(x)|^p \, dx \right)^{\frac{1}{p}} \right) .$$

$$\left(\int_a^b |f(x) + g(x)|^p \, dx \right)^{\frac{1}{q}} .$$

---

[3] A function $f : \mathbb{R} \to \mathbb{R}$ is *convex* if $f(\lambda x_1 + (1 - \lambda) x_2) \leq \lambda f(x_1) + (1 - \lambda) f(x_2)$ for any $\lambda \in [0, 1]$.

$$\therefore \left(\int_a^b |f(x) + g(x)|^p \, dx \right)^{1-\frac{1}{q}} \leq \left(\int_a^b |f(x)|^p \, dx \right)^{\frac{1}{p}} + \left(\int_a^b |g(x)|^p \, dx \right)^{\frac{1}{p}}.$$

$$\therefore \left(\int_a^b |f(x) + g(x)|^p \, dx \right)^{\frac{1}{p}} \leq \left(\int_a^b |f(x)|^p \, dx \right)^{\frac{1}{p}} + \left(\int_a^b |g(x)|^p \, dx \right)^{\frac{1}{p}}.$$

$\square$

With this, we are now completely equipped and ready to define an abstract metric space and look at examples.

2.3 Notion of Metric Space

In the first section, we tried to understand how to calculate distance between two real numbers. Now, our aim is to do the same but in an abstract setting. This means, we want to know the distance between points of any given set, say X. An intuitive reasoning tells us that the procedure seems simple if we know how to calculate the lengths of line segments. But this has a problem! We are provided with a set X, where making these line segments need not be always possible. Making line segments is possible when we have what is called an algebraic structure, in particular two operations that can be called addition and scalar multiplication, and indeed, a set of scalars called a field.[4]

The first thing we wish to do in this section is to avoid all these restrictions. To do so, let us look what the distance between two points means mathematically. Firstly, to find the distance, we need two points (values) which need not be distinct and what we get after measuring the distance is a non-negative real number (our intuition tells us that distance cannot be negative). Therefore, we can look at distance as a function $d : X \times X \to [0, \infty)$. What other intuitive properties should distance satisfy?

1. The distance between two points should be zero only when the two points are identical and for every two distinct points, we should have a positive distance.

---

[4]If we have a vector space, say V, over a field $\mathbb{F}$, then we can easily make "line segments" in the space with the help of two operations "addition" and "scalar multiplication", which are available in the vector space. If we want to make a line segment between two points (vectors), say x and y in V, we form the set $L = \{\lambda \cdot x + (1 - \lambda) \cdot y | \lambda \in [0, 1]\}$. Notice that this is precisely the set containing all the points "between" x and y. Indeed, for $\lambda = 0$, we get y and for $\lambda = 1$, we get x.

2. It should not matter from where we measure the distance. If x and y are two points in the given set X, we would want the distance from x to y and the distance from y to x to remain the same, i.e., $d(x,y) = d(y,x)$.

3. If we take a third stop z while measuring the distance between two points x and y, the distance measured by us cannot be less than the distance between the two points x and y. This will be later called *triangle inequality*.

Based on this intuition, we now make a formal definition of distance, which henceforth we shall call *metric*, and hence a set equipped with metric will be called a *metric space*.

Definition 2.1 (Metric). *Given a set X, a function $d : X \times X \to [0, \infty)$ is called a metric on X if $\forall x, y, z \in X$, the following properties hold.*

1. *$d(x,y) = 0$ if and only if $x = y$.*

2. *$d(x,y) = d(y,x)$.*   *[Symmetry]*

3. *$d(x,y) \le d(x,z) + d(z,y)$.*   *[Triangle inequality]*

A set X equipped with a metric d is called a metric space and is denoted as the tuple (X, d).

Remark.

1. In the definition, X is the universal set. It may or may not have other structures (such as vector space or field structure) in it. In general, throughout the text, we shall assume that the sets under consideration do not have any other structure on it except when mentioned explicitly.

2. Since in the text we shall be mostly dealing with metric spaces and not other type of "spaces", we shall not denote a metric space by the tuple (X, d) but by simply X, except when the context seems ambiguous or there is a need to specify the tuple.

As seen in the definition and discussion above it, rather than abstracting the concept of length of line segment, we have abstracted the concept of distance directly. But, by our experience, distance came as a result of length in $\mathbb{R}$. Since intuitively these distances must mean the same, our original idea of distance between two real numbers should satisfy our definition of distance. This is left as an exercise to the reader.

Exercise 2.3. Prove that the function $d : \mathbb{R} \times \mathbb{R} \to [0, \infty)$ defined as $d(x,y) = |x - y|$ is indeed a metric.

Now, let us look at examples of metric spaces, which we somewhat know by experience and to some extent, we also use them in our daily lives.

Example 2.1 (Euclidean metric). Let us consider a plane. We want to measure distance between two points on the plane and for a while, we assume that we can move freely on the plane. So, intelligent humans such as ourselves would travel along a straight line from one point to another and try to measure the distance. Doing so when we have measuring tapes is easy! But theoretically, we have nothing more than the definition of plane and some of its properties which we can exploit.

By definition, a plane is simply the set $\mathbb{R}^2$ of ordered pairs (x, y) where each entry is a real number. Inherently, when we write (x, y), a notion of axes comes to our mind. To reach the point (x, y) from our "origin" $(0, 0)$, we first need to travel along the X-axis for x units and then parallel to Y-axis for y units.

Now, suppose we want to measure the distance between two given points (x_1, y_1) and (x_2, y_2). Since we are free to move any way we want in $\mathbb{R}^2$, the most natural way is by walking straight to the point (x_2, y_2) from the point (x_1, y_1) and measuring the distance on the way. As discussed earlier, since with $\mathbb{R}^2$, the notion of X- and Y-axes automatically comes to our mind, a way to reach the point (x_2, y_2) from (x_1, y_1) is by first getting to the X-axis by traveling y_1 units, then traveling along the X-axis for $|x_1 - x_2|$ units and then traveling a distance y_2 units parallel to the Y-axis. So, how much distance did we travel? As shown in Figure 2.2, parallel to the X-axis, we moved $|x_1 - x_2|$

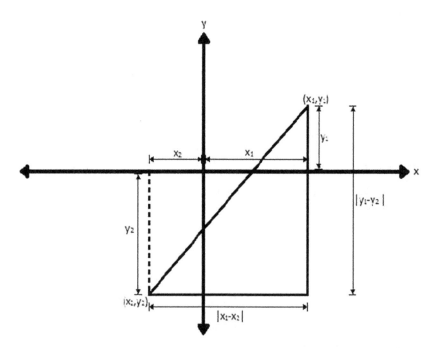

FIGURE 2.2: Measuring distance from (x_1, y_1) to (x_2, y_2) in $\mathbb{R}^2$.

units and parallel to the Y-axis, we moved $|y_1 - y_2|$ units. But initially, we wanted to measure the distance when we traveled along a straight line between the points. Pythagoras tells us that this distance is given by

$$d\left((x_1, y_1), (x_2, y_2)\right) = \sqrt{|x_1 - x_2|^2 + |y_1 - y_2|^2}. \qquad (2.7)$$

Thus, we have a method to find the distance between two points on the plane by the function $d : \mathbb{R}^2 \times \mathbb{R}^2 \to [0, \infty)$, where d is defined as in Equation (2.7). Now, our next job is to check if this definition is really a metric on $\mathbb{R}^2$. First, we see that d is well defined since by definition, we take the non-negative square root of a non-negative number. Next, let us see when can the distance between two points be 0, i.e., when is $d\left((x_1, y_1), (x_2, y_2)\right) = 0$?

$$\sqrt{|x_1 - x_2|^2 + |y_1 - y_2|^2} = 0 \Leftrightarrow |x_1 - x_2|^2 + |y_1 - y_2|^2 = 0,$$
$$\Leftrightarrow |x_1 - x_2|^2 = 0 \text{ and } |y_1 - y_2|^2 = 0,$$
$$\Leftrightarrow |x_1 - x_2| = 0 \text{ and } |y_1 - y_2| = 0,$$
$$\Leftrightarrow x_1 = x_2 \text{ and } y_1 = y_2,$$
$$\Leftrightarrow (x_1, y_1) = (x_2, y_2).$$

The reader should note that in the above proof, we have used many properties from analysis such as sum of two non-negative numbers is zero if and only if each of them is zero, and the equality of ordered pairs. Here, we advise the reader to take a pause and go through the proof to assimilate it!

We have proved that $d\left((x_1, y_1), (x_2, y_2)\right) = 0$ if and only if $(x_1, y_1) = (x_2, y_2)$. It is clear from the very definition of d that $d\left((x_1, y_1), (x_2, y_2)\right) = d\left((x_2, y_2), (x_1, y_1)\right)$, i.e., it does not matter from where we measure the distance. Also, in the Minkowski's inequality [Equation (2.3)] for $n = 2$ and $p = 2$, we get the triangle inequality. Hence, d is indeed a metric. This is called the *Euclidean metric* on the plane.

Just as in the above example, we can define a Euclidean metric in any of the $\mathbb{R}^n$ by simply doing the same procedure for each axis. We do not mention it specifically here since making an illustration for $\mathbb{R}^3$ is difficult on a page and anything beyond $\mathbb{R}^3$ cannot be illustrated through diagrams but can be explained mathematically. Thus, if we are given what is called the n-space, $\mathbb{R}^n$ of n-tuples, we can define the Euclidean metric as $d : \mathbb{R}^n \times \mathbb{R}^n \to [0, \infty)$,

$$d\left(\mathbf{x}, \mathbf{y}\right) = \left(\sum_{i=1}^{n} |x_i - y_i|^2\right)^{\frac{1}{2}}. \qquad (2.8)$$

where $\mathbf{x} = (x_1, x_2, \cdots, x_n)$ and $\mathbf{y} = (y_1, y_2, \cdots, y_n)$ are elements (n-tuples) in $\mathbb{R}^n$.

Exercise 2.4. Prove that the function d as in Equation (2.8) is indeed a metric on $\mathbb{R}^n$.

Remark. Since the way of finding distance which led us to defining the Euclidean metric was to go from one point to another along a straight line, it is also sometimes called the *crow-fly distance*.

At this point, one can ask a question: Is this the only way to define a metric on $\mathbb{R}^n$? Well, the answer is no! Another way to define a metric on $\mathbb{R}^n$ can be found in the method by which we defined the Euclidean metric. We see that in another example where we define a metric again on $\mathbb{R}^2$.

Example 2.2 (Taxi-Cab metric/Manhattan metric). Let us consider again the problem of finding the distance between two points (x_1, y_1) and (x_2, y_2) in $\mathbb{R}^2$. We follow a similar procedure but this time we do not go to Pythagoras for help. Instead, we measure the distance we walk along or parallel to the axes. How much distance did we travel along the X-axis? It was $|x_1 - x_2|$. Similarly, we traveled a distance $|y_1 - y_2|$ parallel to the Y-axis. So, for us the distance we traveled will be

$$d\left((x_1, y_1), (x_2, y_2)\right) = |x_1 - x_2| + |y_1 - y_2|. \tag{2.9}$$

Of course, it is the next task to verify that this function $d : \mathbb{R}^2 \times \mathbb{R}^2 \to [0, \infty)$ is indeed a metric. However, we do not show the proof here. The proof is easy and involves even fewer arguments than the proof for the Euclidean metric. Hence, we shall leave it as an exercise to the reader.

Instead, we make a comment on the name of the metric. Often, it is called the taxi-cab metric or the Manhattan metric. This is because, if one looks at the map of Manhattan, they would observe that to go from one place to another, only paths along two perpendicular roads (which in our case were parallel to the X-axis and the Y-axis) can be taken. So is the case with taxis. They cannot "fly" or go "through" the buildings. They have to follow the roads. Hence, the name of this metric follows.

Exercise 2.5. Verify that the function $d : \mathbb{R}^n \times \mathbb{R}^n \to [0, \infty)$ defined as

$$d\left(\mathbf{x}, \mathbf{y}\right) = \sum_{i=1}^{n} |x_i - y_i|. \tag{2.10}$$

is a metric, where $\mathbf{x} = (x_1, x_2, \cdots, x_n)$ and $\mathbf{y} = (y_1, y_2, \cdots, y_n)$.

Again, these are not the only ways to define distance. Given a $p \geq 1$, we can define distance on $\mathbb{R}^n$ as

$$d\left(\mathbf{x}, \mathbf{y}\right) = \left(\sum_{i=1}^{n} |x_i - y_i|^p\right)^{\frac{1}{p}} \tag{2.11}$$

where $\mathbf{x} = (x_1, x_2, \cdots, x_n)$ and $\mathbf{y} = (y_1, y_2, \cdots, y_n)$. It is left as an exercise to the reader to verify that d is indeed a metric. As a hint, we advise the reader to use Minkowski's inequalities.

We can observe that for $p = 2$, Equation (2.11) gives us our Euclidean distance and for $p = 1$, it gives the Taxi-Cab/Manhattan distance.

Let us take a step ahead and instead of being limited in finite tuples, let us go for infinite tuples. First, let us understand what is even meant by infinite tuples. In analysis, we have seen real-valued sequences. By definition they are just functions which map every element in $\mathbb{N}$ to an element in $\mathbb{R}$ and the nth term of the sequence is denoted by x_n. Clearly, if we change the order of two elements in the sequence, the sequence changes! Hence, we shall refer to these sequences as infinite tuples and the set of all real-valued sequences as $\mathbb{R}^{\mathbb{N}}$. As a matter of notations, we shall denote sequences as $(x_i)_{i=1}^{\infty}$, or $(x_i)_{i \in \mathbb{N}}$, and they will all be elements of $\mathbb{R}^{\mathbb{N}}$.

At a first glance, it seems that we have a structure similar to $\mathbb{R}^n$ apart from the fact that now we have infinite tuples, instead of finite tuples. So, let us try defining the Euclidean metric on $\mathbb{R}^{\mathbb{N}}$. Define $d : \mathbb{R}^{\mathbb{N}} \times \mathbb{R}^{\mathbb{N}} \to [0, \infty)$ as

$$d\left((x_i)_{i=1}^{\infty}, (y_i)_{i=1}^{\infty}\right) = \sqrt{\sum_{i=1}^{\infty} |x_i - y_i|^2}.$$

As usual, our next task will be to verify that d is indeed a metric. For that d has to be well-defined. Clearly, we have taken the non-negative square root and hence d will never give a negative output. But will it give any output in $\mathbb{R}$? Since we are taking non-negative square root of a sum of non-negative real numbers, we are guaranteed that whenever there is an output, it will be a real number. The crucial question that we must ask is: Will we always get an output for any choice of $(x_i)_{i=1}^{\infty}$ and $(y_i)_{i=1}^{\infty}$? Since we are dealing with infinite sums, we must check the convergence of series in play, and we know from analysis that not all series converge. One such series which we know does not converge is $\sum_{i=1}^{\infty} 1$. So, if for some choice of $(x_i)_{i=1}^{\infty}$ and $(y_i)_{i=1}^{\infty}$, we get this series, d will not be well defined.

Well, one such choice is $(x_i)_{i=1}^{\infty} = (1)_{i=1}^{\infty}$, the constant sequence which takes the value 1 everywhere, and $(y_i)_{i=1}^{\infty} = (0)_{i=1}^{\infty}$, the constant sequence which takes the value 0 everywhere. Hence, d is not even well-defined on $\mathbb{R}^{\mathbb{N}}$!

However, based on this observation, we can in fact create sets and then define metric on them. This is shown in the following example.

Example 2.3 (Sequence spaces). Let $p \geq 1$ be given. Define a set $\ell^p = \left\{(x_i)_{i=1}^{\infty} \in \mathbb{R}^{\mathbb{N}} \,\middle|\, \sum_{i=1}^{\infty} |x_i|^p < \infty\right\}$. Here, by $\sum_{i=1}^{\infty} |x_i|^p < \infty$, we mean that the series is convergent. Now, we define $d : \ell^p \times \ell^p \to [0, \infty)$ as

$$d\left((x_i)_{i=1}^{\infty}, (y_i)_{i=1}^{\infty}\right) = \left(\sum_{i=1}^{\infty} |x_i - y_i|^p\right)^{\frac{1}{p}}. \tag{2.12}$$

First, we remember that $f : [0, \infty) \to [0, \infty)$ defined by $f(x) = x^p$ for $p \geq 1$ is a convex function. Thus, we have

$$\left| \frac{x_i}{2} + \left(-\frac{y_i}{2} \right) \right|^p \leq \frac{1}{2} |x_i|^p + \frac{1}{2} |-y_i|^p.$$

$$\therefore \frac{1}{2^p} |x_i - y_i|^p \leq \frac{1}{2} (|x_i|^p + |y_i|^p).$$

$$\therefore |x_i - y_i|^p \leq 2^{p-1} (|x_i|^p + |y_i|^p).$$

Once this is achieved, we make an observation: the sequence $|x_i|^p$ has all terms non-negative and hence the sequence of partial sums $\sum\limits_{i=1}^{n} |x_i|^p$ is non-decreasing. Therefore, by Theorem 2.4, we can say that for every $n \in \mathbb{N}$, $\sum\limits_{i=1}^{n} |x_i|^p \leq \sum\limits_{i=1}^{\infty} |x_i|^p$. Hence, we have

$$\sum_{i=1}^{n} |x_i - y_i|^p \leq 2^{p-1} \sum_{i=1}^{n} (|x_i|^p + |y_i|^p)$$

$$= 2^{p-1} \left(\sum_{i=1}^{n} |x_i|^p + \sum_{i=1}^{n} |y_i|^p \right)$$

$$\leq 2^{p-1} \left(\sum_{i=1}^{\infty} |x_i|^p + \sum_{i=1}^{\infty} |y_i|^p \right).$$

Thus, the set of partial sums $\left\{ \sum\limits_{i=1}^{n} |x_i - y_i|^p \,\middle|\, n \in \mathbb{N} \right\}$ is bounded and by Theorem 2.4, we conclude that the series $\sum\limits_{i=1}^{\infty} |x_i - y_i|^p$ is convergent.

Therefore, d is well-defined. When will $d\left((x_i)_{i=1}^{\infty}, (y_i)_{i=1}^{\infty} \right)$ be 0?

$$\left(\sum_{i=1}^{\infty} |x_i - y_i|^p \right)^{\frac{1}{p}} = 0 \Leftrightarrow \sum_{i=1}^{\infty} |x_i - y_i|^p = 0,$$

$$\Leftrightarrow \sup_{n \in \mathbb{N}} \left\{ \sum_{i=1}^{n} |x_i - y_i|^p \right\} = 0,$$

$$\Leftrightarrow \forall n \in \mathbb{N}, \sum_{i=1}^{n} |x_i - y_i|^p = 0,$$

$$\Leftrightarrow \forall i \in \mathbb{N}, |x_i - y_i| = 0,$$

$$\Leftrightarrow \forall i \in \mathbb{N}, x_i = y_i,$$

$$\Leftrightarrow (x_i)_{i=1}^{\infty} = (y_i)_{i=1}^{\infty}.$$

Thus, we have proved that $d\left((x_i)_{i=1}^{\infty}, (y_i)_{i=1}^{\infty}\right) = 0$ if and only if $(x_i)_{i=1}^{\infty} = (y_i)_{i=1}^{\infty}$. Also, it is clear from the definition of d that $d\left((x_i)_{i=1}^{\infty}, (y_i)_{i=1}^{\infty}\right) = d\left((y_i)_{i=1}^{\infty}, (x_i)_{i=1}^{\infty}\right)$ and Minkowski's inequality [Equation (2.4)] gives us the triangle inequality. Therefore, ℓ^p is a metric space with the metric d defined as above.

In the sequence spaces discussed above, all the sequences in any of the ℓ^p have a special property: all the sequences $(x_i)_{i=1}^{\infty}$ converge to 0 or otherwise the series $\sum_{i=1}^{\infty} |x_i|^p$ would not converge. This seems like a bit of limitation to the sequences we have in our hand to play with! So, instead of sequences which converge to 0, we now consider sequences which are bounded. These may or may not be convergent.

Example 2.4. Define a set

$$\ell^{\infty} = \left\{ (x_i)_{i=1}^{\infty} \in \mathbb{R}^{\mathbb{N}} \mid \exists M > 0 \text{ such that } \forall i \in \mathbb{N} \text{ we have } |x_i| \leq M \right\},$$

namely the set of all bounded sequences. Now, instead of trying to get a distance in terms of series, we find the distance between corresponding values of given sequences. The distance between these two sequences will be the supremum of the individual distances. To find distance between each corresponding value, we use the absolute value function we know from $\mathbb{R}$. The only question that remains to answer is: Will the supremum always exist?

To answer this question, we first observe that for any $i \in \mathbb{N}$, $|x_i - y_i| \leq |x_i| + |y_i|$. Since $(x_i)_{i=1}^{\infty}$ and $(y_i)_{i=1}^{\infty}$ are bounded sequences, $\exists M_1, M_2 > 0$ such that $\forall i \in \mathbb{N}$ we have $|x_i| \leq M_1$ and $|y_i| \leq M_2$. Hence, the set $\{|x_i - y_i| \mid i \in \mathbb{N}\}$ is bounded above by $M_1 + M_2$ and by the LUB axiom in $\mathbb{R}$, we can conclude that the supremum exists.

Therefore, we can now define $d : \ell^{\infty} \times \ell^{\infty} \to [0, \infty)$ as

$$d\left((x_i)_{i=1}^{\infty}, (y_i)_{i=1}^{\infty}\right) = \sup_{i \in \mathbb{N}} \{|x_i - y_i|\} \tag{2.13}$$

Let us now check if this function d forms a metric. We have already shown that d is well-defined. Let us now see when does $d\left((x_i)_{i=1}^{\infty}, (y_i)_{i=1}^{\infty}\right) = 0$.

$$\sup_{i \in \mathbb{N}} \{|x_i - y_i|\} = 0 \Leftrightarrow \forall i \in \mathbb{N}, |x_i - y_i| = 0,$$

$$\Leftrightarrow \forall i \in \mathbb{N}, x_i = y_i,$$

$$\Leftrightarrow (x_i)_{i=1}^{\infty} = (y_i)_{i=1}^{\infty}.$$

Therefore, we have shown that $d\left((x_i)_{i=1}^{\infty}, (y_i)_{i=1}^{\infty}\right) = 0$ if and only if $(x_i)_{i=1}^{\infty} = (y_i)_{i=1}^{\infty}$. From the definition, it follows that $d\left((x_i)_{i=1}^{\infty}, (y_i)_{i=1}^{\infty}\right) = d\left((y_i)_{i=1}^{\infty}, (x_i)_{i=1}^{\infty}\right)$. However, this time, we cannot seek help from Minkowski to prove the triangle inequality! To see if the triangle inequality holds, consider the following for three sequences $(x_i)_{i=1}^\infty, (y_i)_{i=1}^\infty, (z_i)_{i=1}^\infty \in \ell^\infty$.

$$|x_i - y_i| \le |x_i - z_i| + |z_i - y_i|$$

$$\le \sup_{i \in \mathbb{N}} \{|x_i - z_i|\} + \sup_{i \in \mathbb{N}} \{|z_i - y_i|\}$$

$$= d\left((x_i)_{i=1}^\infty, (z_i)_{i=1}^\infty\right) + d\left((z_i)_{i=1}^\infty, (y_i)_{i=1}^\infty\right).$$

This is true for any $i \in \mathbb{N}$. Therefore, the set $\{|x_i - y_i| \,|\, i \in \mathbb{N}\}$ is bounded above by $d\left((x_i)_{i=1}^\infty, (z_i)_{i=1}^\infty\right) + d\left((z_i)_{i=1}^\infty, (y_i)_{i=1}^\infty\right)$. Hence,

$$\sup_{i \in \mathbb{N}} \{|x_i - y_i|\} = d\left((x_i)_{i=1}^\infty, (y_i)_{i=1}^\infty\right) \le d\left((x_i)_{i=1}^\infty, (z_i)_{i=1}^\infty\right) + d\left((z_i)_{i=1}^\infty, (y_i)_{i=1}^\infty\right).$$

Hence, the triangle inequality holds and we conclude that ℓ^∞ is indeed a metric space.

In the example above, we found out the distance between two sequences using the supremum of the distance between corresponding values of each sequence. A natural question to ask at this time is: Why only supremum? We could have as well taken infimum of the set $\{|x_i - y_i| \,|\, i \in \mathbb{N}\}$, which we know exists since the set is bounded below by 0. Let us, for once, try defining the distance $d : \ell^\infty \times \ell^\infty \to [0, \infty)$ as

$$d\left((x_i)_{i=1}^\infty, (y_i)_{i=1}^\infty\right) = \inf_{i \in \mathbb{N}} \{|x_i - y_i|\}.$$

Clearly, this function d is well defined. Let us now see when do we get $d\left((x_i)_{i=1}^\infty, (y_i)_{i=1}^\infty\right) = 0$.

$$\inf_{i \in \mathbb{N}} \{|x_i - y_i|\} = 0 \Rightarrow \forall i \in \mathbb{N}, |x_i - y_i| \ge 0.$$

But, this is something we already know! Infimum of a set being zero is not equivalent to each element in the set being 0. Therefore, it seems that there is some problem with this definition of distance. In what axiom of metric does the problem occur? The problem, as seen from this inspection, is that there may be distinct sequences in ℓ^∞ whose distance may come out to be zero, i.e., $(x_i)_{i=1}^\infty \ne (y_i)_{i=1}^\infty$ but $\inf_{i \in \mathbb{N}} \{|x_i - y_i|\} = 0$. Let us try constructing such sequences.

We know that the sequence $\left(\frac{1}{i}\right)_{i=1}^\infty$ converges to 0 and moreover, $\inf_{i \in \mathbb{N}} \{\frac{1}{i}\} = 0$. Therefore, if we choose $(x_i)_{i=1}^\infty = \left(\frac{1}{i}\right)_{i=1}^\infty$ and $(y_i)_{i=1}^\infty = (0)_{i=1}^\infty$ (the constant sequence which takes the value 0 everywhere) then we get

$$d\left((x_i)_{i=1}^\infty, (y_i)_{i=1}^\infty\right) = \inf_{i \in \mathbb{N}} \left\{\left|\frac{1}{i} - 0\right|\right\}$$

$$= \inf_{i \in \mathbb{N}} \left\{\frac{1}{i}\right\} = 0.$$

Clearly, $\left(\frac{1}{i}\right)_{i=1}^\infty \ne (0)_{i=1}^\infty$. Hence, d cannot form a metric!

Exercise 2.6. Analogous to the metric defined on ℓ^∞, we try to define another metric on $\mathbb{R}^n$ as $d : \mathbb{R}^n \times \mathbb{R}^n \to [0, \infty)$,

$$d(\mathbf{x}, \mathbf{y}) = \max\{|x_i - y_i| \,|\, 1 \le i \le n\}.$$

where $\mathbf{x} = (x_1, x_2, \cdots, x_n)$ and $\mathbf{y} = (y_1, y_2, \cdots, y_n)$. Check if d forms a metric.

Remark. The sequence spaces ℓ^p, all depend on the choice of p. Hence, the metric on ℓ^p is usually denoted by d_p. Even on $\mathbb{R}^n$, this notation can be followed. Therefore, d_1 is the taxi-cab metric and d_2 is the Euclidean metric. Also, as a notation, we shall use d_∞ to denote the supremum or maximum metric.

A natural question one can ask about the d_p metric is: Why should p be at least 1? What would happen is we take $p < 1$? First, let us take the case where $p = 0$. We try to define the metric $d_0 : \mathbb{R}^n \times \mathbb{R}^n \to [0, \infty)$ as

$$d_0(\mathbf{x}, \mathbf{y}) = \left(\sum_{i=1}^n |x_i - y_i|^p \right)^{\frac{1}{p}}.$$

where $\mathbf{x} = (x_1, x_2, \cdots, x_n), \mathbf{y} = (y_1, y_2, \cdots, y_n)$, and $p = 0$. Clearly, this is not defined! Hence, we cannot have a d_0 metric.

Let us try $p < 0$. In that case, if $x_i = y_i$ for some i, then $|x_i - y_i|^p$ will not be defined and overall d_p will not be defined. Now, we are left with $0 < p < 1$. There does not seem any problem with at least the definition of d_p. Also, the two properties that follow while checking that d_p is a metric do not really depend on the choice of p. What does depend on the choice of p are the Minkowski's inequalities, and hence if $0 < p < 1$, our guess is that there will be some problem in the triangle inequality. Therefore, let us try to find three elements $\mathbf{x} = (x_1, x_2, \cdots, x_n), \mathbf{y} = (y_1, y_2, \cdots, y_n)$, and $\mathbf{z} = (z_1, z_2, \cdots, z_n) \in \mathbb{R}^n$ such that $d_p(\mathbf{x}, \mathbf{y}) > d_p(\mathbf{x}, \mathbf{z}) + d_p(\mathbf{y}, \mathbf{z})$.
Consider $\mathbf{x} = (1, 0, \cdots, 0), \mathbf{y} = (0, 1, 0, \cdots, 0)$, and $\mathbf{z} = (0, 0, \cdots, 0)$. Then,

$$d_p(\mathbf{x}, \mathbf{y}) = (|1|^p + |-1|^p)^{\frac{1}{p}}$$
$$= (1^p + 1^p)^{\frac{1}{p}}$$
$$= 2^{\frac{1}{p}}.$$
$$d_p(\mathbf{x}, \mathbf{z}) = (|1|^p)^{\frac{1}{p}} = 1.$$
$$d_p(\mathbf{z}, \mathbf{y}) = (|-1|^p)^{\frac{1}{p}} = 1.$$

Now, since $0 < p < 1$, we have $\dfrac{1}{p} > 1$ and hence $2^{\frac{1}{p}} > 2$. Therefore,

$$d_p(\mathbf{x}, \mathbf{y}) = 2^{\frac{1}{p}} > 2 = d_p(\mathbf{x}, \mathbf{z}) + d_p(\mathbf{z}, \mathbf{y}).$$

However, it seems that our proof works for $\mathbb{R}^n$, where $n \ge 2$. What about $\mathbb{R}$?

First, we observe that for any $p > 0$, $(|x|^p)^{\frac{1}{p}} = |x|$. Hence, for any choice of $p > 0$, on $\mathbb{R}$ we have

$$d_p(x, y) = (|x - y|^p)^{\frac{1}{p}} = |x - y|.$$

Therefore, in $\mathbb{R}$, we have for any p, only one metric, namely, the absolute value.

Note. In a similar manner, one can prove that even for the sequence spaces, we need to have $p \geq 1$. This is left as an exercise to the reader.

Exercise 2.7. Consider the set $M_{m \times n}(\mathbb{R})$ of all $m \times n$ matrices with real entries. Define the d_p metric for $p \geq 1$ on $M_{m \times n}(\mathbb{R})$. **Hint:** Notice that an $m \times n$ matrix contains mn elements in which changing any two distinct elements changes the matrix.

Now, let us define a new metrics on $\mathbb{R}^2$ which does not involve p.

Example 2.5. Suppose in $\mathbb{R}^2$, we have a restriction that to go from one point to another, we need to compulsorily go through the point $(0, 0)$. For notational purpose, we shall denote this point, as $\mathbf{0}$. Suppose we know the distance of each point (x, y) from $\mathbf{0}$. To calculate the distance between the two points (x_1, y_1) and (x_2, y_2), all we need to do is, first reach $\mathbf{0}$ from (x_1, y_1) and then go to (x_2, y_2). The distance will be the total of the distance we would have traveled in the process. Mathematically, we can write $d : \mathbb{R}^2 \times \mathbb{R}^2 \to [0, \infty)$ as

$$d((x_1, y_1), (x_2, y_2)) = \begin{cases} 0, & \text{if } (x_1, y_1) = (x_2, y_2) \\ d'((x_1, y_1), \mathbf{0}) + d'((x_2, y_2), \mathbf{0}), & \text{otherwise} \end{cases}$$

where d' is known to us.

All the axioms of metric, except the triangle inequality, are satisfied automatically as a consequence of its definition. Therefore, our task is to verify the triangle inequality. If $(x_1, y_1) = (x_2, y_2)$, then for any $(x_3, y_3) \in \mathbb{R}^2$, we have $d((x_1, y_1), (x_2, y_2)) = 0 \leq d((x_1, y_1), (x_3, y_3)) + d((x_3, y_3), (x_2, y_2))$.

If $(x_1, y_1) \neq (x_2, y_2)$, then $d((x_1, y_1), (x_2, y_2)) = d'((x_1, y_1), \mathbf{0}) + d'((x_2, y_2), \mathbf{0})$. Since d' is non-negative, for any $(x_3, y_3) \in \mathbb{R}^2$, we have

$$d'((x_1, y_1), \mathbf{0}) \leq d'((x_1, y_1), \mathbf{0}) + d'((x_3, y_3), \mathbf{0}),$$

$$d'((x_2, y_2), \mathbf{0}) \leq d'((x_2, y_2), \mathbf{0}) + d'((x_3, y_3), \mathbf{0}).$$

Hence, we have

$$d((x_1, y_1), (x_2, y_2)) = d'((x_1, y_1), \mathbf{0}) + d'((x_2, y_2), \mathbf{0})$$
$$\leq d'((x_1, y_1), \mathbf{0}) + d'((x_3, y_3), \mathbf{0})$$
$$+ d'((x_2, y_2), \mathbf{0}) + d'((x_3, y_3), \mathbf{0})$$
$$= d((x_1, y_1), (x_3, y_3)) + d((x_2, y_2), (x_3, y_3)).$$

Thus, the triangle inequality holds and d is indeed a metric on $\mathbb{R}^2$.

Exercise 2.8. In Example 2.5, while proving the triangle inequality we inherently assumed that $(x_1, y_1) \neq (x_2, y_2) \neq (x_3, y_3)$. Give a proof for triangle inequality when $(x_1, y_1) = (x_3, y_3)$ or $(x_3, y_3) = (x_2, y_2)$.

Remark. While defining the metric above, we did not need any property of $\mathbb{R}^2$ as such. All we needed was a point from where the distance to every other point is known. Once this is achieved, we can define a metric such as above on any set X. So, in the abstract sense, if we fix a point $x_0 \in X$ from where the distance to every other point in X is known, we can always define a metric as $d : X \times X \to [0, \infty)$,

$$d(x, y) = \begin{cases} 0, & \text{if } x = y. \\ d'(x, x_0) + d'(y, x_0), & \text{otherwise.} \end{cases}$$

In a similar manner, we shall now define metric on an arbitrary set X without knowing the properties of its elements.

Example 2.6 (Discrete metric). Let X be a given set. We define the distance between any two points as 1 if the points are distinct and 0 if they are the same. Such a metric is called *discrete metric*. We write $d : X \times X \to [0, \infty)$ as

$$d(x, y) = \begin{cases} 1, & x \neq y. \\ 0, & x = y. \end{cases} \tag{2.14}$$

Clearly, all the axioms of metric are satisfied by d directly from the definition, except for the triangle inequality. For $x, y, z \in X$, we have we have $d(x, y) \leq 1$. If $x = y$, then $d(x, y) = 0$ and then $d(x, z) + d(y, z) \geq 0 = d(x, y)$. Therefore, we consider $x \neq y$. Then, $d(x, y) = 1$. Now, we have two cases: either $x = z$ (or $y = z$, but not both) or $x \neq y \neq z$. In the first case $d(x, z) + d(y, z) = 1 = d(x, y)$. In the second case, $d(x, z) + d(y, z) = 2 > d(x, y)$. Therefore, for any $x, y, z \in X$, the triangle inequality holds and d forms a metric.

Remark. From Example 2.6, we can say that given any set X, we always have at least one metric on it, namely the discrete metric. Therefore, any set can be a metric space irrespective of whether or not it has algebraic properties as in $\mathbb{R}^n$.

Example 2.7 (Hilbert cube). From the sequence spaces discussed above, let us take into consideration, special type of sequences. We make a set $I^\infty = \left\{ (x_i)_{i=1}^\infty \in \mathbb{R}^\mathbb{N} \middle| \forall i \in \mathbb{N}, 0 \leq x_i \leq \frac{1}{i} \right\}$. This set is called the *Hilbert cube*. We define a metric on I^∞ as $d : I^\infty \times I^\infty \to [0, \infty)$,

$$d\left((x_i)_{i=1}^\infty, (y_i)_{i=1}^\infty\right) = \left(\sum_{i=1}^\infty |x_i - y_i|^2 \right)^{\frac{1}{2}}. \tag{2.15}$$

It seems that we have taken a set and applied the metric from the ℓ^2 metric space. Therefore, it will be enough for us to prove that d is well-defined. All other properties are shown to hold in ℓ^2. Note that d will be well-defined if the series $\sum_{i=1}^{\infty} |x_i - y_i|^2$ converges. We note that from the construction of I^{∞}, we have

$$-\frac{1}{i} \le x_i - y_i \le \frac{1}{i},$$

i.e., $|x_i - y_i| \le \frac{1}{i}$ and hence $|x_i - y_i|^2 \le \frac{1}{i^2}$. From the comparison test, we now know that $\sum_{i=1}^{\infty} |x_i - y_i|^2$ converges and hence d is well defined. Therefore, we have another metric space!

Exercise 2.9. Just as in the above example, can we also borrow metric from any l^p space for $p \ge 1$ (or l^{∞} space) and define a metric on Hilbert cube? Justify.

Exercise 2.10. Consider the set $\mathbb{R}^{\mathbb{N}}$ of all real-valued sequences. If we denote any element in $\mathbb{R}^{\mathbb{N}}$ as $x = (x_i)_{i=1}^{\infty}$, does the following function define a metric?

$$d : \mathbb{R}^{\mathbb{N}} \times \mathbb{R}^{\mathbb{N}} \to \mathbb{R}, \qquad d(x, y) = \sum_{i=1}^{\infty} \frac{1}{2^i} \frac{|x_i - y_i|}{1 + |x_i - y_i|}.$$

Tuples and sequences are not the only spaces we can define metric on. Another useful space is given in the next example.

Example 2.8 (Function space). Consider the space $\mathscr{C}[a, b]$ of all real-valued, continuous functions defined on the interval $[a, b]$. Given any two functions f and g from this space, we wish to find the distance between them (Figure 2.3). A natural thinking would lead us to determining the distance of the values

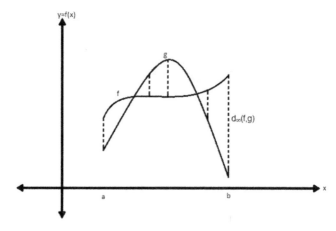

FIGURE 2.3: Physical interpretation of d_{∞}.

of the functions at each point in $[a, b]$ and then using that distance to find the distance between the functions (by taking the supremum). Therefore, we define our metric as $d_\infty : \mathscr{C}[a, b] \times \mathscr{C}[a, b] \to [0, \infty)$, where

$$d_\infty (f, g) = \sup_{x \in [a,b]} \{|f(x) - g(x)|\}. \tag{2.16}$$

Here, the notation d_∞ is borrowed from the notations of the sequence space. The proof that this indeed forms a metric is analogous to that in the sequence spaces and hence is left as an exercise to the reader.

Also, this is not the only way to define a metric on $\mathscr{C}[a, b]$. Another way to do so is by measuring the area between given functions. To get an intuition that this will form a distance, we first observe that if there are two distinct continuous functions, they will have some positive area between them. Hence, by this definition, the distance between two functions can be zero if and only if the functions are identical. Now, measurement of area does not depend from where we start. Therefore, we may define the metric as

$$d_1 (f, g) = \int_a^b |f(x) - g(x)|\ dx. \tag{2.17}$$

To see that the triangle inequality holds, consider the following

$$d_1 (f, g) = \int_a^b |f(x) - g(x)|\ dx$$

$$= \int_a^b |f(x) - h(x) + h(x) - g(x)|\ dx$$

$$\leq \int_a^b |f(x) - h(x)|\ dx + \int_a^b |h(x) - g(x)|\ dx$$

$$= d(f, h) + d(h, g).$$

Hence, d_1 indeed forms a metric (Figure 2.4). Similarly, we can also have for $p > 1$, a metric

$$d_p (f, g) = \left(\int_a^b |f(x) - g(x)|^p\ dx \right)^{\frac{1}{p}}. \tag{2.18}$$

Exercise 2.11. Prove that for $p \geq 1$, the function d_p defined above is indeed a metric. Also prove that d_∞ is a metric.

Finally, we see another example of a metric space which is not trivially constructed.

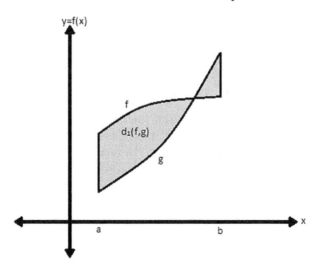

FIGURE 2.4: Physical interpretation of d_1.

Example 2.9 (*p*-adic metric). Consider the set of rational numbers $\mathbb{Q}$ and fix a prime p. Now, any element in $\mathbb{Q}$ is of the form $\frac{a}{b}$, where $a \in \mathbb{Z}$ and $b \in \mathbb{N}$. We now take out p from both the numerator and denominator and cancel any terms that can be canceled. So, any $x \in \mathbb{Q}$ can now be written as $x = p^k \frac{m}{n}$ so that $k \in \mathbb{Z}$ and p neither divides m nor n. Here, this integer k is special to us. That integer tells us "how much" of p do we have in our rational number. We call it the *valuation number* for x and write it as $v_p(x)$.

The reader may note here that the valuation number may not be always non-negative. To see this, consider the rational number $\frac{1}{p}$, where p is the prime number fixed above. Then, the valuation number is -1, which is negative.

Here, we may ask ourselves a question: Can we find the valuation number for each rational number? At the first glance, it does seem so. If we have integers, we can always take out factors and the knowledge of factorization from basic arithmetic does guarantee us that we will stop at some stage. But looking closely to it, that guarantee is only given for natural numbers! So, if we take something which is not a natural number, we may problems. If we take a negative integer, it is just the additive inverse of a natural number so that its natural counterpart does the job to find the valuation number. However, for the number 0, how many p's can we take out? There is no limit to it! So, our valuation number is not really defined for 0. We shall come to this problem later.

The way we started defining the valuation number, our minds may have thought of a natural way to define a metric using them. That way is to simply find the distance between valuation numbers. This means, we can try defining the metric as $d : \mathbb{Q} \times \mathbb{Q} \rightarrow [0, \infty)$, where $d(x, y) = |v_p(x) - v_p(y)|$. But, the rational number 0 creates a problem. Any element of type $(x, 0)$ or $(0, y)$ in

$\mathbb{Q} \times \mathbb{Q}$ cannot have a distance defined! Well, these are just too many elements to manipulate. So, instead we try defining the metric as $d(x,y) = |v_p(x-y)|$. Now, only the elements of type (x,x) in $\mathbb{Q} \times \mathbb{Q}$ do not have distance defined for them.

But, we do know what to do with such elements in case of defining distance. Since (x,x) denotes a pair of identical elements, their distance must be zero! Thus, we now define our metric as

$$d(x,y) = \begin{cases} |v_p(x-y)|, & x \neq y. \\ 0, & x = y. \end{cases}$$

It does seem like things are getting in control! But, as the next step, we must check if this really forms a metric. Clearly, for any input (x,y), the function d gives a non-negative output since the absolute value function gives a non-negative output. Also, for identical elements in $\mathbb{Q}$, the distance is defined to be zero. However, we must check the reverse implication. Whenever the distance is zero, the elements should be identical. Is it always possible? What happens when $x - y$ does not have any p in its representation?

In fact, one can get easily a counterexample, where we take $x = p'$ to be any prime number other than p and $y = 0$. Therefore, $v_p(x - y) = v_p(p') = 0$. However, 0 is not a prime number and hence $x \neq y$. Therefore, the above definition of d does not define a metric!

It seems like all the trivial and natural options are exhausted. Looking at all the functions we might know from a course on analysis, the exponential function is never zero. In fact, it is always positive (when the base is positive). Since prime numbers are positive, we exploit this fact and define the metric as

$$d(x,y) = \begin{cases} p^{-v_p(x-y)}, & x \neq y. \\ 0, & x = y. \end{cases} \tag{2.19}$$

This definition clearly solves all problems of the first two axioms of defining a metric. Note that the second axiom is satisfied since a number and its negative will have the same valuation. This part is left as an exercise to the reader. We now look at the triangle inequality.

Let $x, y \in \mathbb{Q}$ be such that $x \neq 0$, $y \neq 0$, and $x \neq y$. Now, we have the representations $x = p^{k_1} \frac{m_1}{n_1}$ and $y = p^{k_2} \frac{m_2}{n_2}$. Therefore, $x - y = p^{k_1} \frac{m_1}{n_1} - p^{k_2} \frac{m_2}{n_2}$. Therefore, we can take out the minimum power of p and form another representation $p^k \frac{m}{n}$, where $k = \min\{k_1, k_2\}$. However, we cannot assure that p does not divide m or n. However, the usual method of calculating the subtraction of two rational number does tell us that since p does not divide n_1 and p does not divide n_2, it cannot divide their LCM which must be n. Hence, no p's taken out will be canceled off. However, there may be additional p's that could be taken out from the numerator m. Therefore, we make an observation here

$$v_p(x - y) \geq \min\{v_p(x), v_p(y)\}.$$

Therefore, we also have

$$p^{-v_p(x-y)} \leq p^{-\min\{v_p(x), v_p(y)\}}.$$

Now, let $z \in \mathbb{Q}$ be such that $x \neq z$ and $y \neq z$. Therefore, we have

$$x - y = (x - z) + (z - y)$$
$$= p^{\min\{v_p(x-z), v_p(z-y)\}} \frac{m}{n}.$$
$$\therefore d(x, y) \leq p^{\max\{-v_p(x-z), -v_p(z-y)\}}$$
$$= \max \left\{ p^{-v_p(x-z)}, p^{-v_p(z-y)} \right\}$$
$$= \max \{d(x, z), d(y, z)\}.$$

Hence, we have,

$$d(x, y) \leq d(x, z),$$
$$d(x, y) \leq d(y, z),$$
$$\therefore d(x, y) \leq \frac{d(x, z) + d(y, z)}{2} \leq d(x, z) + d(y, z).$$

Note that if $x = y$, $y = z$, or $x = z$, the triangle inequality gets converted to equality. Readers are suggested to check the claim and satisfy themselves.

Hence, we have another way to define metric on $\mathbb{Q}$. This is called the *p-adic metric* and depending on the value of p, we get different metrics.

Exercise 2.12. Define the 2-adic metric on $\mathbb{Q}$ by putting $p = 2$ in the above example. What is the distance between 1 and 3? What is the distance between 1024 and 32? Lastly, what is the distance between $\frac{125}{31}$ and $\frac{1}{31}$?

Exercise 2.13. In the above example, we defined the p-adic metric as $p^{-v_p(x-y)}$. Can we define it as $p^{v_p(x-y)}$? Justify.

Remark. While proving the triangle inequality for the p-adic metric, we in fact proved a stronger inequality

$$d(x, y) \leq \max\{d(x, z), d(y, z)\}.$$

This inequality is sometimes called the *ultra-metric inequality* and the p-adic metric is called an *ultra-metric* or a *super-metric*.

Now, we look at some examples where we can obtain new metrics from previously known metric spaces.

Example 2.10 (Bounded metric). Suppose we have (X, d), a metric space. We define another function $d' : X \times X \to [0, \infty)$ as

$$d'(x, y) = \min\{d(x, y), 1\}. \tag{2.20}$$

Clearly, this function is well-defined since the minimum of non-negative numbers is non-negative. Also, the function takes the value zero if and only if $d(x, y) = 0$ which is possible if and only if $x = y$. Since d is a metric, we have $d'(x, y) = d'(y, x)$. Finally, since $d(x, y) \leq d(x, z) + d(y, z)$, we have,

$$
\begin{aligned}
d'(x, y) = \min\{d(x, y), 1\} &\leq \min\{d(x, z) + d(y, z), 1\} \\
&\leq \min\{d(x, z), 1\} + \min\{d(y, z), 1\} \\
&= d'(x, z) + d'(y, z).
\end{aligned}
$$

Example 2.11 (Subspace). Given a metric space (X, d) and a subset $Y \subseteq X$, we can have the metric space (Y, d) by restricting d to the elements of Y. In this case, the universe now becomes Y and we shall act like we do not know what is outside of it. Clearly, (Y, d) is a metric space since it is given that d satisfies all the axioms. Here, Y is called the *subspace* of X.

Example 2.12 (Product metric). Given two metric spaces (X, d_1) and (Y, d_2), we can define a metric on the Cartesian product $X \times Y$ as $d : (X \times Y) \times (X \times Y) \to [0, \infty)$, where,

$$
d((x_1, y_1), (x_2, y_2)) = \max\{d_1(x_1, x_2), d_2(y_1, y_2)\}. \tag{2.21}
$$

The metric space $(X \times Y, d)$ is called the *product space* of X and Y, and the individual spaces (X and Y) are called the *coordinate spaces*. The proof that this is indeed a metric is left as an exercise to the reader.

Exercise 2.14. Prove that the product metric function defined above is indeed a metric. Can we get some motivation from $\mathbb{R}^n$ to have another metric on $X \times Y$?

2.4 Open Sets

Up until now, we have only been looking at some preliminaries and examples required for developing theory on metric spaces. From this section, we will start developing the theory of metric spaces. To do so, we will apply the same methodology as we have been for all the preceding sections. We take what we know from experience and then abstract its properties so that a similar concept can be defined on any set.

By our experience in the real numbers $\mathbb{R}$ taken as a metric space with the metric defined by absolute value function ($|\cdot|$), we know that open sets are one of the consequences of the absolute value. In $\mathbb{R}$, we define a set to be open, if for any point x in the set, we can get a room completely inside the set. Mathematically, we say that to each x, there is some positive real number ϵ so that the interval $(x - \epsilon, x + \epsilon)$ lies within the given set.

This particular interval $(x - \epsilon, x + \epsilon)$ is of special interest to us. If we can abstract the properties of this particular interval, generalizing open sets to any metric space will be an easy task. First, we see that if any $y \in (x - \epsilon, x + \epsilon)$, then we have the inequality

$$x - \epsilon < y < x + \epsilon.$$

This can also be written as

$$-\epsilon < y - x < \epsilon.$$

Which further can be written as

$$|x - y| < \epsilon.$$

Since the absolute value gives a method of finding the distances between two real numbers, we see that any element in $(x - \epsilon, x + \epsilon)$ is within a distance of ϵ from x. Also, the point x seems to be in the *center* of the interval.

Coming to abstract metric spaces, we want to have a similar concept to this interval. First, note that to define an interval on any set, we would require a partial order. However, it would then mean that any metric space without a partial order specified cannot have open sets! We do not wish this. This is because, originally, the open sets were a consequence of a metric in $\mathbb{R}$ and not its partial order. Therefore, we now take into account the properties discussed above of the interval $(x - \epsilon, x + \epsilon)$ and define a special type of a set.

Definition 2.2 (Open ball). *In a metric space (X, d), we define the open ball, centered at a point $x \in X$ and of radius $r > 0$ as the following set*

$$B(x, r) = \{y \in X | d(x, y) < r\}.$$

We may ask ourselves a question: Can any open ball be empty? To answer it, we first observe that since $r > 0$ and $d(x, x) = 0 < r$, we always have $x \in B(x, r)$. Hence, every open ball is a non-empty set and it always contains its center.

Remark. The name open *ball* is motivated from the structure of the set in the Euclidean space $\mathbb{R}^2$ and $\mathbb{R}^3$. This is emphasized in the example that follows.

Exercise 2.15. Justify that the interval $(x - \epsilon, x + \epsilon)$ is indeed an open ball about $x \in \mathbb{R}$ of radius ϵ in the metric space $(\mathbb{R}, |\cdot|)$. Is every interval (a, b) in $\mathbb{R}$ where $a < b$ also an open ball?

In the examples that follow, we try to see the structure of open balls in various metric spaces.

Example 2.13 (Open balls in the plane). Consider the set $\mathbb{R}^2$. We have seen many ways of defining a metric on $\mathbb{R}^2$. We shall now see what an open ball looks like here. First, we take the metric in $\mathbb{R}^2$ we are used to with, namely, the Euclidean metric d_2. Instead of defining an arbitrary open ball, we confine ourselves to defining an open ball centered at $(0,0)$ and of radius 1. This is called the *unit open ball*. To make the notations short, we shall use $\mathbf{0}$ to denote the point $(0,0)$. Then,

$$B\left(\mathbf{0},1\right) = \left\{(x,y) \in \mathbb{R}^2 \big| d_2\left((x,y),\mathbf{0}\right) < 1\right\}.$$

The condition $d_2\left((x,y),\mathbf{0}\right) < 1$ gives us $\sqrt{x^2 + y^2} < 1$, which is same as $x^2 + y^2 < 1$. But, we know that $x^2 + y^2 = 1$ defines the unit circle centered at origin. Therefore, $B\left(\mathbf{0},1\right)$ (in this case) gives the set of all points inside the circle (excluding the circle).

This is where the term "open ball" gets its motivation. As seen in Figure 2.5, the open unit ball is everything inside the circle of radius 1 centered at origin. Similarly, for an arbitrary open ball in the Euclidean metric, the shape of the "boundary"[5] remains the same: a circle!

Since this open ball was formed by the metric d_2, we denote it as $B_2\left(\mathbf{0},1\right)$. Since for $p \geq 1$, we can define metric on $\mathbb{R}^2$ and we denote that metric as d_p, we shall denote the open ball constructed using d_p as $B_p\left(\mathbf{x},r\right)$, where $\mathbf{x} = (x,y) \in \mathbb{R}^2$.

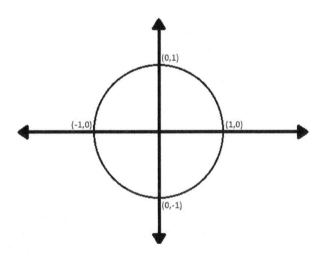

FIGURE 2.5: A circle in the Euclidean plane used in determining the open unit ball.

---

[5]We are cheating here by using the word "boundary". Since we have not yet defined what boundary means in our context, it is not really appropriate to use it. However, we believe that the reader will get a geometric sense of the boundary from the condition used to construct the circle. In the text that follows, we shall eventually define the boundary of any set and this shall become clear. However, until then, the readers are advised to take the geometric intuition of the boundary and currently forget about the definition they might have gone through in a course of analysis.

Let us consider the d_1 metric defined as in Equation (2.9). An open unit ball in this metric is the set,

$$B_1(\mathbf{0}, 1) = \left\{ (x, y) \in \mathbb{R}^2 | d_1((x, y), \mathbf{0}) < 1 \right\}.$$

We shall first see what $d_1((x, y), \mathbf{0}) = 1$ can lead us to. We have,

$$|x| + |y| = 1.$$

Depending on the values of x and y, we can have one of the following four cases:

Case I: $x \geq 0$ and $y \geq 0$. Then, $|x| = x$ and $|y| = y$. Therefore, the description above gives the line

$$x + y = 1.$$

This line has slope -1 and a y-intercept 1.

Case II: $x \geq 0$ and $y < 0$. Then, $|x| = x$ and $|y| = -y$. Therefore, we have the line

$$x - y = 1.$$

This line has a slope 1 and a y-intercept -1.

Case III: $x < 0$ and $y \geq 0$. Then, $|x| = -x$ and $|y| = y$. Therefore, we get the line

$$-x + y = 1.$$

This line has a slope 1 and a y-intercept 1.

Case IV: $x < 0$ and $y < 0$. Then $|x| = -x$ and $|y| = -y$. Thus, we get

$$-x - y = 1.$$

This line has a slop -1 and a y-intercept -1.

Hence, the condition $d_1((x, y), \mathbf{0}) = 1$ gives us the union of the four line segments stated above and the condition $d_1((x, y), \mathbf{0}) < 1$ gives us everything inside the union of these line segments excluding the lines themselves. Geometrically, the shape of $B_1(\mathbf{0}, 1)$ is as shown in Figure 2.6.

Now, we look at the metric d_∞. To construct an open unit ball in the d_∞ metric, we will require the condition $d_\infty((x, y), \mathbf{0}) < 1$. However, we first look at $d_\infty((x, y), \mathbf{0}) = 1$. This gives us $\max\{|x|, |y|\} = 1$. Therefore, we have two cases, either $|x| = 1$ or $|y| = 1$. This gives us the union of lines $x = 1$, $x = -1$, $y = 1$, and $y = -1$ as can be seen in Figure 2.7. Then, the open unit ball,

$$B_\infty(\mathbf{0}, 1) = \left\{ (x, y) \in \mathbb{R}^2 | d_\infty((x, y), \mathbf{0}) < 1 \right\}$$

is everything inside the union of lines excluding the lines themselves.

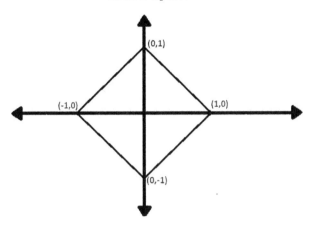

FIGURE 2.6: A unit "circle" in $\mathbb{R}^2$ with the Manhattan metric, used to construct open unit ball.

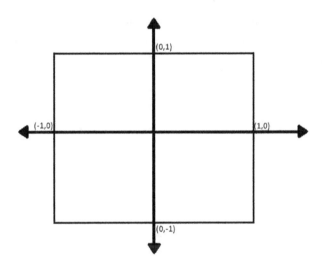

FIGURE 2.7: A unit "circle" in $\mathbb{R}^2$ with d_∞, used to determine the open unit ball.

Similarly, we can have any open unit ball $B_p(\mathbf{0}, 1)$. However, for values of p other than those discussed above, making the curves is difficult and hence omitted from the text.

We can keep making open balls in any metric space we have. In particular, we can also make open balls, similar to those discussed above, in the 3 dimensional space $\mathbb{R}^3$ and even in the sequence spaces. However, visualizing them may not be that easy, especially for $\mathbb{R}^n$ where $n \geq 4$ and the sequence spaces ℓ^p and $\mathbb{R}^\mathbb{N}$.

Exercise 2.16. Describe the open balls $B_1(\mathbf{0},1)$, $B_2(\mathbf{0},1)$, and $B_\infty(\mathbf{0},1)$ in $\mathbb{R}^3$.

We now move one step ahead and look at open unit ball in the function space $\mathscr{C}[a,b]$ with different metrics.

Example 2.14 (Open balls in function space). Consider the set $\mathscr{C}[a,b]$ equipped with the metric d_∞. We shall make the open unit ball

$$B_\infty(\mathbf{0},1) = \{f \in \mathscr{C}[a,b] \,|\, d_\infty(f,\mathbf{0}) < 1\}.$$

First, we look at the condition $d_\infty(f,\mathbf{0}) = 1$. This gives us $\sup_{x \in [a,b]} \{|f(x)|\} = 1$. From the definition of supremum, we know that $\forall x \in [a,b]$, we will have $|f(x)| \le 1$. Hence, any function in the region specified in the figure will satisfy the criterion $d_\infty(f,\mathbf{0}) = 1$. Now, the criterion $d_\infty(f,\mathbf{0}) < 1$ will be satisfied by those functions that do not touch (or cross) the constant functions which take value 1 and -1 everywhere in the interval $[a,b]$.

The region where the functions in the unit open ball lie can be seen in Figure 2.8.

Similarly, we can look at the open unit ball

$$B_1(\mathbf{0},1) = \{f \in \mathscr{C}[a,b] \,|\, d_1(f,\mathbf{0}) < 1\}$$

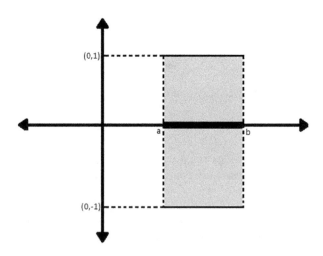

FIGURE 2.8: The region in XY plane where the functions inside the unit open ball lie with d_∞.

constructed from the metric d_1 on $\mathscr{C}[a,b]$. The condition $d_1(f,0) < 1$ leads us to

$$\int_a^b |f(x)| \ dx < 1.$$

Physically this means that the function f has an area less than 1. Although at first glance it may appear that such functions lie within some region as in the previous case, it is not true! We can have a triangle with a very small base and correspondingly large height so that it has the same area as a triangle with unit base and unit height. By this we mean to say that if we fix a number $r < 1$, then there can be infinitely many functions with increasing heights so that the integral involved is exactly r and hence all those functions lie in the open unit ball. This can be seen in Figure 2.9.

We have been looking at open balls in some special metric spaces where we knew how the elements behave. Now, we shall look at open balls in the discrete metric space.

Example 2.15 (Open balls in the discrete space). Consider the discrete metric space from Example 2.6. We do not have an origin to make a open unit ball. However, the metric function d does not take many values in $[0, \infty)$. For distinct points, it takes the value 1 and for identical points, it takes the value 0. Therefore, if we fix any point $x \in X$, where X is the set where the discrete metric is defined, and fix a radius $r \leq 1$, then, the open ball $B(x,r) = \{y \in X | d(x,y) < r\}$ will contain only x. This is because for any

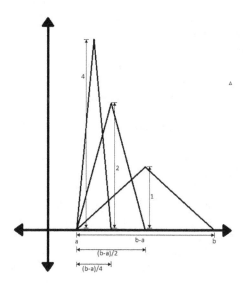

FIGURE 2.9: Functions in $\mathscr{C}[0,1]$ with the same area $\frac{b-a}{2}$ but different "heights".

$y \neq x$, we have $d(x,y) = 1 \not< r$ and hence $y \notin B(x,r)$. Similarly, if we have $r > 1$, then $\forall y \in X$, $d(x,y) \leq 1 < r$. Hence, $B(x,r) = X$.

Therefore, in the discrete metric space, about any point we have only two possible open balls: One which contains only the center and other, the whole space X. We may write this as

$$B(x,r) = \begin{cases} \{x\}, & r \leq 1. \\ X, & r > 1. \end{cases}$$

Now that we have seen sufficient examples of open balls in various metric spaces, we can now move to defining open sets in an arbitrary metric space. However, it is necessary to comment upon why we looked at so many examples of open balls (and their structures) in various metric spaces. In the beginning of this section, we mentioned that to abstract the properties of an open set in $\mathbb{R}$, we would need to abstract the properties of the interval $(x - \epsilon, x + \epsilon)$ that arose due to the definition of "openness". In the process we had to define, in an abstract metric space, an open ball. Therefore, we shall now make use of this concept (open ball) to define open sets in any metric space. The formal definition of open sets is as follows.

Definition 2.3 (Open set). *A set $G \subseteq X$ is said to be open in the metric space (X, d), if $\forall x \in G$, $\exists r > 0$ such that $B(x,r) \subseteq G$.*

Now, we are ready to find open sets in a metric space. Given any non-empty set X, we always have at least two subsets available with us $\emptyset$ and X itself. Therefore, it is natural to ask whether these are open or not. First, we look at X. As discussed in Chapter 1, whenever we are given a universal set, we shall act like we do not know what is outside this universe. Therefore, for any point $x \in X$, we have $B(x,1) \subseteq X$, which means X is open.

Looking at $\emptyset$, we claim that it is open. Suppose someone wants to prove us wrong. Then, the person will have to prove that $\emptyset$ is not open. To do so, they will have to negate the definition of open set and obtain the statement:

$$\exists x \in \emptyset \text{ such that } \forall r > 0, B(x,r) \not\subseteq \emptyset.$$

This means the person will have to get an element x from the empty set, $\emptyset$, which is not possible. Since the person cannot do this, we win and $\emptyset$ is open!

Another natural question to ask is: If we know open sets in a metric space, what about their union and intersection? In particular, is the union (finite as well as arbitrary) of open sets again open? Is the intersection (finite and arbitrary) of open sets again open? Let us answer these questions one by one.

Let G_1, G_2 be two given open sets in a metric space (X, d). Consider their union $G_1 \cup G_2$ and let $x \in G_1 \cup G_2$. This means that $x \in G_1$ or $x \in G_2$. In either case, since both G_1 and G_2 are open, $\exists r > 0$ such that $B(x,r) \subseteq G_1$ or $B(x,r) \subseteq G_2$. Therefore, we get $B(x,r) \subseteq G_1 \cup G_2$ and since the choice of x was arbitrary, we may conclude that $G_1 \cup G_2$ is open.

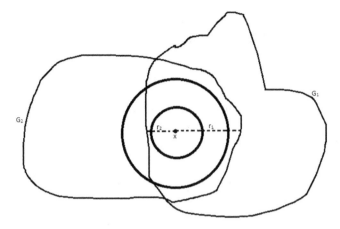

FIGURE 2.10: Intersection of two open sets.

Similarly, if we take $x \in G_1 \cap G_2$, then it would mean that $x \in G_1$ and $x \in G_2$. Since G_1 and G_2 are both open, $\exists r_1 > 0$ such that $B(x, r_1) \subseteq G_1$ and $\exists r_2 > 0$ such that $B(x, r_2) \subseteq G_2$. A similar scenario is depicted in Figure 2.10. Now, if we consider $r = \min\{r_1, r_2\}$ and take the ball $B(x, r)$, then we will have $\forall y \in B(x, r)$, $d(x, y) < r = \min\{r_1, r_2\} \le r_1$ and hence $B(x, r) \subseteq B(x, r_1)$. Similarly, we also have $\forall y \in B(x, r)$, $d(x, y) < r = \min\{r_1, r_2\} \le r_2$ and hence $B(x, r) \subseteq B(x, r_2)$. Therefore, $B(x, r) \subseteq G_1$ and $B(x, r) \subseteq G_2$. Hence, $\forall x \in G_1 \cap G_2$, $\exists r > 0$ (given by $r = \min\{r_1, r_2\}$ as discussed above) such that $B(x, r) \subseteq G_1 \cap G_2$. Therefore, $G_1 \cap G_2$ is open.

Exercise 2.17. Prove that $\bigcup_{i=1}^{n} G_i$ and $\bigcap_{i=1}^{n} G_i$ are open, given that G_i are open sets for $i = 1, 2, \cdots, n$.

We shall now see arbitrary unions and intersections of open sets. Let $\mathscr{F} = \{G_\lambda \subseteq X | \lambda \in \Lambda\}$ be a family of open sets indexed by Λ. It is easy to prove that the union $\bigcup_{\lambda \in \Lambda} G_\lambda$ is open. We exploit the fact that any element in the union is contained in at least one of the G_λ and the corresponding open ball is again inside the union. The complete proof is left as an exercise to the reader. However, we look at arbitrary intersections $\bigcap_{\lambda \in \Lambda} G_\lambda$. If we take any element $x \in \bigcap_{\lambda \in \Lambda} G_\lambda$, then $\forall \lambda \in \Lambda$, $\exists r_\lambda > 0$ such that $B(x, r_\lambda) \subseteq G_\lambda$. If we are to adopt the same strategy as we used in the case of intersection of two open sets, we would now have to find $r = \inf_{\lambda \in \Lambda} \{r_\lambda\}$. However, we note that this need not be always positive but can also be zero. Therefore, we may not get any positive r for which $B(x, r) \subseteq \bigcap_{\lambda \in \Lambda} G_\lambda$.

Keeping this in mind, we now construct a counterexample.

FIGURE 2.11: A pictorical representation of the family $\mathscr{F}$.

Example 2.16. Consider the family $\mathscr{F} = \left\{\left(-\frac{1}{n}, \frac{1}{n}\right) \subseteq \mathbb{R} \,\middle|\, n \in \mathbb{N}\right\}$ indexed by $\mathbb{N}$, in the metric space $(\mathbb{R}, |\cdot|)$. This is a family of nested intervals, in which each interval is centered at 0 as seen in Figure 2.11. Therefore, we know that $0 \in \bigcap_{n \in \mathbb{N}} \left(-\frac{1}{n}, \frac{1}{n}\right)$.

Now, let $x \in \bigcap_{n \in \mathbb{N}} \left(-\frac{1}{n}, \frac{1}{n}\right)$. Then, we have $\forall n \in \mathbb{N}$, $-\frac{1}{n} < x < \frac{1}{n}$ which we may write as $|x| < \frac{1}{n}$. Let $\epsilon > 0$ be arbitrary. Then, by the Archimedean property[6] of $\mathbb{N}$ in $\mathbb{R}$, $\exists n_0 \in \mathbb{N}$ such that $n_0\epsilon > 1$, i.e., $\frac{1}{n_0} < \epsilon$. Hence, from above, we have $|x| < \frac{1}{n_0} < \epsilon$. Since ϵ was arbitrary, we can conclude that $\forall \epsilon > 0$, $|x| < \epsilon$ and hence $x = 0$.

This gives us the following result:

$$\bigcap_{n \in \mathbb{N}} \left(-\frac{1}{n}, \frac{1}{n}\right) = \{0\}.$$

Now, since this set has only one element, if we can find an $r > 0$ such that $B(0, r) \subseteq \{0\}$, then it will be proved that $\{0\}$ is open. So, let us see what happens to $B(0, r)$ for $r > 0$. We know that $B(0, r) = (-r, r)$. Therefore, $\frac{r}{2} \in B(0, r)$ for each $r > 0$. Since $r \neq 0$, $\frac{r}{2} \neq 0$ and hence $B(0, r) \not\subseteq \{0\}$. This is true for any $r > 0$ and therefore, $\{0\}$ is not an open set.

This example proves that an arbitrary intersection of open sets need not be open.

To see where the problem occurs in finding the infimum (from where we got an idea of construct the counterexample), consider the point 0 which is in each of $\left(-\frac{1}{n}, \frac{1}{n}\right)$. For every set $\left(-\frac{1}{n}, \frac{1}{n}\right)$, the corresponding open ball is $B\left(0, \frac{1}{n}\right)$, i.e., $r_n = \frac{1}{n}$. Therefore, when we find the infimum, we get $\inf_{n \in \mathbb{N}} \left\{\frac{1}{n}\right\} = 0 \not> 0$.

Exercise 2.18. Prove that arbitrary union of open sets is open.

We summarize all of the above discussion as the following theorem.

---

[6]Archimedean property says that if $x, y \in \mathbb{R}$ are given with $x > 0$, then $\exists n \in \mathbb{N}$ such that $nx > y$. Indeed, this has a philosophical meaning, which we would like to mention here! If we have some goal in our lives (y), all we need to do is take some positive steps (of size x) toward it and one day (after taking n such steps) we will reach the goal (in fact, exceed it).

Theorem 2.8. *In a metric space (X, d), the following statements hold:*

1. *The whole space X and the empty set $\emptyset$ are open.*

2. *Arbitrary union of open sets is open.*

3. *Finite intersection of open sets is open. However, arbitrary intersection of open sets is not necessarily open.*

In defining an open set, we made use of an open ball. As the name suggests, we expect an open ball to be open. This is dealt with in the following theorem.

Theorem 2.9. *Any open ball in a metric space is an open set.*

Proof. Let (X, d) be a metric space and let $B(x, r)$ be any open ball in X. Let $y \in B(x, r)$. Then, we know that $d(x, y) < r$.
   We define $B(y, r')$, for $r' = r - d(x, y)$.
   Let $z \in B(y, r')$. Then, from the triangle inequality we have

$$
\begin{aligned}
d(x, z) &\leq d(x, y) + d(y, z) \\
&< d(x, y) + r - d(x, y) \\
&= r.
\end{aligned}
$$

Hence, $z \in B(x, r)$ so that $B(y, r') \subseteq B(x, r)$. Therefore, $B(x, r)$ is open. This can be seen in Figure 2.12. $\qquad \square$

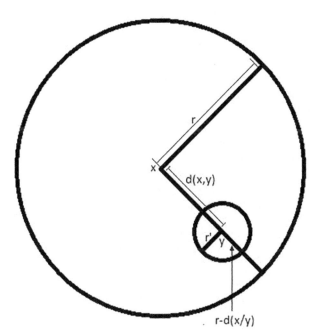

FIGURE 2.12: A diagrammatic representation of an open ball.

Let us get back to finding open sets in some metric spaces. First, let us look at the discrete space (X, d). Consider any subset $S \subseteq X$ and let $x \in S$. Then, the open ball $B(x, 1)$ contains only its center as discussed in Example 2.15. Therefore, $\{x\} = B(x, 1) \subseteq S$. Hence, S is an open set. Since S was arbitrary, we can conclude that every subset of X is open when X is equipped with the discrete metric.

Let us get back to $\mathbb{R}$ and its known subsets. We shall try to answer which of $\mathbb{N}$, $\mathbb{Z}$, $\mathbb{Q}$ is open in $\mathbb{R}$. First, looking at $\mathbb{N}$, let $n \in \mathbb{N}$ and $r > 0$. Then, $\min\left\{n + \frac{1}{2}, n + \frac{r}{2}\right\} \in B(n, r)$. If $\min\left\{n + \frac{1}{2}, n + \frac{r}{2}\right\} = n + \frac{1}{2} \in B(n, r)$, since $n + \frac{1}{2} \notin \mathbb{N}$, we have $B(n, r) \not\subseteq \mathbb{N}$. On the other hand, if $\min\left\{n + \frac{1}{2}, n + \frac{r}{2}\right\} = n + \frac{r}{2} \in B(n, r)$, then $n + \frac{r}{2} \leq n + \frac{1}{2}$ and since $r > 0$, $n + \frac{r}{2} > n$. Therefore, $n + \frac{r}{2} \notin \mathbb{N}$ and again, $B(n, r) \not\subseteq \mathbb{N}$. Hence, $\mathbb{N}$ is not open in $\mathbb{R}$.

Exercise 2.19. Check if $\mathbb{Z}$ is open in $\mathbb{R}$.

For $\mathbb{Q}$, we shall use the property that both rationals and irrationals are dense in $\mathbb{R}$. This means, given any two real numbers x and y such that $x < y$, we can always have a rational number r such that $x < r < y$ and an irrational number s such that $x < s < y$. Therefore, if $q \in \mathbb{Q}$ and $r > 0$, we have $q + \frac{r}{2} \in B(q, r)$ and $q < q + \frac{r}{2}$. From the density of $\mathbb{Q}^c$ in $\mathbb{R}$ we have an element $s \in \mathbb{Q}^c$ such that $q < s < q + \frac{r}{2}$ so that $s \in B(q, r)$ and hence $B(q, r) \not\subseteq \mathbb{Q}$. Therefore, we may conclude that $\mathbb{Q}$ is not open in $\mathbb{R}$. Similarly, we can prove that $\mathbb{Q}^c$ is not open in $\mathbb{R}$.

The question is: What are the open sets in $\mathbb{R}$? We know that any interval of the form (a, b) for $a < b$, being an open ball (Exercise 2.15), is open and hence an arbitrary union of open intervals is also open. But can there be any set which is not a union of intervals and still open?

To see this, consider an open set G in $\mathbb{R}$ and let $x \in G$. Then, since G is open, there is at least one open interval, say J, such that $x \in J \subseteq G$. Define J_x to be the union of all such intervals which contain x and are contained in G. Now, let $a, b \in J_x$ such that $a < b$. Also, let $c \in \mathbb{R}$ such that $a < c < b$. Since J_x is the union of intervals, a and b lie in some interval. If they lie in the same interval, then c also lies in that interval and hence lies in J_x. However, if a lies in an interval J_1 and b lies in an interval J_2 where $J_1 \neq J_2$, then we would like to show that c still belongs to J_x. To see this, we first observe that $x \in J_1$ and $x \in J_2$. We have the following cases:

Case I: $x \leq a < b$. In this case, since J_2 is an interval, $x \leq a < c < b$ holds and hence $c \in J_2$. Since J_x is the union of intervals which contain x and contained in G, $c \in J_x$.

Case II: $a < x < b$. From the argument above, $c \in J_x$.

Case III: $a < x = b$. Then, since J_1 is an interval and $a < c < b$, we have $c \in J_1$ and hence $c \in J_x$.

Case IV: $a < b < x$. Again since J_1 is an interval and $a < c < b$, we have $c \in J_1$ and hence $c \in J_x$.

Therefore, in any case for $a < c < b$ with $a, b \in J_x$ we have $c \in J_x$ and hence J_x is an interval.

Therefore, $G = \bigcup_{x \in G} J_x$, i.e., G can be written as a union of (open) intervals. Now, consider a relation $\sim$ defined on G as $x \sim y$ if and only if $J_x = J_y$. Clearly, the relation is an equivalence relation so that the intervals J_x and J_y are either identical or completely disjoint. Therefore, $G = \bigcup_{i \in I} J_i$ for some index set I. Now, let $r_i \in J_i$ be a fixed rational number. Define a map $f : I \to \mathbb{Q}$ as $f(i) = r_i$. Then for $i \neq j$, we have $r_i \neq r_j$ since $J_i \neq J_j$. Hence, f is one-to-one and I is countable. Therefore we have proved a much stronger and useful result which is stated as the following theorem.

Theorem 2.10. *Any open set in $\mathbb{R}$ is a countable union of open intervals.*

Let us move to the Euclidean plane $\mathbb{R}^2$. Consider any set of the form $\{(x, 0) \in \mathbb{R}^2 | a < x < b\}$. If we drop the second coordinate, it is just the interval (a, b) from $\mathbb{R}$. Is this open in $\mathbb{R}^2$ equipped with the d_2 metric? To see if it is open, consider any open ball of radius $r > 0$ centered at any point x in the set considered. The point $\left(x, \frac{r}{2}\right)$ lies in the ball $B(x, r)$ but not in the set. Therefore, the set $\{(x, 0) \in \mathbb{R}^2 | a < x < b\}$ is not open. The situation is represented in Figure 2.13.

Therefore we see that taking intervals and adding another coordinate will not give us open set in $\mathbb{R}^2$. Looking at the definition of $\mathbb{R}^2$, we realize that it is but the Cartesian product of $\mathbb{R}$ with itself. So, what would happen if we take Cartesian product of two intervals? Will $(a, b) \times (c, d)$ be open in $\mathbb{R}^2$?

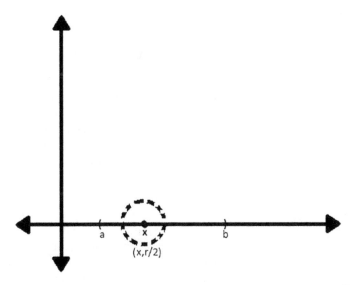

FIGURE 2.13: An interval on the X axis is not open in $\mathbb{R}^2$.

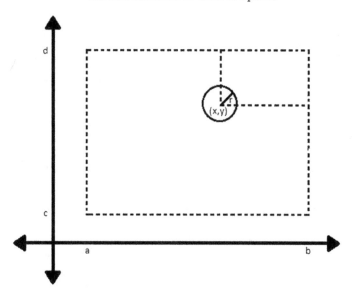

FIGURE 2.14: Cartesian product of open intervals is open in $\mathbb{R}^2$.

Let (x, y) be any point in the set $(a, b) \times (c, d)$. Consider the following real numbers: $r_1 = |b - x|, r_2 = |a - x|, r_3 = |c - y|, r_4 = |d - y|$. If we choose $r = \min \{r_1, r_2, r_3, r_4\}$ and make an open ball $B((x, y), r)$, then we expect it to be completely contained in the set $(a, b) \times (c, d)$ as can be seen in Figure 2.14. We leave the text book proof of this fact to the reader.

Just as in the case of $\mathbb{R}$, we would like to have open sets in any metric space to be more intimately related to open balls. This is shown in the following theorem.

Theorem 2.11. *In any metric space (X, d), a set $G \subseteq X$ is open if and only if it is a union of open balls.*

Proof. First we see that a union of open balls is open since open balls are open sets. We shall prove the converse. For the same, we will apply the same strategy used to prove that any open set in $\mathbb{R}$ is the union of open intervals.

Let $G \subseteq X$ be open. Then, for each $x \in G$, there is an open ball $B(x, r_x) \subseteq G$. Also, we know that $x \in B(x, r)$. Therefore, we can write $G = \bigcup_{x \in G} B(x, r_x)$, which completes the proof! $\qquad\square$

Proceeding a step ahead, can we say that open sets are a countable union of open balls, like in the case of $\mathbb{R}$? Well, looking at the proof for the case of $\mathbb{R}$, we observe that there was a set $\mathbb{Q}$ which was countable and had a property, "density", which enabled the open sets in $\mathbb{R}$ to be a countable union of open intervals (balls). Since in an arbitrary metric space, we may not have such a set, in general we would like to answer the question negatively.

Example 2.17. Consider a discrete metric space X, where X is uncountable. Fix a point $x \in X$. We know that $X \setminus \{x\}$ is open and uncountable. Also, in a discrete space, there are only two types of open balls available, singletons and the whole space. Since $X \setminus \{x\} \neq X$, the only way to write the set as a union of open balls is

$$\bigcup_{y \in X \setminus \{x\}} \{y\}.$$

Clearly, this is a union of family indexed by an uncountable set.

However, later in this text, we shall try to see some special metric spaces where we may get open sets as a union of countably many open balls.

Before we close this section, we point at a very special property enjoyed by metric spaces. Suppose there are two distinct points x and y in a metric space (X, d). Then, by the definition of metric, we have $d(x, y) > 0$. If we construct open balls around each of these points with radius $\frac{1}{2} d(x, y)$, then these two open balls will contain the center but not the other point, i.e., $B\left(x, \frac{1}{2} d(x, y)\right)$ will contain x but not y and similarly $B\left(y, \frac{1}{2} d(x, y)\right)$ will contain y but not x. In fact, these two open balls do not intersect at all! This is true for any two points and is called the *Hausdorff property*. A formal definition is as follows.

Definition 2.4 (Hausdorff property). *In a metric space X, $\forall x, y \in X$, $\exists \epsilon_1, \epsilon_2 > 0$ such that $B(x, \epsilon_1) \cap B(y, \epsilon_2) = \emptyset$. This property of metric spaces is called Hausdorff property.*

We have been looking at open sets for quite a while now. Therefore, it is of importance that we mention here about the collection of all open sets. Since in the further study of metric spaces, open sets are going to play a major role, we will give a name to this collection. Formally, the collection of all open sets is called a *topology* on X, and is often denoted by $\mathscr{T}$. Notice that it is a collection of subsets of X which satisfies the following properties:

1. $\emptyset, X \in \mathscr{T}$.

2. For any family $\mathscr{F} = \{G_\lambda \in \mathscr{T} \mid \lambda \in \Lambda\}$, we have $\bigcup_{\lambda \in \Lambda} G_\lambda \in \mathscr{T}$.

3. For $G_1, G_2 \in \mathscr{T}$, we have $G_1 \cap G_2 \in \mathscr{T}$.

We introduced this technicality here (although it will not be used much in the study of metric spaces), because often when we study subspaces of metric spaces and their product spaces, we do not say "Open sets in subspace/product space". Rather we simply say *subspace topology* and *product topology*. We will shortly see what is exactly the subspace and product topology, i.e., what are all the open sets in a subspace and a product space. The important part of this study will be to see if we can have the open sets in subspace and product space in terms of open sets of the original metric spaces.

2.4.1   Subspace topology

Here, we prove a major result for finding open sets in a subspace of a metric space by using the open sets at hand.

Theorem 2.12. *Let (X, d) be a metric space and (S, d) its subspace. A set G is open in S if and only if there is an open set G' in X such that $G = G' \cap S$.*

Proof. We shall use two notations: $B_X(x, r)$ and $B_S(x, r)$ to denote the open balls centered at x and of radius r, in the space X and S respectively. First, let G be open in S. We know that,

$$G = \bigcup_{x \in G} B_S(x, r_x),$$

for suitable choices of $r_x > 0$. Now, the definition says

$$B_S(x, r_x) = \{y \in S \,|\, d(x, y) < r_x\},$$

and

$$B_X(x, r_x) = \{y \in X \,|\, d(x, y) < r_x\}.$$

It is clear from these definitions that $B_S(x, r_x) = B_X(x, r_x) \cap S$. Hence, we have

$$G = \bigcup_{x \in G} (B_X(x, r_x) \cap S) = \left(\bigcup_{x \in G} B_X(x, r_x) \right) \cap S.$$

This proves the first part of the result.

Conversely, let $G = G' \cap S$, where G' is open in X. We need to show that G is open in S. Consider $x \in G$. Then, $x \in G'$ and $x \in S$. Since G' is open in X, there is some $r > 0$ such that $B_X(x, r) \subseteq G'$. Also, we have seen that $B_S(x, r) = B_X(x, r) \cap S \subseteq G' \cap S = G$. Hence, G is open in S. This is shown in Figure 2.15. □

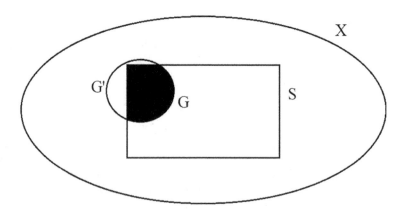

FIGURE 2.15: Open set in subspace topology.

This result directly gives us all the open sets in the subspace topology. Indeed, all we have to do is take an open set in the whole space and intersect it with our subspace. Whatever is common is the new open set.

Exercise 2.20. Consider $(0, 1]$ as a subspace of $(\mathbb{R}, |\cdot|)$. Is $\left(\frac{1}{2}, 1\right]$ open in this subspace? What about $\left[\frac{1}{2}, 1\right]$?

2.4.2 Product topology

Just like the previous section on subspace topology, here we will see what are the open sets in the product space. However, we shall deal with the result for the product of two metric spaces and leave it to the reader for product of n metric spaces.

Theorem 2.13. *Let (X, d_X) and (Y, d_Y) be metric spaces. Then, a set G is open in the product space $(X \times Y, d)$ if and only if $G = G_1 \times G_2$, where G_1 is open in X and G_2 is open in Y.*

Proof. First, let G be an open set in $X \times Y$. Then,

$$G = \bigcup_{(x,y) \in G} B_{X \times Y}\left((x, y), r\right).$$

Now, consider $B_X\left(X, r_x\right)$ and $B_Y\left(y, r_y\right)$. For any point $(x', y') \in X \times Y$, we have

$$d\left((x, y), (x', y')\right) = \max\left\{d_X\left(x, x'\right), d_Y\left(y, y'\right)\right\} < r \Leftrightarrow d_X\left(x, x'\right) < r$$
$$\text{and } d_Y\left(y, y'\right) < r.$$

That is to say

$$B_{X \times Y}\left((x, y), r\right) = B_X\left(x, r\right) \times B_Y\left(y, r\right).$$

Hence, we have

$$G = \bigcup_{(x,y) \in G}\left(B_X\left(x, r\right) \times B_Y\left(y, r\right)\right) = \left(\cup B_X\left(x, r\right)\right) \times \left(\cup B_Y\left(y, r\right)\right).$$

This completes the first part of the proof.

Now, let $G = G_1 \times G_2 \subseteq X \times Y$, where G_1 is open in X and G_2 is open in Y. Let $(x, y) \in G$. Then, $x \in G_1$ and $y \in G_2$. Since G_1 and G_2 are open in the respective spaces, we have r_x and $r_y > 0$ such that $B_X\left(x, r_x\right) \subseteq G_1$ and $B_Y\left(y, r_y\right) \subseteq G_2$. Then, consider $r = \min\left\{r_x, r_y\right\}$. We have,

$$B_{X \times Y}\left((x, y), r\right) = B_X\left(x, r\right) \times B_Y\left(y, r\right) \subseteq G_1 \times G_2 = G.$$

Hence, G is open. $\square$

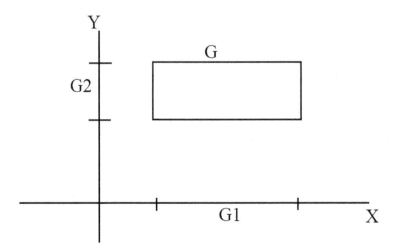

FIGURE 2.16: Open set in product topology.

Thus, a typical open set in $X \times Y$ is a product of open sets in X and Y as shown in Figure 2.16.

Exercise 2.21. Find the open sets in the product space $X_1 \times X_2 \times X_3 \times \cdots \times X_n$. Can we generalize it to arbitrary products?

2.5   Closed Sets

As an immediate consequence of defining open sets, it is now time to define closed sets in any metric space. In a course on analysis, one may have studied various definitions of closed sets in $\mathbb{R}$ which may include terminologies such as limit point, sequences, convergence, and others. However, we avoid using those terminologies since we have not yet defined them. Therefore, we make the most natural definition of closed sets as follows:

Definition 2.5 (Closed set). *A set $S \subseteq X$ in a metric space (X, d) is closed if its complement is open.*

Remark. One may argue that a more natural definition for closed sets would be those sets which are not open. Although this seems intuitive and natural, we must take care that sets are not like doors: they may be open and closed at the same time and can also be neither open nor closed. This is dealt with in the text that follows.

As we have been proceeding in Section 2.4, we first see the two trivial subsets of any metric space: the empty set, $\emptyset$, and the whole space, X. Since $X^c = \emptyset$ and $\emptyset^c = X$, and both these sets are open, from the definition of closed sets we can conclude that both these sets are also closed. We also have other properties of closed sets analogous to open sets. We state them and leave the proof as an exercise to the reader.

Theorem 2.14. *In a metric space (X, d), the following statements are true:*

1. *The empty set, $\emptyset$, and the whole space X are closed.*

2. *Finite union of closed sets is closed. However, arbitrary union of closed sets may not be closed.*

3. *Arbitrary intersection of closed sets is closed.*

To look at a counter example to show that arbitrary union of closed sets may not be closed, consider the family $\mathscr{F} = \left\{ \left[\frac{1}{n}, 1\right] \subseteq \mathbb{R} \,\middle|\, n \in \mathbb{N} \right\}$ of closed intervals. First, our claim is that $\bigcup_{n \in \mathbb{N}} \left[\frac{1}{n}, 1\right] = (0, 1]$. To prove this claim, we consider $x \in \bigcup_{n \in \mathbb{N}} \left[\frac{1}{n}, 1\right]$. Therefore, we have $0 < \frac{1}{n_0} \leq x \leq 1$ for some $n_0 \in \mathbb{N}$, i.e., $x \in (0, 1]$ so that $\bigcup_{n \in \mathbb{N}} \left[\frac{1}{n}, 1\right] \subseteq (0, 1]$. Conversely, if $x \in (0, 1]$, then we have $x > 0$ and $x \leq 1$. By Archimedean property of $\mathbb{N}$ in $\mathbb{R}$, $\exists n_0 \in \mathbb{N}$ such that $n_0 x > 1$, i.e., $x > \frac{1}{n_0}$. Hence, $x \in \left[\frac{1}{n_0}, 1\right]$ and we get $\bigcup_{n \in \mathbb{N}} \left[\frac{1}{n}, 1\right] = (0, 1]$. We advise the reader to draw appropriate diagrams for a better understanding.

Now, $(0, 1]^c = (-\infty, 0] \cup (1, \infty)$. If we consider the point $0 \in (0, 1]^c$, for any $r > 0$ we always have $\min\left\{\frac{1}{2}, \frac{r}{2}\right\} \in (0, 1]$ so that for any $r > 0$, $B(0, r) \not\subseteq (0, 1]^c$. Hence, $(0, 1]^c$ is not open and therefore $(0, 1]$ is not closed. This proves that arbitrary union of closed sets is not necessarily closed.

Exercise 2.22. Prove parts 2 and 3 of Theorem 2.14 which are not discussed above.

Now, we shall look at known subsets of $\mathbb{R}$ and check if they are closed. First, we look at $\mathbb{N}$. Take any element x in $\mathbb{N}^c$. If $x < 0$, then $x + 1 < 1$ so that $B(x, 1) \subseteq \mathbb{N}^c$. If $0 < x < 1$, we can choose $r = \min\left\{\frac{x}{2}, \frac{1-x}{2}\right\}$ so that $B(x, r) \subseteq \mathbb{N}^c$. Similarly, if $x > 1$, then $\exists n_0 \in \mathbb{N}$ such that $n_0 < x < n_0 + 1$. Therefore, if we choose $r = \min\left\{\frac{x-n_0}{2}, \frac{n_0+1-x}{2}\right\}$, we have $B(x, r) \subseteq \mathbb{N}^c$. Therefore, $\mathbb{N}^c$ is open and hence $\mathbb{N}$ is closed. This working is depicted in Figure 2.17.

FIGURE 2.17: Method of construction of open balls around each point in $\mathbb{N}^c$.

Exercise 2.23. Is $\mathbb{Z}$ closed in $\mathbb{R}$?

We know that neither $\mathbb{Q}$ nor $\mathbb{Q}^c$ are open in $\mathbb{R}$. Therefore, we can also conclude that these two sets are not closed as well.

Example 2.18. Let us try to find closed sets in a discrete metric space. We know any subset of a discrete metric space is open. Therefore, if we take any subset of the space and then consider its complement, it is again a subset of the original space and hence open. Therefore, each set in a discrete metric space is closed.

Exercise 2.24. Check if the following sets are closed in $\mathbb{R}^2$ equipped with the Euclidean metric.

1. $S = \{(x, y) \in \mathbb{R}^2 | xy = 0\}$.

2. The set of points on the unit circle $x^2 + y^2 = 1$.

3. $S = \{(x, y) \in \mathbb{R}^2 | x \in \mathbb{Q}\}$.

What can you say about the sets when $\mathbb{R}^2$ is equipped with the Manhattan metric, d_1, and when it is equipped with the maximum metric, d_∞?

Before ending this section, we make an observation about finite sets in a metric space.

Theorem 2.15. *A finite set in a metric space is always closed.*

Proof. Let (X, d) be a metric space and $S = \{x_1, x_2, \cdots, x_n\}$ be a finite set in X. Consider any point $x \in S^c$. We know the distances of each of the point of S from x. Let $r_i = d(x, x_i)$ for $i = 1, 2, \cdots, n$. If we choose $r = \min\{r_1, r_2, \cdots, r_n\}$ and construct an open ball $B(x, r)$, we expect that $B(x, r) \subseteq S^c$. This is shown in Figure 2.18. Suppose that this is not true. Then, $\exists y \in B(x, r)$ such that $y \notin S^c$. But, $y \notin S^c$ means that $y \in S$ and hence $y = x_{i_0}$ for some $i_0 \in \{1, 2, \cdots, n\}$.

Now, we also have $r_{i_0} = d(x, x_{i_0}) < r = \min\{r_1, r_2, \cdots, r_n\}$. However, this is not possible since $r = \min\{r_1, r_2, \cdots, r_n\} \leq r_{i_0}$. Therefore our assumption is wrong and $B(x, r) \subseteq S^c$. Hence, S is closed. $\qquad\square$

Exercise 2.25. Prove that the set $\left\{n + \frac{1}{n} \middle| n \in \mathbb{N} \setminus \{1\}\right\}$ is closed in $\mathbb{R}$, equipped with the usual metric. **Hint:** Draw a picture of the set on the real line and then try proving that the complement is open.

Exercise 2.26. What are the closed sets in subspace topology and product topology, in terms of the closed sets of the original metric spaces in hand?

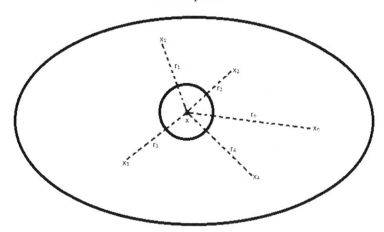

FIGURE 2.18: Method of construction of open balls in the complement of a finite set in any metric space.

2.6 Interior, Exterior, and Boundary Points

In the previous two sections, we tried to characterize sets in a metric space based on certain properties. In this section, given a set, we shall try to characterize the points of the metric space with respect to the given set. First, we look at the motivation of open sets. We motivated the definition of open sets by saying that every element in a set should have some "space" inside the set. That space provided for each point was given through open balls.

These points are special to us. Since their personal open ball space lies completely inside the set, we shall call them *interior points*. Similarly, suppose there is a point in the metric space whose personal open ball space lies completely outside a given set. Such points are called *exterior points*. Finally, there may be points for which every personal open ball space is neither completely inside nor completely outside the given set. Such points will be called *boundary points*. We now make the formal definitions of these concepts.

Definition 2.6 (Interior point). *A point $x \in X$ is called an interior point of a set $S \subseteq X$ if $\exists \epsilon > 0$ such that $B(x, \epsilon) \subseteq S$.*

Definition 2.7 (Exterior point). *A point $x \in X$ is called an exterior point of a set $S \subseteq X$ if $\exists \epsilon > 0$ such that $B(x, \epsilon) \subseteq S^c$.*

Definition 2.8 (Boundary point). *A point $x \in X$ is called a boundary point of a set $S \subseteq X$ if $\forall \epsilon > 0$, $B(x, \epsilon) \cap S \neq \emptyset$ and $B(x, \epsilon) \cap S^c \neq \emptyset$.*

Note. If the formal definition of boundary point seems intimidating (or confusing), we can think of it like every open ball constructed around the point

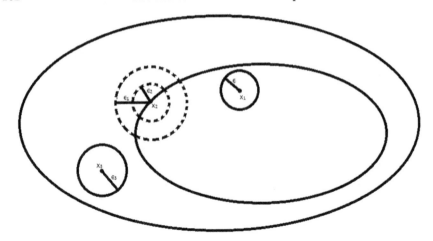

FIGURE 2.19: Pictorical representation of interior point (x_1), Exterior point (x_3) and a boundary point (x_2) in a metric space.

x has points from both the given set and outside it. Therefore, x is neither properly interior nor exterior.

A pictorical representation of interior, exterior, and boundary points is given in Figure 2.19.

Example 2.19. Consider any metric space (X, d). Earlier, we have constructed open balls around some points. Let $x_0 \in X$ be fixed and consider an open ball

$$B(x_0, 1) = \{y \in X | d(y, x_0) < 1\}.$$

Then, we can easily see that every point in $B(x_0, 1)$ is an interior point.

The following exercises will help the reader in actually finding these special types of points for known sets.

Exercise 2.27. Consider the following sets in $\mathbb{R}^2$ equipped with the Euclidean metric. Determine the interior, exterior, and boundary points for each of the sets:

1. $\{(0, y) \in \mathbb{R}^2 | y \in \mathbb{R}\}$.

2. $\mathbb{Q} \times \mathbb{Q} = \{(x, y) \in \mathbb{R}^2 | x, y \in \mathbb{Q}\}$.

3. $\{(x, y) \in \mathbb{R}^2 | xy \neq 0\}$.

Hint: Think geometrically for each of these sets before starting to write a rigorous proof.

Exercise 2.28. Find out the interior, exterior, and boundary points of the following subsets of $\mathbb{R}$.

1. $(a, b) \subseteq \mathbb{R}$ where $-\infty < a < b < \infty$.

2. $\mathbb{N}$.

3. $\mathbb{Q}^c$.

Since we have defined these special types of points, we can now start collecting them to form sets. As a matter of notation, the collection of interior points of a set S, called the interior of the set, will be denoted by $\text{Int}(S)$ or S^o. Similarly, the set of boundary points, called the boundary of the set, will be denoted by ∂S and the set of all exterior points will be denoted by $\text{Ext}(S)$, and will be called *exterior* of S.

The first observation we make is that the interior of any set is open. To prove it, assume any point $x \in S^o$. Then, $\exists \epsilon > 0$ such that $B(x, \epsilon) \subseteq S$. Now, since open ball is an open set, $\forall y \in B(x, \epsilon)$, $\exists \delta > 0$ such that $B(y, \delta) \subseteq B(x, \epsilon) \subseteq S$. This gives us that y is also an interior point of S. Since y was arbitrary, we can conclude that $B(x, \epsilon) \subseteq S^o$ and S^o is open.

Also, suppose for any given set S, let there by an open set R such that $S^o \subseteq R \subseteq S$. Suppose we assume the strict inclusion $S^o \subset R$. This would mean that $\exists x \in R$ and $x \notin S^o$. However, since R is open, $\exists \epsilon > 0$ such that $B(x, \epsilon) \subseteq R \subseteq S$ so that x is an interior point of S. This is a contradiction! Therefore, the strict inclusion cannot hold. What does this mean? It means that there is no open set "between" a set and its interior. Mathematically speaking, the interior of a set is the largest open set contained in the given set. We summarize this as a theorem.

Theorem 2.16. *In a metric space (X, d), given a set $S \subseteq X$, the interior of S is the largest open set contained in S.*

For an open set, every point is its interior point. This follows trivially from the definition of open set. Conversely, if every point of a set is its interior point, then to each point we can have a personal open ball space inside the set. Therefore, we have a characterization of open sets in terms of its interior points as follows:

Theorem 2.17. *In a metric space (X, d), a set $G \subseteq X$ is open if and only if $G = G^o$.*

Exercise 2.29. Give an example of a non-empty set in a metric space with empty interior. What can you say about its boundary?

On similar lines, we can observe that the exterior of a set is actually the interior of its complement and hence open. But, what can we say about the boundary? Is the boundary of a set open? Clearly, it is not. This is because the open ball about any point on the boundary also intersects the exterior.

A next natural question we can ask is: Is the boundary of the set closed? We see that any point in the complement of the boundary of a set is either an interior point or an exterior point, i.e., $(\partial S)^c = \text{Int}(A) \cup \text{Ext}(A)$. Since both interior and exterior are open sets, we can conclude that ∂S is a closed set. From the same arguments we gave to conclude that boundary is not open, we can also conclude that boundary of a set and that of its complement are the same. The textbook proof of this result is left as an exercise to the reader.

Exercise 2.30. In a metric space (X, d), for any set $S \subseteq X$, prove that $\partial S = \partial S^c$.

Note. ∂S^c and $(\partial S)^c$ are, in general, two different sets. The first denotes the boundary of the complement of S while the latter denotes the complement of the boundary of S.

2.7 Limit and Cluster Points

Consider visiting a mathematics convention where all great mathematicians of our time have been invited. An interesting phenomenon that is bound to occur is that once any of the great mathematicians enters the convention, everyone tries to get as close to the person as possible (maybe for taking an autograph or a photograph). On the other hand, people such as ourselves, who are not so famous in the community, can have our own space where no one bothers us. Based on this phenomenon, we will be defining two special types of points for a given set in a metric space: *limit point* and *cluster point*.

Definition 2.9 (Limit point). *In a metric space (X, d), a point $x \in X$ is called a limit point of a set $S \subseteq X$ if $\forall \epsilon > 0$, $B(x, \epsilon) \cap S \neq \emptyset$.*

Definition 2.10 (Cluster points). *In a metric space (X, d), a point $x \in X$ is called a cluster point of a set $S \subseteq X$ if $\forall \epsilon > 0$, $B(x, \epsilon) \cap (S \setminus \{x\}) \neq \emptyset$.*

In the situation described above, the famous mathematicians invited to the convention are the cluster points of the set of people in the convention while all the non-famous mathematicians (such as ourselves) are limit points. It is to be noted that many books do not keep a difference between limit and cluster points. The definition of cluster points which we have given is the definition of limit points in various books. Such books then name our limit points as adherent points. If the reader has gone through a course on analysis, then he or she might have come across adherent points and limit points in $\mathbb{R}$. However, throughout this book we shall make sure to keep a difference between limit points and cluster points and will not name any point in the set as an adherent point. The cluster points can also be named as *accumulation points* or *condensation points*.

Note that the main difference in limit and cluster points from the definition we gave above is that every open ball around a limit point can contain itself as the only member in the intersection. On the other hand, this is not allowed for cluster points. Every open ball constructed around a cluster point must have at least one point of the set other than itself.

Based on the definition, we can make the following observations immediately, which we state as a theorem. The textbook proof for this is left to the reader.

Theorem 2.18.

1. *Every point of the set is a limit point.*

2. *Every cluster point of a set is a limit point.*

Exercise 2.31. Prove Theorem 2.18

An intuition may tell the reader, at this point, that the points on the boundary of a set are cluster points. This thinking is natural and is in fact encouraged. However, whatever we think to be true needs to be proved with mathematical rigor. So, let us first see why do we think that boundary of a set consists of cluster points.

Firstly, the definition of boundary of a set mentions that every open ball constructed about a boundary point has a non-empty intersection with the given set and its complement. This is similar to our definition of cluster point. However, there is a small glitch! What if the boundary point is included in the set? There may be a possibility of getting an open ball which contains no point other than itself in the intersection. To illustrate this, let us look at an example.

We have seen that $\mathbb{N}$ has an empty interior and in fact, every point in $\mathbb{N}$ is a boundary point. So, if we take any $n \in \mathbb{N}$ and construct the open ball $B\left(n, \frac{1}{2}\right) = \left(n - \frac{1}{2}, n + \frac{1}{2}\right)$, then we get $\left(n - \frac{1}{2}, n + \frac{1}{2}\right) \cap \mathbb{N} = \{n\}$. Therefore, $\mathbb{N}$ is an example where the boundary points are not cluster points. In fact, we have also proved that no point in $\mathbb{N}$ is a cluster point. Let us move a step ahead and ask whether $\mathbb{N}$ has any real cluster point.

To see this, let $r \in \mathbb{R}$. Then, we have one of the two cases $r < 1$ or $r > 1$. The case where $r = 1$ is already handled since we know that 1 is not a cluster point of $\mathbb{N}$. If $r < 1$, then we construct an open ball $B\left(r, \frac{1-r}{2}\right)$ which has no natural number inside it. Similarly, if $r > 1$ and r is a natural number, then we know that r is not a cluster point. Therefore, we consider the case where $r > 1$ and r is not a natural number. Therefore, $\exists n \in \mathbb{N}$ such that $n < r < n + 1$. Now, we construct the open ball $B\left(r, \frac{1}{2}\min\{r - n, n - r + 1\}\right)$ which contains no natural number. Therefore, we can conclude that $\mathbb{N}$ has no cluster points.

Previously we made an observation that every point in a given set is a limit point. Is the converse true? Does every limit point of a set always lie in the set? The answer to this is no! A simple example is the unit open ball in $\mathbb{R}^2$. All points on the unit circle are limit points of this open ball but none of them is inside the open ball. Therefore, in general, we may say that limit points may not always lie in the set. But an interesting question that arises is: What would happen if all limit points lie in the set?

In this case, any point outside the set will have a personal open ball space which does not intersect with our set. This is because, if not, then that point will again be a limit point and would contradict the hypothesis we assumed. This would further imply that the complement of the set is open and hence the given set is closed.

Here, we have proved that if a set contains all of its limit points, then it is closed. One may now ask the converse question: If a set is closed, will it contain all of its limit points? Suppose not! Then, there is at least one limit point outside the given closed set and hence is in its complement. However, the complement is open and therefore this limit point has a personal open ball space with no intersection with the given set. This contradicts the definition of limit point.

Therefore, we have characterized closed sets in terms of limit points as the following theorem mentions.

Theorem 2.19. *A set F in a metric space is closed if and only if it contains all of its limit points.*

We leave the textbook proof as an exercise to the reader.

Exercise 2.32. Prove Theorem 2.19.

As a next step, let us collect all the limit points of a given set S and make another set, say $\bar{S}$. This set will be called *closure* of S and contains all of its limit points. As the name suggests, we expect $\bar{S}$ to be closed. So, let x be a limit point of $\bar{S}$. Then, $\forall \epsilon > 0$, $B(x, \epsilon) \cap \bar{S} \neq \emptyset$. Hence, $\exists y \in B(x, \epsilon) \cap \bar{S}$. This means that y is a limit point of S and also is in the open ball $B(x, \epsilon)$. Hence, $\exists \delta > 0$ such that $B(y, \delta) \subseteq B(x, \epsilon)$ and also, $B(y, \delta) \cap S \neq \emptyset$. This proves that x is a limit point of S and therefore, $\bar{S}$ contains all of its limit points. From the above characterization (Theorem 2.19), we can conclude that closure of a set is a closed set.

Now, just as we did in the case of interior of a set, let $S \subseteq R \subseteq \bar{S}$. Can the strict inclusion $R \subset \bar{S}$ hold when R is closed? Suppose it does hold. Then, there is at least one point x in $\bar{S}$ which is not in R. Since R is closed, it must contain all of its limit points and hence x is not a limit point of R. Therefore, we have an $\epsilon > 0$ such that $B(x, \epsilon) \cap R = \emptyset$, which further implies $B(x, \epsilon) \cap S = \emptyset$. Thus, we reach at a conclusion that x is not a limit point of S, which is a contradiction to the fact that $x \in \bar{S}$. Therefore, there is no closed set "between" S and its closure. We summarize this in the following theorem.

Theorem 2.20. *The closure of a set S is the smallest closed set containing S.*

Let us look at the boundary of a set once again. We have already mentioned that the boundary of a set contains limit points of the set. Can it contain any interior point? The answer is no! It is clear from the definition of boundary and interior itself. Therefore, for any set S, we can conclude that $\partial S \subseteq \bar{S} \setminus S^o$. The real question that needs to be asked is: Are they equal? Or do we have examples where the inclusion is strict?

Let us try to prove the equality. If we get stuck in the proof, we shall get an idea of constructing an example for strict inequality. Let $x \in \bar{S} \setminus S^o$. Then, x is a limit point of S and we have $\forall \epsilon > 0$, $B(x, \epsilon) \cap S \neq \emptyset$. If for some $\epsilon > 0$ we get $B(x, \epsilon) \cap S^c = \emptyset$, then x would be an interior point of S. This would be a contradiction to our hypothesis. Therefore, $\forall \epsilon > 0$, we get $B(x, \epsilon) \cap S \neq \emptyset$ and $B(x, \epsilon) \cap S^c \neq \emptyset$. Hence, $\partial S = \bar{S} \setminus S^o$. Note that many authors consider this as the standard definition of the boundary, and it can be found in many books on metric spaces.

Before closing this section, we define another special type of set which will be interesting to study once we have developed the notion of sequences.

Definition 2.11 (Dense sets). *A set S in a metric space X is said to be dense if every point in X is a limit point of S.*

An immediate consequence of this definition is that every metric space is dense in itself.

Let us look at the metric space we know by experience, $(\mathbb{R}, d)$, where d is given by the absolute value function. Now, what are the dense sets in $\mathbb{R}$? Those who have taken a course in analysis will know that $\mathbb{Q}$ and $\mathbb{Q}^c$ are both dense. The set of rational numbers is countable, while the set of irrational numbers is not. Similarly, if we look at $(\mathbb{R}^n, d_2)$, then the set $\mathbb{Q}^n$ is dense in $\mathbb{R}^n$ and is *countable*. Therefore, it seems that these metric spaces are special. Based on this idea, we now make the following definition of what is called a separable metric space.

Definition 2.12 (Separable). *A metric space (X, d) is separable if it has a countable dense subset.*

Note that since every metric space is dense in itself, if X is countable, then (X, d) is separable.

Example 2.20. Consider the metric space ℓ^p for $p \geq 1$ and consider the set $S = \{(r_i)_{i=1}^{\infty} \in \ell^p | r_i \in \mathbb{Q}$, where $r_i = 0$ for $i > n$ for $n \in \mathbb{N}\}$. This means the set is a collection of all rational sequences which, after finitely many terms, take the value 0. Let $x = (x_i)_{i=1}^{\infty}$ be any point in ℓ^p. Since $\mathbb{Q}$ is dense in $\mathbb{R}$, $\forall \epsilon > 0$, there is a sequence $r = \left(r_i^{(\epsilon)}\right)_{i=1}^{\infty} \in S$ such that $d(x, r) < \epsilon$. This is because $(x_i)_{i=1}^{\infty} \in \ell^p$ means $\sum_{i=1}^{\infty} |x_i|^p$ converges. Consequently, $\sum_{i=n+1}^{\infty} |x_i|^p$ also converges to that $\sum_{i=n+1}^{\infty} |x_i|^p < \frac{\epsilon}{2}$ for some $n \in \mathbb{N}$. From the density of rationals, we can also have $\sum_{i=1}^{n} |x_i - r_i| < \frac{\epsilon}{2}$. Now, combining these two we get the estimate for $d(x, r)$ as mentioned above. This shows that S is dense in ℓ^p. Also, S is countable (why?). Hence, ℓ^p is separable metric space for each $p \geq 1$.

Let us now look at another example of a separable metric space which is not so trivial. If the reader feels that the proof is not understandable, the reader is advised to visit this proof after studying continuity, uniform continuity, and density in $\mathbb{R}$. The proof uses all three concepts together.

Example 2.21. Consider the set $\mathscr{C}[0, 1]$ equipped with the metric d_∞ (sup metric). Let $f \in \mathscr{C}[0, 1]$. Then, it is easy to see that f is uniformly continuous[7] on $[0, 1]$.

Therefore, $\forall \epsilon > 0$, $\exists n_0 \in \mathbb{N}$ such that $\forall x, y \in [0, 1]$ with $|x - y| < \frac{1}{n_0}$, we have $|f(x) - f(y)| < \epsilon$. This is just a consequence of the definition of uniform continuity and we have done nothing new. Now, we define a polygonal function $g : [0, 1] \to \mathbb{R}$ as follows:

(i) $\forall k \in \{0, 1, \cdots, n_0\}$, $g\left(\frac{k}{n_0}\right) = f\left(\frac{k}{n_0}\right)$.

(ii) The function g is linear in each of the subintervals $\left(\frac{k}{n_0}, \frac{k+1}{n_0}\right)$ for $k \in \{0, 1, \cdots, n_0 - 1\}$.

Construction of g is represented in Figure 2.20.

By this definition of g, it is easy to check that $d_\infty(f, g) < \frac{\epsilon}{2}$. We leave this computation as an exercise to the reader and move forward. We make another function h such that $h\left(\frac{k}{n_0}\right)$ is rational and $d(h, g) < \frac{\epsilon}{2}$. Now, by simple application of triangle inequality we get $d(f, h) < \epsilon$, thereby making the set of all such functions h dense in $\mathscr{C}[0, 1]$. Also, this set is countable (why?) so that $\mathscr{C}[0, 1]$ is separable.

There is another proof of this fact that $\mathscr{C}[0, 1]$ is separable. It uses a famous result called the *Weierstraß approximation theorem*, which states that any continuous function on a closed and bounded interval can be approximated by a sequence of polynomials which converge uniformly to the given function. Since it uses too many technical terms, we skipped the approach in this text. However, we encourage the reader to search up on the proof involving Weierstraß approximation of functions.

---

[7]The readers who are not familiar with uniform continuity are advised to postpone this example until Chapter 6, where we have formally defined the concept and developed some theory on it. Otherwise, one may refer to any Real Analysis book for reference on uniform continuity of real-valued functions.

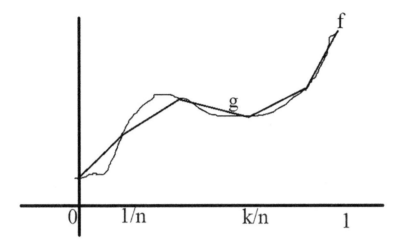

FIGURE 2.20: Approximating a continuous function with a polygonal function.

Exercise 2.33. Prove that the sets considered to prove that ℓ^p and $\mathscr{C}[0,1]$ are separable are indeed countable.

Exercise 2.34. What can you say about the separability of an uncountable set equipped with discrete metric?

Now, let us look at a question posed by us in the section on open sets. We asked: Is every open set (in any metric space) a countable union of open balls? Indeed, we answered the question negatively as in Example 2.17. We would now like to see when can the question be answered affirmatively.

Theorem 2.21. *In a separable metric space, every open set is a countable union of open balls.*

Proof. Let X be a separable metric space and let Y be its countable dense subset. Let G be an open set in X. Since Y is dense, the points of G can be made as close to the points of Y as we please. Therefore, if we consider the set $G \cap Y$, and around each point in this set, we construct an open ball of appropriate radius, we expect to get the whole union as G. First, we observe that

$$\bigcup_{y \in G \cap Y} B(y, r_y) \subseteq G,$$

for appropriate $r_y > 0$ because G is open. Now, we would like the other subsethood relation.

Let $x \in G$. Then, there is a $r_x > 0$ such that $B(x, r_x) \subseteq G$. Let $y \in Y$ be such that $d(x, y) < \frac{r_x}{2}$ and choose $r_y > 0$ such that $d(x, y) < r_y < \frac{r_x}{2}$.

Clearly, now $x \in B(y, r_y) \subseteq \bigcup_{y \in G \cap Y} B(y, r_y)$. Hence, the desired equality is obtained. $\square$

Clearly, since the discrete space defined on an uncountable set is not separable (Exercise 2.34), there are open sets which cannot be written as a countable union of open balls. One can also ask the converse question: If every open set in a metric space is a countable union of open balls, is the metric space separable? To answer this question, we would require the knowledge of continuum hypothesis and a few technical terms beyond the scope of this text. Hence, we do not discuss the details here. However, we give the reader a hint that if one assumes the continuum hypothesis to be true, then our question is answered positively.

2.8   Bounded Sets

The notion of bounded sets in $\mathbb{R}$ must be intuitively clear to almost every reader. However, we review it once again before we can abstract its properties. In $\mathbb{R}$, we call a set S to be bounded above if there is a real number $u \in \mathbb{R}$ such that $\forall x \in S$ we get $x \leq u$. Similarly, we call the set S to be bounded below if $\exists l \in \mathbb{R}$ such that $\forall x \in S$ we get $l \leq x$. Overall, a set is said to be bounded if $\exists M > 0$ such that $\forall x \in S$ we have $|x| \leq M$.

All these three notions of boundedness are not really useful in an abstract metric space. This is because the first two notions use partial order on $\mathbb{R}$, which may not be available in every metric space. On the other hand, the third notion uses *norm* instead of metric and we have decided in the beginning of this chapter not to talk about norms. However, the third notion does give to us an idea to define a bounded set in arbitrary metric space.

To do so, let us first look geometrically in $\mathbb{R}^2$ which is easy to visualize. Let us look at $B(0, 1)$, the open unit ball in the Euclidean plane $\mathbb{R}^2$. Geometric intuition tells us that this set is bounded. Why do we have this intuition? This is because of a simple yet a non-trivial reason. As a part of natural processing, we start looking at distances between points in a given set. If the distance goes on increasing for certain points, our intuition tells us that this set must be unbounded. We shall use this natural intuition to define bounded sets.

Definition 2.13 (Bounded sets). *A set S in a metric space X is said to be bounded if the set $\{d(x, y) \,|\, x, y \in S\}$ is bounded in $\mathbb{R}$.*

Remark. One may argue that in the definition, we may have excluded the empty set. Since intuitively, the empty set seems bounded, we would like it to satisfy the definition of our bounded sets. We observe that for the empty set, the corresponding set $\{d\,(x,y)\,|x,y\in\emptyset\}$ is itself empty. From analysis, we know that $\emptyset$ is bounded in $\mathbb{R}^8$, and therefore, $\emptyset$ is bounded in any arbitrary metric space X.

An immediate consequence of this definition is the concept of *diameter* of a set. Since for a bounded set in $\mathbb{R}$, supremum exists (by the LUB axiom), we can exploit this property to get the following definition.

Definition 2.14 (Diameter). *The diameter of a bounded set S in a metric space X is defined as*

$$diam\,(S) = \sup\{d\,(x,y)\,|x,y\in S\}.$$

Note that the diameter of a set is the measure of how far two points in the set can be. However, it is not guaranteed that the set will contain points x and y such that $d\,(x,y)$ is the diameter of the set. A classic example is the unit open ball in $\mathbb{R}^2$, equipped with the Euclidean metric. A simple computation would lead us to the fact that the diameter of the unit open ball is 2. However, for any two points in the open ball, the distance is always strictly less than 2.

Also, the name diameter is misleading in a sense that one may think $B\,(x,r)$ has the diameter $2r$. But, this is not always the case. Consider the discrete metric space and an open ball $B\,(x,1)$ about any point $x\in X$. Then, we have $B\,(x,1) = \{x\}$ and diam $(B\,(x,1)) = 0$.

However, consider the following for any two points $y,z\in B\,(x,r)$ (for an arbitrary metric space).

$$d\,(y,z) \leq d\,(y,x) + d\,(x,z)$$
$$< r + r = 2r.$$

Therefore, the set $\{d\,(y,z)\,|y,z\in B\,(x,r)\}$ is bounded above by $2r$ and hence diam $(B\,(x,r)) \leq 2r$. That is to say that the diameter of an open ball never exceeds twice its radius.

We now characterize bounded sets as follows.

Theorem 2.22. *A set S in a metric space X is bounded if and only if $\exists x_0 \in X$ and $r > 0$ such that $S \subseteq B\,(x_0,r)$.*

Proof. First, consider that S is bounded. Therefore, the set $\{d\,(x,y)\,|x,y\in S\}$ is bounded in $\mathbb{R}$. Let $r = $ diam (S). For a fixed $x_0 \in S$, consider the open ball $B\,(x_0, r+1)$. Now, if $y \in S$, then $d\,(x_0,y) \leq r < r + 1$. Hence, $y \in B\,(x_0, r+1)$, i.e., $S \subseteq B\,(x_0,r)$.

---

[8]Notice that if we want to prove that $\emptyset$ is bounded, all we have to do is produce a real number $M > 0$ such that $\forall x \in \emptyset$, $|x| \leq M$. We claim that $M = 1$. If someone wants to prove us wrong, they will have to find an element $x \in \emptyset$ such that $|x| > 1$, which is not possible. Therefore, we win and the empty set, $\emptyset$, is bounded in $\mathbb{R}$.

Conversely, let there be a point $x_0 \in X$ and a positive number r such that $S \subseteq B(x_0, r)$. Now, for any two points $x, y \in S$ we have

$$d(x, y) \leq d(x, x_0) + d(x_0, y) < r + r = 2r.$$

Hence, the set $\{d(x, y) \, | \, x, y \in S\}$ is bounded in $\mathbb{R}$ and consequently S is bounded in X. □

This is sometimes used as an equivalent definition of boundedness of sets in a metric space. We leave some simple results regarding the diameter of a set as an exercise to the reader.

Exercise 2.35. Let A and B be non-empty subsets of a metric space (X, d). Prove the following:

1. $\text{diam}(A) = 0$ if and only if A is a singleton set.

2. If $A \subseteq B$, then $\text{diam}(A) \leq \text{diam}(B)$.

3. For each $x \in A$ and $y \in B$, $d(x, y) \leq \text{diam}(A \cup B)$.

4. If $A \cap B \neq \emptyset$, then $\text{diam}(A \cup B) \leq \text{diam}(A) + \text{diam}(B)$.

2.9 Distance Between Sets

Until now, we tried to measure the distance between two points in a metric space. Now, we may take a step forward and ask ourselves: How far is a point from a given set? This means we wish to find out the distance between a given point and a given set. The first thing we should ask ourselves is: Is this question meaningful? In other words, is there any known method to calculate the distance of a point from a set? Well, there isn't! But, there is an intuition. Since we are provided with a metric function, we know the distances between any two points. Now, for notational purposes, let $x \in X$ be the given point and $A \subseteq X$ be the given set, where (X, d) is a metric space. The first thing that comes to our mind is let us find out all the distances of the point x with each point in A. Then, by intuition, the distance of x from A would be the distance between x and the closest point to x in A. In mathematical terms, the closest point (or distance) will be given by infimum. Therefore, we now make the formal definition.

Definition 2.15 (Distance between a point and a set). *The distance between a point $x \in X$ and a set $A \subseteq X$ of a metric space X is given by*

$$d(x, A) = \inf \{d(x, y) \, | \, y \in A\}.$$

Note. One may question that instead of infimum, why could we not use minimum? One must observe that there could be infinitely many points in the set and therefore, the minimum may not always exist. However, since the distance between points is bounded below by 0, the set used in Definition 2.15 is bounded below and the infimum always exists. In fact, we can also comment that the distance between the point and the set is non-negative, as per our intuition.

Let us, as an example, try to find the distance between a point and a set in $\mathbb{R}$, with the usual metric.

Example 2.22. Consider $\mathbb{R}$ with the usual metric and the set $[0, 1]$. Let us find the distance of $2 \in \mathbb{R}$ from the set $[0, 1]$. First, we observe that for any point $0 \le x \le 1$, $1 \le |2 - x| = 2 - x \le 2$. Therefore, the set $\{d(2, x) \,|\, x \in [0, 1]\}$ is bounded below by 1. Now, let $\epsilon > 0$ and consider $x = \max\{0, 1 - \frac{\epsilon}{2}\}$. Therefore, we have

$$|2 - x| = 2 - x$$
$$\le 2 - \left(1 - \frac{\epsilon}{2}\right)$$
$$= 1 + \frac{\epsilon}{2}$$
$$< 1 + \epsilon.$$

Hence, $1 + \epsilon$ is not a lower bound for the set $\{d(2, x) \,|\, x \in [0, 1]\}$ and we have $d(2, [0, 1]) = 1$.

Let us now ask ourselves a question: What would happen if the point belongs to the set? Clearly, if $x \in A$, then $d(x, x) = 0 \in \{d(x, y) \,|\, y \in A\}$. Also, this set is bounded below by 0, which suggests that the infimum must be 0. Therefore, we have the following result.

Theorem 2.23. If $x \in A$, then $d(x, A) = 0$.

One may ask the converse question, Does $d(x, A) = 0$ imply $x \in A$? To see this, let us consider the following example.

Example 2.23. Again, let us consider $\mathbb{R}$ with the usual metric and the set $A = (0, 1)$. Let us find $d(0, A)$. First, we see that the sequence $\left(\frac{1}{n}\right)_{n \in \mathbb{N}}$ converges to 0, which means that some points of $(0, 1)$ can be made as close to 0 as we like. Now, consider $\epsilon > 0$. Then, we know that $\exists n_0 \in \mathbb{N}$ such that $\frac{1}{n_0} < \epsilon$. Hence, if we choose $0 < x < \frac{1}{n_0}$, then we have $d(0, x) = x < \epsilon$ and therefore $\inf\{d(0, x) \,|\, x \in (0, 1)\} = 0$.

The above example shows us that it is possible for the distance between a point and a set to be 0 even if the point does not belong to the set. Therefore, the converse question asked above is negatively answered.

In the above example, we observe that 0 was an accumulation point of $(0, 1)$. Therefore, it is natural to ask: Is the distance between a set and its accumulation point (which does not belong to the set) 0? Let us try to answer this question affirmatively.

Consider a set $A \subseteq X$ of a metric space (X, d) and its accumulation point $l \notin A$. By the definition, we know that $\forall \epsilon > 0$, $\exists x \in A$ such that $d(l, x) < \epsilon$. Therefore, $\epsilon > 0$ is not a lower bound for the set, $\{d(l, x) | x \in A\}$ and at the same time, the set is bounded below by 0. Hence, $d(l, A) = \inf \{d(l, x) | x \in A\} = 0$.

Conversely, suppose that $d(l, A) = 0$ for some $l \notin A$. Then, $\forall \epsilon > 0$, $\exists d(l, x) \in \{d(l, x) | x \in A\}$ such that $d(l, x) < \epsilon$. This further means that $\forall \epsilon > 0$, $\exists x \in A$ such that $d(l, x) < \epsilon$, i.e., l is an accumulation point of A. Therefore, we have proved the following.

Theorem 2.24. *Consider a point $l \in X$ and a set $A \subseteq X$, where (X, d) is a metric space such that $l \notin A$. Then, $d(l, A) = 0$ if and only if l is an accumulation point of A.*

Now, we move another step ahead and define the distance between two sets in a metric space. The basic intuition remains the same and therefore, we do not provide all the details as above.

Definition 2.16 (Distance between sets). *Given two sets A and B in a metric space (X, d), the distance between A and B is given by*

$$d(A, B) = \inf \{d(x, y) | x \in A, y \in B\}.$$

Exercise 2.36. It is clear that if $A \cap B \neq \emptyset$, then $d(A, B) = 0$. Is the converse true? Justify. What can you say when $d(A, B) = 0$ but $A \cap B = \emptyset$?

As first time readers, we would have guessed that if A and B are two disjoint open sets with a common accumulation point, then their distance can be zero. However, this is not restricted to disjoint open sets alone. The next example shows the same from disjoint closed sets.

Example 2.24. Consider the metric space $(\mathbb{R}, |\cdot|)$ and the two sets $\mathbb{N}$ and $A = \{n + \frac{1}{n} | n \in \mathbb{N} \setminus \{1\}\}$. Clearly, $\mathbb{N}$ and A are closed in $\mathbb{R}$. Also, $\mathbb{N} \cap A = \emptyset$, since all elements of A are rational numbers. We know that the set $\{|m - n - \frac{1}{n}| \, | m \in \mathbb{N}, n + \frac{1}{n} \in A\}$ is bounded below by 0.

Let $\epsilon > 0$. By Archimedean property, $\exists n \in \mathbb{N}$ such that $\frac{1}{n} < \epsilon$. Then, for $n \in \mathbb{N}$ and $n + \frac{1}{n} \in A$, we have

$$\left| n - n - \frac{1}{n} \right| = \frac{1}{n} < \epsilon.$$

Hence, $\epsilon > 0$ is not a lower bound for the set, and we get $d(\mathbb{N}, A) = 0$.

Exercise 2.37. If A and B are non-empty subsets, of a metric space (X, d), then prove the following:

1. $\text{diam}(A \cup B) \leq \text{diam}(A) + \text{diam}(B) + d(A, B)$.

2. $\forall x, y \in X$, $d(x, A) \leq d(x, y) + d(y, A)$.

2.10 Equivalent Metrics

In earlier sections, we saw that on the same set X, we can define as many metrics as we want and as a consequence, each yields a collection of open sets. The question remains whether these open sets defined by different metrics are truly different. In this section, we try to answer this question. But, before that, we first look at an example from the very common metric space $\mathbb{R}^2$, for which we have gained quite a lot of experience and which is also easy to visualize.

Example 2.25. Consider the set $\mathbb{R}^2$ equipped with two different metrics: d_1 and d_2, given by Equations (2.9) and (2.7), respectively.

Now, consider the open balls $B_1\left(\mathbf{0}, 1\right)$ and $B_2\left(\mathbf{0}, 1\right)$, the open unit balls in d_1 and d_2 metrics, respectively.

$$B_1\left(\mathbf{0}, 1\right) = \left\{(x, y) \in \mathbb{R}^2 \middle| |x| + |y| < 1\right\},$$

$$B_2\left(\mathbf{0}, 1\right) = \left\{(x, y) \in \mathbb{R}^2 \middle| \sqrt{x^2 + y^2} < 1\right\}.$$

If we draw the open balls (boundary), it will look as in Figure 2.21. As is visible, $B_1\left(\mathbf{0}, 1\right) \subseteq B_2\left(\mathbf{0}, 1\right)$. However, we need to prove such an observation. First, notice that

$$(|x| + |y|)^2 = x^2 + y^2 + 2\,|x|\,|y| \geq x^2 + y^2 \geq 0.$$

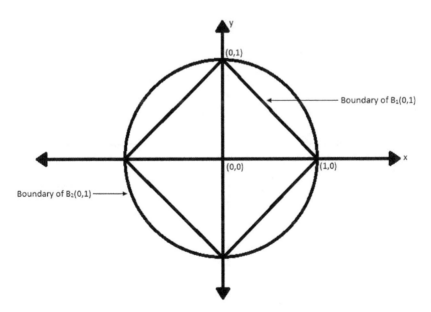

FIGURE 2.21: d_1 open unit ball "fits" inside the d_2 open unit ball.

Hence, we have

$$\sqrt{x^2 + y^2} \le |x| + |y|.$$

Therefore, if we consider $(x, y) \in B_1(\mathbf{0}, 1)$, then we have the following

$$\sqrt{x^2 + y^2} \le |x| + |y| < 1,$$

which further means $(x, y) \in B_2(\mathbf{0}, 1)$, i.e., $B_1(\mathbf{0}, 1) \subseteq B_2(\mathbf{0}, 1)$.

Now, consider another open ball with respect to d_2,

$$B_2\left(\mathbf{0}, \frac{1}{\sqrt{2}}\right) = \left\{(x, y) \in \mathbb{R}^2 \,\middle|\, \sqrt{x^2 + y^2} < \frac{1}{\sqrt{2}}\right\}.$$

The situation now looks as shown in Figure 2.22.

Notice that $(|x| - |y|)^2 \ge 0$ leads us to

$$x^2 + y^2 \ge 2|x||y|,$$

which further gives

$$x^2 + y^2 \ge \frac{1}{2}(|x| + |y|)^2.$$

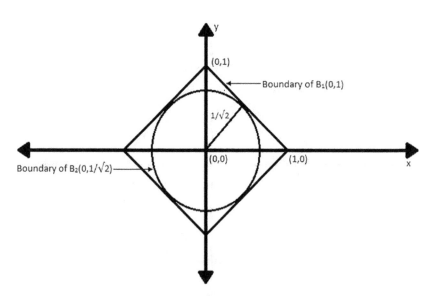

FIGURE 2.22: There is a d_2 open ball which "fits" inside the d_1 open unit ball.

Therefore, for any $(x, y) \in B_2\left(\mathbf{0}, \frac{1}{\sqrt{2}}\right)$, we have

$$|x| + |y| \leq \sqrt{2}\sqrt{x^2 + y^2} < 1.$$

Hence, $B_2\left(\mathbf{0}, \frac{1}{\sqrt{2}}\right) \subseteq B_1\left(\mathbf{0}, 1\right)$.

What did we observe in the example? We saw that for the two metrics d_1 and d_2 defined on $\mathbb{R}^2$, in each of the open unit balls, we can have another smaller open ball with the same center. Also, in $\mathbb{R}^2$, we are equipped with the two operations of addition and scaling, so that this holds true not only for the open balls around origin, but for any open ball about any point in $\mathbb{R}^2$ and of any positive radius. To put this rigorously, we have the following amazing property of the two metrics d_1 and d_2 in $\mathbb{R}^2$:

$$\forall \mathbf{x} \in \mathbb{R}^2 \text{ and } \forall r > 0, \exists r' > 0 \text{ such that } B_1\left(\mathbf{x}, r'\right) \subseteq B_2\left(\mathbf{x}, r\right),$$

and

$$\forall \mathbf{x} \in \mathbb{R}^2 \text{ and } \forall r > 0, \exists r'' > 0 \text{ such that } B_2\left(\mathbf{x}, r''\right) \subseteq B_1\left(\mathbf{x}, r\right).$$

One may ask: Why is such a property important? This means we now have to comment upon the consequences of this property. Let us look at open sets defined with respect to the two metrics d_1 and d_2. Let U be an open set with respect to d_1. Now, take any point $\mathbf{x} \in U$. Since U is open, there is a d_1-open ball around this which is completely contained in U. By the translation and scaling properties, we can take this open ball to origin and make it the unit open ball. From the above example, we see that there is a smaller d_2-open ball inside this one and again by translation and scaling, we can take the whole setting back to our set U. Therefore, in U, about every point $\mathbf{x}$, there is also a d_2-open ball completely contained in U. This means that U is open in d_2. Similarly, we can say that if U is open in d_2, then it is open in d_1. However, the arbitrary choice of U forces us to conclude that the open sets defined by d_1 are exactly the same as those defined by d_2. We leave the textbook proof of the above explanation as an exercise to the reader.

Exercise 2.38. Prove that the open sets defined with respect to d_1 are exactly the same as those defined with respect to d_2 in the set $\mathbb{R}^2$.

The next question one can ask is: Can we generalize this to more metrics and more sets? Since the arguments required for answering this question affirmatively, at least in some cases, is essentially the same as above, we leave it as exercises to the reader.

Exercise 2.39. Prove that the open sets in $\mathbb{R}^2$ defined with respect to d_2 and d_∞ are exactly the same.

Exercise 2.40. Prove that the metrics d_{p_1} and d_{p_2}, where $p_1, p_2 \geq 1$ and $p_1 \neq p_2$ defined on $\mathbb{R}^n$ as

$$d_p(\mathbf{x}, \mathbf{y}) = \left(\sum_{i=1}^{n} |x_i - y_i|^p \right)^{\frac{1}{p}}$$

for $p \geq 1$ and $\mathbf{x} = (x_i)_{i=1}^{n}$, $\mathbf{y} = (y_i)_{i=1}^{n} \in \mathbb{R}^n$ define the same collection of open sets.

Based on these observations, we now define what is called *equivalent metrics*.

Definition 2.17 (Equivalent metric). *If d_1, d_2 are two metrics defined on the same set X, they are said to be equivalent if for all open sets U with respect to metric d_1, U is also open with respect to d_2 and vice-versa.*

Remark. From Exercise 2.40, we can conclude that all p-metrics on $\mathbb{R}^n$ are equivalent. This further means that to study $(\mathbb{R}^n, d_p)$ for $p \geq 1$, it is sufficient to study only one of them. All others will have the same properties.[9]

In all the above description, we saw that if we can have open balls (with respect to different metrics) inside one another, then metrics can be equivalent. However, they were just examples! Can we have such a result for any two metrics? And, can we answer the converse question also? First, let us answer the converse question, i.e., if there are two equivalent metrics, then can we have open balls inside one another? Let X be the given set and d_1, d_2 be two equivalent metrics defined on X.

Consider $B_1(x, r)$, the d_1-open ball centered at $x \in X$ and of radius $r > 0$. Since the metrics are equivalent, $B_1(x, r)$ is also open with respect to d_2, i.e., $\exists r' > 0$ such that $B_2(x, r') \subseteq B_1(x, r)$. Similarly, we can argue for $B_2(x, r)$. Therefore, for equivalent metrics, we can have the open balls inside each other.

Now, suppose that we can have these balls inside one another, i.e., if d_1, d_2 are two metrics on the same set X, then

$$\forall x \in X \text{ and } \forall r > 0, \exists r' > 0 \text{ such that } B_1(x, r') \subseteq B_2(x, r),$$

and

$$\forall x \in X \text{ and } \forall r > 0, \exists r'' > 0 \text{ such that } B_2(x, r'') \subseteq B_1(x, r).$$

---

[9]This will be clear when we study further properties like convergence, continuity, and compactness further in the text.

Let U be an open set with respect to d_1 and let $x \in U$. Then, $\exists r > 0$ such that $B_1(x, r) \subseteq U$. Also, $\exists r' > 0$, such that $B_2(x, r') \subseteq B(x, r) \subseteq U$. Therefore, U is also open with respect to d_2. Similarly if U is open with respect to d_2, from the above reasoning, it is also open with respect to d_1.

Hence, we have the following result.

Theorem 2.25. *Two metrics d_1 and d_2, defined on the same set X, are equivalent if and only if*

$$\forall x \in X \text{ and } \forall r > 0, \exists r' > 0 \text{ such that } B_1(x, r') \subseteq B_2(x, r),$$

and

$$\forall x \in X \text{ and } \forall r > 0, \exists r'' > 0 \text{ such that } B_2(x, r'') \subseteq B_1(x, r)$$

where $B_1(x, r)$ denotes the open ball centered at $x \in X$ of radius r, with respect to the metric d_1 and $B_2(x, r)$ denotes the open ball centered at $x \in X$ of radius r, with respect to metric d_2.

Exercise 2.41. Is the discrete metric equivalent to the metric d_1, each defined on $\mathbb{R}^2$?

In Exercise 2.40, the reader must have proved that any two p-metrics on $\mathbb{R}^n$ are equivalent. Moving a step further, we may ask: Are the ℓ^p metrics equivalent? The question is natural in the sense that the metric is defined with the motivation from $\mathbb{R}^n$, except that we have certain additional conditions of series convergence. First, we see that for two metrics to be equivalent, they must be defined on the same set. Therefore, we first check if two sets ℓ^{p_1} and ℓ^{p_2} are same when $p_1 \neq p_2$. With the usual notation, ℓ^p is defined to be the set of all sequences $(x_i)_{i=1}^{\infty}$ such that the series $\sum_{i=1}^{\infty} |x_i|^p$ converges.

Now, consider the two sets ℓ^{p_1} and ℓ^{p_2}, where $p_1 \neq p_2$, and without loss of generality let us assume that $p_1 > p_2$. Consider the sequence $\left(\frac{1}{n^{p_2}}\right)_{n \in \mathbb{N}}$. Since $\frac{p_1}{p_2} > 1$, the series $\sum_{n=1}^{\infty} \frac{1}{n^{\frac{p_1}{p_2}}}$ converges, while the series $\sum_{n=1}^{\infty} \frac{1}{n}$ is not convergent. Therefore, $\left(\frac{1}{n^{p_2}}\right)_{n \in \mathbb{N}} \in \ell^{p_1}$ but $\left(\frac{1}{n^{p_2}}\right)_{n \in \mathbb{N}} \notin \ell^{p_2}$. Hence, $\ell^{p_1} \neq \ell^{p_2}$. Therefore, we cannot talk about the equivalence of the metrics d_p for ℓ^p.

Finally, we consider another set of interest, and of importance, $\mathscr{C}[a, b]$ for $a < b$. Recall that it is the set of all continuous real valued functions defined on the interval $[a, b]$. We can define the metrics on $\mathscr{C}[a, b]$ in the following way:

$$d_p(f, g) = \left(\int_a^b |f(x) - g(x)|^p \, dx \right)^{\frac{1}{p}},$$

for $p \geq 1$. And,

$$d_\infty(f, g) = \sup_{x \in [a,b]} \{|f(x) - g(x)|\}.$$

We want to see if these metrics are equivalent. We shall see the easy case, d_1 and d_∞.

Observe that

$$d_1(f, g) = \int_a^b |f(x) - g(x)| \, dx$$

$$\leq (b - a) \sup_{x \in [a,b]} \{|f(x) - g(x)|\}$$

$$= (b - a) d_\infty(f, g).$$

Based on our observation, it seems that we can have a d_∞-open ball inside the d_1 open unit ball. Therefore, consider $B_1(\mathbf{0}, 1) = \left\{ f \in \mathscr{C}[a, b] \,\middle|\, \int_a^b |f(x)| \, dx < 1 \right\}$. Construct the open ball

$$B_\infty\left(\mathbf{0}, \frac{1}{b-a}\right) = \left\{ f \in \mathscr{C}[a, b] \,\middle|\, \sup_{x \in [a,b]} \{|f(x)|\} < \frac{1}{b-a} \right\}.$$

Now, for any $f \in B_\infty\left(\mathbf{0}, \frac{1}{b-a}\right)$, we have

$$d_1(\mathbf{0}, f) \leq (b - a) d_\infty(\mathbf{0}, f) < 1.$$

Hence, $f \in B_1(\mathbf{0}, 1)$ and $B_\infty\left(\mathbf{0}, \frac{1}{b-a}\right) \subseteq B_1(\mathbf{0}, 1)$. Since $\mathscr{C}[a, b]$ is equipped with the operations of addition and scaling, this holds true for any open ball centered at any point $f \in \mathscr{C}[a, b]$ and of any radius $r > 0$.

Can we have the converse, i.e., can we construct an open ball with respect to d_1 inside the open unit ball with respect to d_∞? First, observe that both d_1 and d_∞ have some geometrical meaning. While d_1 gives the area between two functions, d_∞ gives the maximum distance between the two functions. Therefore, the functions in $B_\infty(\mathbf{0}, 1)$ can never attain the value 1 in their domain $[a, b]$.[10] This is shown in Figure 2.7. On the other hand, we can have a function with large maximum value but area as small as we please. Therefore, our intuition tells us that the converse question is answered negatively. The next step is to prove it!

---

[10]This is because continuous functions on closed and bounded intervals attain their bounds.

Let $\delta > 0$. Define $B_1(\mathbf{0}, \delta)$. If $\delta > \frac{b-a}{2}$, consider the function $f : [a, b] \to \mathbb{R}$ defined as

$$f(x) = \begin{cases} \dfrac{2}{b-a}(x-a), & a \le x < \dfrac{a+b}{2} \\ -\dfrac{2}{b-a}(x-b), & \dfrac{a+b}{2} \le x \le b. \end{cases}$$

The construction of this function is shown in Figure 2.23.

It is easy to check that f is continuous and $\sup\limits_{x \in [a,b]} \{|f(x)|\} = 1$. Also,

$$\int_a^b |f(x)| \, dx = \int_a^{\frac{a+b}{2}} \left(\frac{2}{b-a}(x-a) \right) dx + \int_{\frac{a+b}{2}}^b \left(-\frac{2}{b-a}(x-b) \right) dx$$

$$= \frac{1}{b-a} \left[(x-a)^2 \right]_a^{\frac{a+b}{2}} - \frac{1}{b-a} \left[(x-b)^2 \right]_{\frac{a+b}{2}}^b$$

$$= \frac{1}{b-a} \left(\frac{b-a}{2} \right)^2 + \frac{1}{b-a} \left(\frac{b-a}{2} \right)^2$$

$$= \frac{b-a}{2}$$

$$< \delta.$$

Therefore, $f \in B_1(\mathbf{0}, \delta)$ but $f \notin B_\infty(\mathbf{0}, 1)$.

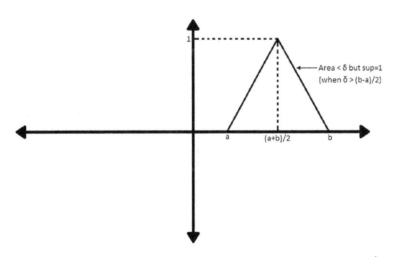

FIGURE 2.23: Constructing a counterexample for the case $\delta > \frac{b-a}{2}$.

Similarly, if $\delta \leq \dfrac{b-a}{2}$, then consider the function $f : [a,b] \to \mathbb{R}$ defined as

$$
f(x) = \begin{cases}
\dfrac{2}{\delta}(x-a), & a \leq x < a + \dfrac{\delta}{2}. \\[2mm]
-\dfrac{2}{\delta}(x-a-\delta), & a + \dfrac{\delta}{2} \leq x < a + \delta. \\[2mm]
0, & a + \delta \leq x \leq b.
\end{cases}
$$

The construction of this function is shown in Figure 2.24.

It is again easy to check that f is continuous and $\displaystyle\sup_{x \in [a,b]} \{|f(x)|\} = 1$. Also,

$$
\begin{aligned}
\int_a^b |f(x)|\, dx &= \int_a^{a+\frac{\delta}{2}} \left(\frac{2}{\delta}(x-a)\right) dx + \int_{a+\frac{\delta}{2}}^{a+\delta} \left(-\frac{2}{\delta}(x-a-\delta)\right) dx \\
&= \frac{1}{\delta}\left[(x-a)^2\right]_a^{a+\frac{\delta}{2}} - \frac{1}{\delta}\left[(x-a-\delta)^2\right]_{a+\frac{\delta}{2}}^{a+\delta} \\
&= \frac{\delta}{2} \\
&< \delta.
\end{aligned}
$$

Hence, $f \in B_1(\mathbf{0}, \delta)$ but $f \notin B_\infty(\mathbf{0}, 1)$. Therefore, we cannot have a d_1-open ball inside the d_∞-open unit ball for the set $\mathscr{C}[a,b]$, as suggested by our intuition. Hence, the two metrics d_1 and d_∞ defined on $\mathscr{C}[a,b]$ are not equivalent.

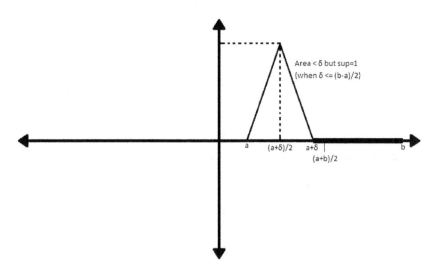

FIGURE 2.24: Constructing a counterexample for the case $\delta \leq \dfrac{b-a}{2}$.

Similar comments can be made about other d_p metrics on $\mathscr{C}[a, b]$. However, it is beyond the scope of this text and therefore we omit it. We do encourage the reader to dig up on this subject. Before we end this section, we now comment on certain abstract metrics. Until now, we were very particular about the sets and the metrics defined on them. Now, we take an arbitrary metric space (X, d) and then try to comment on the equivalence of other metrics defined on it.

Example 2.26 (Bounded metric). Consider a metric space (X, d). Consider the bounded metric $\delta : X \times X \to \mathbb{R}$.

$$\delta(x, y) = \min\{d(x, y), 1\}.$$

Now, let U be an open set with respect to the metric d. Then, $\forall x \in U, \exists r > 0$ such that $B_d(x, r) \subseteq U$. We can very well assume, without disturbing any concept, that $r < 1$. In this case, $B_\delta(x, r) = B_d(x, r) \subseteq U$, and as a consequence, U is also open with respect to δ.

Similarly, if U is an open set with respect to the metric δ, we consider the open ball $B_\delta(x, r) \subseteq U$ for an arbitrary $x \in U$. Choose an $r' > 0$ such that $r' < r$. Then, for any $x, y \in B_d(x, r')$, we have $\delta(x, y) \leq r' < r$. Hence, U is open with respect to d.

Finally, we can conclude that the two metrics d and δ are equivalent.

Remark. Although the two metrics in Example 2.26 are equivalent, notice that X is bounded with respect to metric δ but may not be bounded in metric d. Hence, equivalent metrics do not preserve bounded sets. This is because bounded sets are not defined in terms of open sets. Since equivalent metrics define the same open sets, all the properties defined in terms of open sets (or for that matter, closed sets) are preserved by equivalent metrics.

Problem Set

1. Consider the metric space $\mathbb{R}^2$ equipped with d_2. Which of the following sets are open?

   (a) $\{(x, y) \in \mathbb{R}^2 | xy \neq 0\}$.
   (b) $\{(x, y) \in \mathbb{R}^2 | x \notin \mathbb{Z}, y \notin \mathbb{Z}\}$.
   (c) $\{(x, y) \in \mathbb{R}^2 | x^2 + y^2 \neq 1\}$.

2. What are all the open sets in a finite metric space?

3. Consider the set $\mathscr{B}[0, 1]$ of all bounded real-valued functions defined on $[0, 1]$. Clearly, the set of continuous functions $\mathscr{C}[0, 1]$ is a subset of this set. Let $\mathscr{B}[0, 1]$ be equipped with the d_∞ metric. Is $\mathscr{C}[0, 1]$ open in $\mathscr{B}[0, 1]$?

4. Let $f : [0, \infty) \to [0, \infty)$ be a continuous function with the following properties:

   (a) $f(t) = 0$ if and only if $t = 0$.

   (b) f is non-decreasing, i.e., $x \leq y \Rightarrow f(x) \leq f(y)$.

   (c) f is subadditive, i.e., $f(x + y) \leq f(x) + f(y)$.

   Prove that if (X, d) is a metric space then $(X, f \circ d)$ is also a metric space. Are these metrics equivalent?

5. Consider a plane in $\mathbb{R}^3$ with a general equation, $ax + by + cz = d$. Is it open? Is it closed? Here, let $\mathbb{R}^3$ be equipped with the Euclidean metric.

6. For two sets $A, B \in \mathbb{R}^n$, we define the addition of these sets as

$$A + B = \{\mathbf{x} + \mathbf{y} | \mathbf{x} \in A, \mathbf{y} \in B\},$$

   where $\mathbf{x}$ and $\mathbf{y}$ are n-tuples. Now, if A and B are open, is $A + B$ open? If A and B are closed, is $A + B$ closed?

7. What is the relation between the closures of $A \cup B$ and $A \cap B$ in terms of $\bar{A}$ and $\bar{B}$?

8. Show that every point in an open set of $\mathbb{R}^n$ (equipped with the Euclidean metric) is a cluster point.

9. Let A be a subset of a metric space X. Then, show that a point $x \in X$ is a cluster point of A if and only if every open set containing x contains infinitely many points of A.

10. Consider the set $M(n, \mathbb{R})$ equipped with the metric defined as follows:

    For a matrix A, consider $A^T A$. Then, the distance from $\mathbf{0}$, the zero matrix, will be given by the Euclidean distance of $A^T A$ from $\mathbf{0}$ for $\mathbb{R}^{n^2}$. Investigate if the following sets are bounded.

    (a) The set of all orthogonal matrices.[11] This set is often denoted by $O(n, \mathbb{R})$.

    (b) The set $SL(n, \mathbb{R})$ of all matrices with determinant 1. This is often termed as *special linear group of order* n.

11. Does there exists a finite set which is dense in $\mathbb{R}$?

12. What can we say about a metric space with a finite dense set?

13. If in a metric space X, the only dense set is X itself, what can we comment upon the metric structure or topology (i.e., the collection of open sets) of this space?

---

[11]A matrix A is orthogonal if $AA^T = A^T A = I$, where A^T denotes the transpose of A.

Biographical Notes

William Henry Young (20 October, 1863 to 7 July, 1942) was an English mathematician. Young was educated at City of London School and Peterhouse, Cambridge. He worked on measure theory, Fourier series, differential calculus, among other fields, and made contributions to the study of functions of several complex variables.

Otto Ludwig Hölder (22 December, 1859 to 29 August, 1937) was a German mathematician born in Stuttgart. Hölder first studied at the Polytechnikum (which today is the University of Stuttgart) and then in 1877 went to Berlin where he was a student of Leopold Kronecker, Karl Weierstrass, and Ernst Kummer. He is noted for many theorems including: Hölder's inequality, the Jordan-Hölder theorem, the theorem stating that every linearly ordered group that satisfies an Archimedean property is isomorphic to a subgroup of the additive group of real numbers, the classification of simple groups of order up to 200, the anomalous outer automorphisms of the symmetric group S_6, and Hölder's theorem, which implies that the Gamma function satisfies no algebraic differential equation. Another idea related to his name is the Hölder condition (or Hölder continuity) which is used in many areas of analysis, including the theories of partial differential equations and function spaces.

Hermann Minkowski (22 June, 1864 to 12 January, 1909) was a German mathematician and professor at Königsberg, Zürich, and Göttingen. He created and developed the geometry of numbers and used geometrical methods to solve problems in number theory, mathematical physics, and the theory of relativity. Minkowski is perhaps best known for his work in relativity, in which he showed in 1907 that his former student Albert Einstein's special theory of relativity (1905) could be understood geometrically as a theory of four-dimensional spacetime, since known as the "Minkowski spacetime".

Euclid (300 BC), sometimes called Euclid of Alexandria to distinguish him from Euclid of Megara, was a Greek mathematician, often referred to as the "founder of geometry" or the "father of geometry". He was active in Alexandria during the reign of Ptolemy I (323-283 BC). His Elements is one of the most influential works in the history of mathematics, serving as the main textbook for teaching mathematics (especially geometry) from the time of its publication until the late nineteenth or early twentieth century. In the Elements, Euclid deduced the theorems of what is now called Euclidean geometry from a small set of axioms. Euclid also wrote works on perspective, conic sections, spherical geometry, number theory, and mathematical rigor. The "Euclidean distance" is named in his honor since the Eulidean geometry pops up when we measure distances in $\mathbb{R}^2$ by the "Euclidean distance".

Henri Leon Lebesgue (28 June, 1875 to 26 July, 1941) was a French mathematician known for his theory of integration, which was a generalization of the seventeenth-century concept of integration–summing the area between an axis and the curve of a function defined for that axis. His theory was published originally in his dissertation *Intégrale, longueur, aire* ("Integral, length, area") at the University of Nancy during 1902. The notation ℓ^p was made in his honor after the notation L^p which denotes the space of Lebesgue integrable functions in some sense.

David Hilbert (23 January, 1862 to 14 February, 1943) was a German mathematician and one of the most influential and universal mathematicians of the nineteenth and early twentieth centuries. Hilbert discovered and developed a broad range of fundamental ideas in many areas, including invariant theory, calculus of variations, commutative algebra, algebraic number theory, the foundations of geometry, spectral theory of operators and its application to integral equations, mathematical physics, and foundations of mathematics (particularly proof theory).

Felix Hausdorff (8 November, 1868 to 26 January, 1942) was a German mathematician who is considered to be one of the founders of modern topology and who contributed significantly to set theory, descriptive set theory, measure theory, function theory, and functional analysis. The term "Hausdorff property" is named in honor for his work in topological spaces and their separation axioms.

Chapter 3

Complete Metric Spaces

Functions with domain as natural numbers $\mathbb{N}$, sequences as they are called, play a crucial role in analysis. Be it defining the continuity of a function or the compactness of a set, sequences provide a very natural way to quantify these rather abstract concepts. In this chapter, we take a quick glance at the properties of sequences and study in detail a special type of sequences known as Cauchy sequences and how they relate to complete metric space. The main aim of this chapter is to look at the process of making a metric space complete. Finally, we prove the Baire category theorem, which is a powerful tool used to provide non-constructive proofs to some rather not-so-straightforward results in analysis.

3.1 Sequences

To begin with, we give the formal definition of sequences which is surprisingly straightforward for such an elegant concept. For any given set X, we define:

Definition 3.1 (Sequence). *A sequence in X is a function from $\mathbb{N}$, the set of all natural numbers, to X.*

Note that this definition is not universal, far from it. But among all the definitions used, this one is the more natural way of doing so. The reader may have experience with "real sequences", which are sequences in $\mathbb{R}$. We will look at a few of those but before that, we must look at the customary notations for sequences.

Instead of looking at sequences as functions, we look at them as a "list" of elements of X. The list need not necessarily be exhaustive, i.e., the function need not be surjective and indeed it rarely is, nor non-repetitive. Thus the notation for sequence follows: We use x_n to denote the nth entry in the list. More formally, a sequence defined by the function $f : \mathbb{N} \to X$ is denoted as $(x_n)_{n \in \mathbb{N}}$ where $x_n = f(n)$. The use of the subscript $n \in \mathbb{N}$ will become clear when we deal with subsequences. We first look at a few examples.

Example 3.1. We look at a few real sequences and the corresponding notations

1. The sequence defined by the identity function $I : \mathbb{N} \to \mathbb{N}$, $I(n) = n$ can be "listed" as $(1, 2, 3, 4, \cdots)$ and is denoted by $(n)_{n \in \mathbb{N}}$.

2. The sequence $\left(1, \frac{1}{2}, \frac{1}{3}, \frac{1}{4}, \cdots\right)$ is denoted by $\left(\frac{1}{n}\right)_{n \in \mathbb{N}}$.

3. Sequences can also be characterized by iterative schemes, in which the next term depends on the previous terms of the sequence in some way. The most notable of such examples is the sequence of Fibonacci numbers $(1, 1, 2, 3, 5, 8, \cdots)$ denoted by $(F_n)_{n \in \mathbb{N}}$. It follows the iterative scheme $F_n = F_{n-1} + F_{n-2}$, for $n \geq 2$ with $F_1 = F_2 = 1$.

4. The sequence of successive rational approximations of $\sqrt{2}$ given by

$$(1, 1.4, 1.41, 1.414, 1.4142, 1.41421, \cdots).$$

While real sequences are important in their own right, sequences go far beyond real numbers. We have sequences of functions, sequences of sequences, sequences of sets, and so on. It thus becomes imperative that we dedicate a few examples to such sequences.

Example 3.2.

1. $\left(1, x, x^2, x^3, x^4, x^5, \cdots\right)$ is a sequence of polynomials.

2. $(\sin x, \cos x, \sin 2x, \cos 2x, \sin 3x, \cos 3x, \cdots)$ is a sequence of trigonometric functions.[1]

3. $\left((-1, 1), \left(\frac{-1}{2}, \frac{1}{2}\right), \left(\frac{-1}{3}, \frac{1}{3}\right), \left(\frac{-1}{4}, \frac{1}{4}\right), \left(\frac{-1}{5}, \frac{1}{5}\right), \cdots\right)$ is a sequence in $\mathbb{R}^2$.

4. Another example is the sequence of functions $f_n : \ell^\infty \to \mathbb{R}$ defined by $f_n(\mathbf{x}) = \xi_n$ where $\mathbf{x} = (\xi_k)_{k \in \mathbb{N}}$.

3.1.1 Subsequences

Sometimes, it is useful to have a way to "take out" some entries from a "list". Subsequences play that role for sequences. They are defined as follows:

Definition 3.2 (Subsequence). *A subsequence for a given sequence* $(x_n)_{n \in \mathbb{N}}$ *is the sequence* $\left(x_{\phi(k)}\right)_{k \in \mathbb{N}}$ *where* $\phi : \mathbb{N} \to \mathbb{N}$ *is a strictly increasing function.*[2]

---

[1]Note that although, the two sequences in parts 1 and 2 are sequences of functions, at a particular $x \in \mathbb{R}$, we get a real-valued sequence.

[2]Note that a function $f : \mathbb{R} \to \mathbb{R}$ is strictly increasing if for $x_1 < x_2$ we have $f(x_1) < f(x_2)$. The same definition works for any subset of $\mathbb{R}$ as the domain (and codomain) of the function f.

It is usually denoted by $(x_{n_k})_{k \in \mathbb{N}}$ where it is assumed that $n_k \in \mathbb{N}$ strictly increases with k. The subscript is thus used to show the "running index". The definition looks overloaded at first, but is rather simpler if we continue to look at it from the "list" analogy. What we're basically doing is traversing the list one element at a time and "picking" the elements we want in the subsequence and noting down their indices with the help of ϕ without disturbing their relative positions.

Example 3.3. Notice that the function $\phi(n) = 2n$ gives the subsequence $(2n)_{n \in \mathbb{N}}$ of the sequence $(n)_{n \in \mathbb{N}}$.

The reader may verify that $\left(\frac{1}{n^2}\right)_{n \in \mathbb{N}}$ is a subsequence of $\left(\frac{1}{n}\right)_{n \in \mathbb{N}}$.

Exercise 3.1. The Fibonacci sequence does not form a subsequence of $(n)_{n \in \mathbb{N}}$. Justify.

Exercise 3.2. Find the function ϕ which proves that $\left(\frac{1}{n^2}\right)_{n \in \mathbb{N}}$ is a subsequence of $\left(\frac{1}{n}\right)_{n \in \mathbb{N}}$, as per the notations of Definition 3.2.

3.2 Convergence of Sequence

Take a look at the following sequences:

- $\left(1, \dfrac{1}{2}, \dfrac{1}{3}, \dfrac{1}{4}, \cdots\right)$

- $(1, e^{-1}, e^{-2}, e^{-3}, \cdots)$

- $\left(\sin\left(\dfrac{1}{n}\right)\right)_{n \in \mathbb{N}}$

They all seem to "tend" to 0. As the sequence progresses, the value of each term gets closer and closer to 0. This tendency of some sequences to go close to a particular value is characterized by convergence. First, to talk about getting "close" to a particular value, we need to have a notion of distance. The definition of convergence then easily follows:

Definition 3.3 (Convergence). *The sequence $(x_n)_{n \in \mathbb{N}}$ converges to a limit x in a metric space (X, d) if*

$$\forall \epsilon > 0, \exists n_0 \in \mathbb{N} \text{ such that } \forall n \geq n_0 \text{ we have } d(x_n, x) < \epsilon.$$

This is often written as

$$d(x_n, x) \to 0 \text{ as } n \to \infty.$$

Example 3.4. The sequence $\left(\sin\left(\frac{1}{n}\right)\right)_{n\in\mathbb{N}}$ as we already know, converges to 0 in the usual metric on $\mathbb{R}$.

To prove this we note that $\sin x < x$, for $x > 0$. Thus for given $\epsilon > 0$ from the Archimedean property, there is an $n_0 \in \mathbb{N}$ such that $n_0 > \frac{1}{\epsilon}$ and for $n \geq n_0$, we have

$$\left|\sin\frac{1}{n} - 0\right| < \frac{1}{n} \leq \frac{1}{n_0} < \epsilon.$$

The definition of convergence puts no restriction on the number of limits a sequence can have. So, we wonder what happens if a sequence is convergent to two distinct limits. It turns out that in a metric space the triangle inequality prevents a sequence from having more than one limit.

Theorem 3.1. *In a metric space* (X, d), *a convergent sequence converges to a unique limit.*

Proof. Let $(x_n)_{n\in\mathbb{N}}$ be a convergent sequence such that it converges to both x and y in X. Thus, by definition of limit, we have

$$\forall \epsilon > 0, \exists n_0 \in \mathbb{N} \text{ such that } \forall n \geq n_0, d(x_n, x) < \frac{\epsilon}{2},$$

and

$$\forall \epsilon > 0, \exists n_1 \in \mathbb{N} \text{ such that } \forall n \geq n_1, d(x_n, y) < \frac{\epsilon}{2}.$$

The triangle inequality gives

$$0 \leq d(x, y) \leq d(x, x_n) + d(x_n, y), \ \forall n \in \mathbb{N}.$$

Taking $n > \max\{n_0, n_1\}$, we get

$$d(x, y) \leq d(x_n, x) + d(y_n, y) < \epsilon.$$

Thus, $d(x, y) = 0$ so that $x = y$ and the limit of convergence is unique. $\square$

We now wish to ask whether the convergence of a subsequence necessarily implies the convergence of a sequence. The answer is no! Consider the sequence $(1, 0, 1, 0, 1, 0, \cdots)$. Its subsequences $(1, 1, 1, 1, \cdots)$ and $(0, 0, 0, 0, \cdots)$ are both convergent; however, the sequence is not. In fact, a sequence with subsequences convergent to distinct limits is not convergent. We leave the proof of this fact as an exercise to the reader (see Problem Set).

As is apparent from the definition, the convergence of a sequence very much depends on the metric space it is considered in. Given two metrics on the same set, a sequence may converge in one but not in another. We now look at a few examples to illustrate this concept.

Example 3.5. On the space of all continuous real-valued functions, $\mathscr{C}[0,1]$, we define two metrics: the metric d_∞ as $d_\infty(f,g) = \sup\limits_{x \in [0,1]} \{|f(x) - g(x)|\}$

and the metric d_1 as $d_1(f,g) = \int\limits_0^1 |f(x) - g(x)|\, dx$.

Consider the sequence $f_n(x) = e^{-nx}$. This sequence converges point-wise[3] to 0 at each point except $x = 0$, where it converges to 1. But the function that is non-zero at only one point is discontinuous and thus, does not belong to $\mathscr{C}[0,1]$.

First, we look at the convergence of the sequence to the constant function $\mathbf{0} : \mathbb{R} \to \mathbb{R}$, defined as $\mathbf{0}(x) = 0$, $\forall x \in [0,1]$.

$$d_1(f_n, 0) = \int\limits_0^1 |f_n(x) - 0|\, dx = \int\limits_0^1 e^{-nx}\, dx = \frac{1 - e^{-n}}{n} \to 0 \text{ as } n \to \infty$$

and

$$d_\infty(f_n, 0) = \sup\limits_{x \in [0,1]} \{|f_n(x) - 0|\} = \sup\limits_{x \in [0,1]} \{e^{-nx}\} = 1.$$

Since the limit of convergence is unique, f_n converges to $\mathbf{0}$ in $\mathscr{C}[0,1]$ with the metric d_1. As for the convergence of the sequence with metric d_∞, we have seen that it is not convergent to $\mathbf{0}$. How do we know that f_n does not converge to any other function? To see this, consider the following.

If f is not identically 0, it must be non-zero throughout an open interval say (c,d) and thus, $|f(x)| \geq M$ $\forall x \in (c,d)$ for some $M > 0$. Also since for any fixed $x \in (c,d)$, $\lim\limits_{n \to \infty} e^{-nx} = 0$, $\exists N_1 > 0$ such that $|e^{-nx}| < M$, $\forall x \in (c,d)$.

Now,

$$\left|f(x) - e^{-nx}\right| \geq \left||f(x)| - |e^{-nx}|\right|,$$

which for $n > N_1$ gives

$$\left|f(x) - e^{-nx}\right| \geq |f(x)| - |e^{-nx}|.$$

Therefore,

$$\sup\limits_{x \in (c,d)} \{|f(x) - e^{-nx}|\} \geq M - e^{-nc}, \ \forall n > N_1$$

which gives

$$\lim\limits_{n \to \infty} \sup\limits_{x \in (c,d)} \{|f(x) - e^{-nx}|\} \geq M.$$

---

[3] Recall that a sequence of real-valued functions $(f_n)_{n \in \mathbb{N}}$ is said to converge *point-wise* to a limit function $f : \mathbb{R} \to \mathbb{R}$ if for each x in the domain of f_n, the real-valued sequence $(f_n(x))_{n \in \mathbb{N}}$ converges to $f(x)$.

Therefore, we have

$$\lim_{n\to\infty} d_\infty(f_n, f) = \lim_{n\to\infty} \sup_{x\in[0,1]} \{|f(x) - f_n(x)|\} \geq M > 0.$$

Thus, f_n does not converge to any other function in $\mathscr{C}[0,1]$.

As one can note from the above examples, the major drawback of the definition of convergence is that one has to know the limit of convergence to prove the convergence of the sequence. We would like to try to formulate a new definition of convergence that does not depend on the limit. A crucial observation would be the fact that for terms of a sequence to come "closer" to a particular limit, they have to come close to each other. Thus, we try to look at sequences whose terms come close to each other as we move to higher indices. Unfortunately, such sequences are not always convergent, as we will shortly see. Such sequences are of great importance nevertheless. They are called *Cauchy sequences*.

Definition 3.4 (Cauchy sequences). *A sequence $(x_n)_{n\in\mathbb{N}}$ in metric space (X, d) is said to be a Cauchy sequence if $\forall \epsilon > 0, \exists n_0 \in \mathbb{N}$ so that $\forall m, n \geq n_0$, $d(x_m, x_n) < \epsilon$.*

Or equivalently, the sequence is Cauchy if $\forall \epsilon > 0, \exists n_0 \in \mathbb{N}$ so that $\forall n > n_0$ and $\forall m \in \mathbb{N}$, $d(x_{m+n}, x_n) < \epsilon$.

The latter definition is equivalent to saying $\lim_{n\to\infty} d(x_{m+n}, x_n) = 0$. We will use this definition a lot in the proceeding sections.

Exercise 3.3. In a metric space (X, d), prove that if sequences $(x_n)_{n\in\mathbb{N}}$ and $(y_n)_{n\in\mathbb{N}}$ are Cauchy, then the real-valued sequence $(d(x_n, y_n))_{n\in\mathbb{N}}$ is also Cauchy.

Exercise 3.4. Prove that every convergent sequence is Cauchy.

The converse of Exercise 3.4 is not true. The most interesting example is that of rational approximations of irrational numbers. Consider, $\sqrt{2}$ for instance. It can be approximated recursively by

$$a_{n+1} = \frac{a_n}{2} + \frac{1}{a_n}, \text{ for any positive } a_1 \in \mathbb{Q} \text{ and } n \in \mathbb{N}$$

or any other approximation technique. After obtaining approximation up to any n^{th} decimal place, the error in approximation can be made less than 10^{-n} for any $n \in \mathbb{N}$. The sequence of such approximations say $(a_n)_{n\in\mathbb{N}}$ is Cauchy since for $m, n \in \mathbb{N}$, we have

$$|a_{m+n} - a_n| < \left|a_{m+n} - \sqrt{2}\right| + \left|\sqrt{2} - a_n\right| < 10^{-n} + 10^{-(m+n)} \to 0 \text{ as } n \to \infty.$$

However, the sequence of said rational approximation $(a_n)_{n \in \mathbb{N}}$ is not convergent **in** $\mathbb{Q}$ since there is no $\sqrt{2}$ in $\mathbb{Q}$ for it to converge to. Note that the sequence here has a "tendency" to converge, in the sense that it seems to converge. It just doesn't have a point to converge to. In a certain sense, we can say that our set is not "complete": it does not contain the points to which a lot of sequences tend to converge to.

Another example of an "incomplete" space is given below.

Example 3.6. Consider the space $\mathscr{C}[0,1]$ with the metric d_1 defined by

$$d_1(f,g) = \int_0^1 |f(x) - g(x)| \, dx.$$

Consider the following sequence of functions in $\mathscr{C}[0,1]$, for $n \geq 2$

$$f_n(x) = \begin{cases} 0, & 0 \leq x < \frac{1}{2} - \frac{1}{n}. \\ \frac{n}{2}\left(x - \left(\frac{1}{2} - \frac{1}{n}\right)\right), & \frac{1}{2} - \frac{1}{n} < x < \frac{1}{2} + \frac{1}{n}. \\ 1, & \frac{1}{2} + \frac{1}{n} < x \leq 1. \end{cases}$$

One such function is shown in Figure 3.1.

To prove that this sequence is Cauchy, consider another function

$$f(x) = \begin{cases} 0, & 0 \leq x \leq \frac{1}{2}. \\ 1, & \frac{1}{2} < x \leq 1. \end{cases}$$

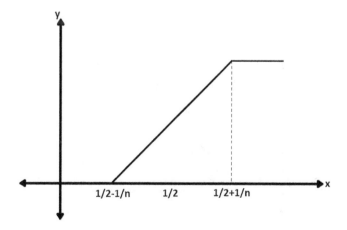

FIGURE 3.1: One of the functions from our sequence.

Note that $f \notin \mathscr{C}[0,1]$; however, it is integrable. Then we have

$$d_1\left(f_{m+n}, f_n\right) = \int_0^1 \left|f_{m+n}\left(x\right) - f_n\left(x\right)\right| \, dx$$

$$\leq \int_0^1 \left|f_{m+n}\left(x\right) - f\left(x\right)\right| \, dx + \int_0^1 \left|f_n\left(x\right) - f\left(x\right)\right| \, dx.$$

Now note that the value of integral is given by the area of the shaded portion as shown in Figure 3.2.

This gives

$$d_1\left(x_{m+n}, x_n\right) \leq \frac{1}{2n} + \frac{1}{2\left(m+n\right)} \to 0 \text{ as } n \to \infty.$$

If we assume that the sequence converges to, say, $g \in C[0,1]$, then we must have

$$d_1\left(f_n, g\right) = \int_0^1 \left|f_n\left(x\right) - g\left(t\right)\right| \, dx$$

$$= \int_0^{\frac{1}{2}-\frac{1}{n}} \left|g\left(x\right)\right| \, dx + \int_{\frac{1}{2}-\frac{1}{n}}^{\frac{1}{2}+\frac{1}{n}} \left|f_n\left(x\right) - g\left(x\right)\right| \, dx + \int_{\frac{1}{2}+\frac{1}{n}}^1 \left|1 - g\left(x\right)\right| \, dx.$$

Since each integral is non-negative, each must tend to 0 as $n \to \infty$.

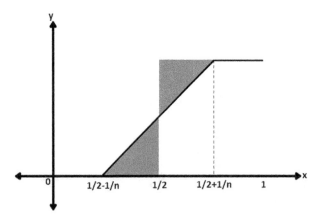

FIGURE 3.2: The area between f_n and f.

By properties of the Riemann integral, we know that the middle integral tends to 0 as $n \to \infty$. Finally,

$$\lim_{n \to \infty} \int_0^{\frac{1}{2} - \frac{1}{n}} |g(x)| \, dx = 0 \implies g(x) = 0, \text{ for } 0 \le x < \frac{1}{2}.$$

$$\lim_{n \to \infty} \int_{\frac{1}{2} + \frac{1}{n}}^{1} |g(x)| \, dx = 0 \implies g(x) = 1, \text{ for } \frac{1}{2} < x \le 1.$$

This is contradictory to the assumption that $g \in \mathscr{C}[0,1]$. Thus the sequence is not convergent **in** $\mathscr{C}[0,1]$.

Now, we provide the reader with a few results on sequences and their convergence, and relate convergence with Cauchy sequences.

Theorem 3.2. *A Cauchy sequence is convergent if it has a convergent subsequence.*

Proof. Let $(x_n)_{n \in \mathbb{N}}$ be a Cauchy sequence in a metric space (X, d) such that a subsequence $(x_{m_k})_{k \in \mathbb{N}}$ converges to $x \in X$. Thus, $\forall \epsilon > 0$, $\exists N_1 \in \mathbb{N}$ such that

$$d(x_{m_k}, x) < \frac{\epsilon}{2}, \ \forall m_k > N_1.$$

Also since $(x_n)_{n \in \mathbb{N}}$ is Cauchy, $\exists N_2 \in \mathbb{N}$ such that

$$d(x_{m_k}, x_n) < \frac{\epsilon}{2} \ \forall m_k, n > N_2.$$

Therefore from triangle inequality, we obtain

$$d(x_n, x) \le d(x_n, x_{m_k}) + d(x_{m_k}, x).$$

Picking any $m_k > \max\{N_1, N_2\}$, we get

$$d(x_n, x) < \epsilon \ \forall n > N_2.$$

Thus, $(x_n)_{n \in \mathbb{N}}$ converges to x. □

We now take a slight detour to look at how convergence works in the more complex setting of a product space. Briefly, the convergence of a sequence in a product space is equivalent to the convergence of each of the coordinate sequences. Before we proceed to the theorem, we take a quick look at the notation we will be using for the theorem.

Given the metric spaces $(X_1, d_1), (X_2, d_2), (X_3, d_3), \cdots, (X_m, d_m)$, we define a metric d on the Cartesian product $X = X_1 \times X_2 \times \cdots \times X_m$ by

$$d\left((x_i)_{i=1}^m, (y_i)_{i=1}^m\right) = \max_{1 \leq i \leq m} \{d_i(x_i, y_i)\}. \tag{3.1}$$

Theorem 3.3. *In the product space (X, d) (defined as above), the sequence $(\mathbf{x}_n)_{n \in \mathbb{N}}$ where $\mathbf{x}_n = \left(x_1^{(n)}, x_2^{(n)}, \cdots x_m^{(n)}\right)$ is convergent iff the sequence $\left(x_k^{(n)}\right)_{n \in \mathbb{N}}$ is convergent in X_k for $k = 1, 2, \cdots, m$.*

Proof. Consider a sequence $(\mathbf{x}_n)_{n \in \mathbb{N}}$ in X where $\mathbf{x}_n = (x_1^{(n)}, x_2^{(n)}, \cdots x_m^{(n)})$.

If the sequence $(\mathbf{x}_n)_{n \in \mathbb{N}}$ converges to $\mathbf{x} \in X$ where $\mathbf{x} = (x_1, x_2, \cdots, x_m)$, then for any $k = 1, 2, \cdots, m$, we have

$$d_k\left(x_k^{(n)}, x_k\right) \leq \max_{r=1,2,\cdots,m}\left\{d_r\left(x_r^{(n)}, x_r\right)\right\} = d(\mathbf{x}_n, \mathbf{x}) \to 0 \text{ as } n \to \infty.$$

Thus, the corresponding coordinate sequences converge in the respective coordinate spaces.

Conversely, if the coordinate sequences each converge, i.e., $(x_k^{(n)})_{n \in \mathbb{N}}$ converges to $x_k \in X_k$ for each $k = 1, 2, \cdots, m$, then $\forall \epsilon > 0$, we have $N_1, N_2, \cdots, N_m \in \mathbb{N}$ so that

$$d_k\left(x_k^{(n)}, x_k\right) < \epsilon, \text{ for } n \geq N_k \text{ and for each } k = 1, 2, \cdots, m.$$

We pick $N = \max\{N_1, N_2, \cdots, N_m\}$ so that

$$d(\mathbf{x}_n, \mathbf{x}) = \max_{k=1,2,\cdots,m}\left\{d_k\left(x_k^{(n)}, x_k\right)\right\} < \epsilon, \forall n \geq N$$

where $\mathbf{x} = (x_1, x_2, \cdots, x_m) \in X$. Thus, the sequence $(\mathbf{x}_n)_{n \in \mathbb{N}}$ converges to $\mathbf{x}$ in X. $\qquad \square$

Before ending the section, let us look at the range of a sequence $(x_n)_{n \in \mathbb{N}}$, i.e., the set

$$R = \{x_n | n \in \mathbb{N}\}.$$

Suppose that the sequence is convergent to $x \in X$. Then, by the definition of convergence, for each $\epsilon > 0$, there is at least one $x_{n_0} \in R$ such that $x_{n_0} \in B(x, \epsilon)$. This is equivalent to saying that $B(x, \epsilon) \cap R \neq \emptyset$ for every $\epsilon > 0$ so that x is a limit point of R.

However, a question can be asked in a converse manner: If S is a set and x is its limit point, does there exist a sequence in S that converges to x? To answer this question, we "attack" the definition of limit point. Clearly, for each

$\epsilon > 0$, $\exists x_\epsilon \in S$ such that $x_\epsilon \in B(x, \epsilon)$. Hence, in particular, for each $n \in \mathbb{N}$ we have $x_n \in B(x, \frac{1}{n})$. Therefore, we have a sequence $(x_n)_{n \in \mathbb{N}}$ in S that seems to get "close" to x. To prove this, however, consider any $\epsilon > 0$. Then, by the Archimedean property, $\exists n_0 \in \mathbb{N}$ such that $n_0 \epsilon > 1$. Also, for $n_1 < n_2$ we have $B(x, \frac{1}{n_2}) \subseteq B(x, \frac{1}{n_1})$ so that for all $n > n_0$ we have $d(x_n, x) < \frac{1}{n} < \frac{1}{n_0} < \epsilon$. Hence, the sequence indeed converges to x. We summarize this discussion as a theorem.

Theorem 3.4. *In a metric space (X, d), a point $x \in X$ is a limit point of $S \subseteq X$ if and only if there is a sequence $(x_n)_{n \in \mathbb{N}}$ that converges to x.*

3.3 Complete Metric Spaces

In the previous section, the concept of Cauchy sequences was introduced. We also saw a metric space where a Cauchy sequence was not convergent, due to some "missing" points. A natural question to ask is: What would happen if a metric space does not "lack" any point (for convergence of Cauchy sequences)? We call such spaces complete. The formal definition follows:

Definition 3.5 (Complete metric space). *A metric space (X, d) is said to be complete if every Cauchy sequence is convergent.*

Let us look at a few examples of complete metric spaces.

Example 3.7. The simplest example of a complete metric space is $\mathbb{R}$ with the usual metric ($|\cdot|$). To prove that it is complete, one needs to recall the least upper bound property of real numbers. It states that every subset of $\mathbb{R}$ that has an upper bound has a least upper bound (also known as the supremum). The property, in turn, also implies the existence of a greatest lower bound (also known as the infimum) if a subset of $\mathbb{R}$ has a lower bound.

Consider a real sequence $(x_n)_{n \in \mathbb{N}}$ that is Cauchy.
For $\epsilon_n = \frac{1}{2^n}$, $\exists N_n \in \mathbb{N}$ such that

$$|x_m - x_{N_n}| < \epsilon_n, \; \forall m > N_n.$$

$$\therefore x_{N_n} - \epsilon_n < x_m < x_{N_n} + \epsilon_n, \; \forall m > N_n.$$

Thus $(x_{N_n} - \epsilon_n)$ is a lower bound for the set $\{x_m | m > N_n\}$ and $(x_{N_n} + \epsilon_n)$ is its upper bound. We define the sequences $\alpha_k = \sup_{n \geq k} \{x_n\}$ and $\beta_k = \inf_{n \geq k} \{x_n\}$.

Therefore by definition of supremum and infimum, we obtain

$$x_{N_n} - \epsilon_n \leq \beta_{N_n} \leq x_m \leq \alpha_{N_n} \leq x_{N_n} + \epsilon_n, \; \forall m > N_n.$$

This in turn gives, $\alpha_{N_n} - \beta_{N_n} < 2\epsilon_n \to 0$ as $n \to \infty$ which leads us to $\lim\limits_{n\to\infty} \alpha_{N_n} = \lim\limits_{n\to\infty} \beta_{N_n}$. Thus, by sandwich theorem the limit $\lim\limits_{n\to\infty} x_n$ exists. In other words, the sequence $(x_n)_{n\in\mathbb{N}}$ converges in $\mathbb{R}$.

The assumption that $\mathbb{R}$ has least upper bound property is equivalent to saying that $\mathbb{R}$ is complete. Hence, the property is also known as "completeness axiom".[4]

Example 3.8. The space $\mathbb{R}^m$ equipped with the Euclidean metric is complete for any $m \in \mathbb{N}$.

Let $(\mathbf{x}_n)_{n\in\mathbb{N}}$ be any Cauchy sequence in $\mathbb{R}^m$ and let $\mathbf{x}_n = (x_1^{(n)}, x_2^{(n)}, \cdots x_m^{(n)}) = (x_k^{(n)})_{k=1}^m$. The superscript (n) denotes the terms of the sequence and the subscript k the coordinate.

$(\mathbf{x}_n)_{n\in\mathbb{N}}$ is any Cauchy sequence implies $\forall \epsilon > 0$, $\exists N_1 > 0$ such that

$$d_2(x_n, x_r) = \sqrt{\sum_{k=1}^m \left| x_k^{(n)} - x_k^{(r)} \right|^2} < \epsilon, \ \forall n, r > N_1.$$

Thus for a fixed k_0, we have

$$\left| x_{k_0}^{(n)} - x_{k_0}^{(r)} \right| \leq \sqrt{\sum_{k=1}^m \left| x_k^{(n)} - x_k^{(r)} \right|^2} < \epsilon, \ \forall n, r > N_1.$$

This means the sequence $\left(x_{k_0}^{(n)} \right)_{n\in\mathbb{N}}$ is Cauchy for $k_0 = 1, 2, \cdots, m$ and thus convergent in $\mathbb{R}$. Let $\lim\limits_{n\to\infty} x_{k_0}^{(n)} = x_{k_0}$ and $\mathbf{x} = (x_1, x_2, \cdots, x_m)$.

We shall prove that $(\mathbf{x}_n)_{n\in\mathbb{N}}$ converges to $\mathbf{x}$ in $\mathbb{R}^m$

$\forall \epsilon > 0$, $\exists M_1, M_2, \cdots, M_m \in \mathbb{N}$ such that for $k = 1, 2, \cdots, m$

$$\left| x_k^{(n)} - x_k \right| < \frac{\epsilon}{\sqrt{m}}, \ \forall n > M_k.$$

Finally, we have,

$$d_2(\mathbf{x}_n, \mathbf{x}) = \sqrt{\sum_{k=1}^m \left| x_k^{(n)} - x_k \right|^2} < \sqrt{\sum_{k=1}^m \left| \frac{\epsilon}{\sqrt{m}} \right|^2} = \epsilon, \ \forall n > \max_{1\leq i\leq m} \{M_i\}.$$

Thus, the sequence converges to $\mathbf{x}$.

---

[4]The names least upper bound "property" and completeness "axiom" are largely due to historical reason and both act as axioms in this context.

Example 3.9. The sequence space ℓ^p is complete for $1 \le p < \infty$.

Consider a Cauchy sequence $(\mathbf{x}_n)_{n \in \mathbb{N}}$ in ℓ^p. Note that for each value of n, $\mathbf{x}_n$ itself is a real sequence. With this in mind, we denote $\mathbf{x}_n = \left(x_k^{(n)} \right)_{k \in \mathbb{N}}$ where $x_k^{(n)} \in \mathbb{R}$.

Then, $\forall \epsilon > 0$, $\exists N_1 \in \mathbb{N}$ such that

$$d_p\left(\mathbf{x}_{m+n}, \mathbf{x}_n\right) = \left(\sum_{k=1}^{\infty} \left| x_k^{(m+n)} - x_k^{(n)} \right|^p \right)^{\frac{1}{p}} < \epsilon, \ \forall n > N_1 \text{ and } m \in \mathbb{N}. \tag{3.2}$$

$$\therefore \left| x_k^{(m+n)} - x_k^{(n)} \right|^p < \epsilon^p, \ \forall m, k \in \mathbb{N}, n > N_1.$$

Therefore, for a fixed k_0, the sequence $\left(x_{k_0}^{(n)} \right)_{n \in \mathbb{N}}$ is Cauchy in $\mathbb{R}$ and hence convergent.

Let $\lim_{n \to \infty} x_{k_0}^{(n)} = x_{k_0}$.

We claim that $\mathbf{x}_n \to \mathbf{x} \in \ell^p$ where $\mathbf{x} = (x_k)_{k \in \mathbb{N}}$. We begin with proving that $\mathbf{x} \in \ell^p$.

For some fixed $r \in \mathbb{N}$, Equation (3.2) implies

$$\left(\sum_{k=1}^{r} \left| x_k^{(m+n)} - x_k^{(n)} \right|^p \right)^{\frac{1}{p}} < \epsilon, \ \forall n > N_1, m \in \mathbb{N}.$$

As $m \to \infty$, we obtain

$$\left(\sum_{k=1}^{r} \left| x_k - x_k^{(n)} \right|^p \right)^{\frac{1}{p}} < \epsilon, \ \forall n > N_1.$$

Since the right side of the preceding inequality is independent of r, by Theorem 2.4, we get

$$\left(\sum_{k=1}^{\infty} \left| x_k - x_k^{(n)} \right|^p \right)^{\frac{1}{p}} \le \epsilon, \ \forall n > N_1. \tag{3.3}$$

Thus by definition of ℓ^p, we have $(\mathbf{x} - \mathbf{x}_n)_{n \in \mathbb{N}} \in \ell^p$. Therefore, $\mathbf{x} = (\mathbf{x} - \mathbf{x}_n) + \mathbf{x}_n \in \ell^p$. Finally, Equation (3.3) gives $x_n \to x$.

Exercise 3.5. Prove that if $\mathbf{x}, \mathbf{y} \in \ell^p$, then $\alpha \cdot \mathbf{x} + \beta \cdot \mathbf{y} \in \ell^p$ for $\alpha, \beta \in \mathbb{R}$. Here $+$ and $\cdot$ are two algebraic operations of addition and scaling defined for any $\mathbf{x} = (x_i)_{i \in \mathbb{N}}, \mathbf{y} = (y_i)_{i \in \mathbb{N}} \in \ell^p$ as

$$\mathbf{x} + \mathbf{y} = (x_i + y_i)_{i \in \mathbb{N}},$$
$$\alpha \cdot \mathbf{x} = (\alpha x_i)_{i \in \mathbb{N}}.$$

This will confirm the second-last line of the proof of Example 3.9.

Exercise 3.6. Is ℓ^∞ complete?

Example 3.10. Consider the space of continuous real-valued functions $\mathscr{C}[a, b]$ equipped with the metric d_∞.

Consider a Cauchy sequence $(f_n)_{n \in \mathbb{N}}$ in $\mathscr{C}[a, b]$. Then, $\forall \epsilon > 0$, $\exists N_1 \in \mathbb{N}$ so that

$$\sup_{x \in [a,b]} \{|f_{m+n}(x) - f_n(x)|\} < \epsilon, \ \forall n > N_1 \text{ and } m, \in \mathbb{N}.$$

$$\therefore |f_{m+n}(x) - f_n(x)| < \epsilon, \ \forall n > N_1, \forall x \in [a, b] \text{ and } m, \in \mathbb{N}.$$

Thus, for a fixed x_0, the sequence $(f_n(x_0))_{n \in \mathbb{N}}$ is Cauchy in $\mathbb{R}$ and hence convergent. We define a function $f : [a, b] \to \mathbb{R}$ by

$$f(x) = \lim_{n \to \infty} f_n(x), \ \forall x \in [a, b].$$

By definition of the limit, there exists $N_2 \in \mathbb{N}$ such that

$$|f_n(x_0) - f(x_0)| < \frac{\epsilon}{3}, \ \forall n > N_2.$$

Since $f_n \in \mathscr{C}[a, b]$, there exists $\delta > 0$ such that,

$$|f_n(x) - f_n(x_0)| < \frac{\epsilon}{3}, \ \text{whenever } |x - x_0| < \delta.$$

Thus for $n = \max\{N_1, N_2, N_3\} + 1$ and $|x - x_0| < \delta$ we have

$$|f(x) - f(x_0)| \le |f(x) - f_n(x)| + |f_n(x) - f_n(x_0)| + |f_n(x_0) - f(x_0)| < \epsilon.$$

Since the choice of ϵ and x_0 was arbitrary, we have $f \in \mathscr{C}[a, b]$. Finally, since f_n is a Cauchy sequence,

$$d_\infty(f_{m+n}, f_n) < \epsilon, \ \forall n > N_1, m \in \mathbb{N}.$$

$$\therefore \sup_{x \in [a,b]} \{|f_{m+n}(x) - f_n(x)|\} < \epsilon.$$

Taking $m \to \infty$, we obtain

$$\sup_{x \in [a,b]} \{|f(x) - f_n(x)|\} < \epsilon.$$

Thus, the sequence $(f_n)_{n \in \mathbb{N}}$ converges to f and we conclude that $\mathscr{C}[a, b]$, is complete.

Exercise 3.7. Discuss all the Cauchy sequences in a discrete space. What can we say about its completeness?

Before ending this section, we would like to expose the reader to a few theorems involving complete metric spaces.

Theorem 3.5. *For a subspace Y of (X, d), if Y is complete then Y is closed in X.*

Proof. Let $x \in X$ be a limit point of Y. Then there is a sequence $(y_n)_{n \in \mathbb{N}}$ in Y such that $\lim_{n \to \infty} d(y_n, x) = 0$. Since every convergent sequence is Cauchy, the sequence $(y_n)_{n \in \mathbb{N}}$ is Cauchy in X. From the definition of Cauchy sequences, we have $\forall \epsilon > 0$, $\exists N_1 \in \mathbb{N}$ so that (in Y)

$$d(y_n, y_m) < \epsilon, \ \forall m, n > N_1.$$

Thus the sequence is Cauchy in Y as well. Now since Y is complete, the sequence converges in Y, i.e., $\exists y \in Y$ such that $\lim_{n \to \infty} d(y_n, y) = 0$. Again by definition of subspace, we have $\lim_{n \to \infty} d(y_n, y) = 0$ in X and thus the sequence converges to y in X as well. But since the point of convergence is unique in a metric space, we get $x = y$ and hence the limit point $x \in Y$ implying Y is closed. $\qquad\square$

Theorem 3.6. *For a subspace Y of a complete metric space (X, d), if Y is closed then Y is complete.*

Proof. Let, $(y_n)_{n \in \mathbb{N}}$ be any Cauchy sequence in Y. Thus, from reasoning similar to Theorem 3.5 it is Cauchy in X as well. Since X is complete, the sequence must converge to a point in X, say x. Thus, x is a limit point of Y and Y being closed implies $x \in Y$. Finally, from the definition of subspace, $\lim_{n \to \infty} d(y_n, x) = 0$ in Y, i.e., the sequence $(y_n)_{n \in \mathbb{N}}$ converges in Y and therefore, Y is complete. $\qquad\square$

3.4 Completion of Metric Spaces

We have seen the definition of Cauchy sequences. Less rigorously, we can say that Cauchy sequences are ones in which the elements "come close" to each other as we move to higher indices. Recall the completeness axiom of a real number system. It ensures that every Cauchy sequence is convergent in $\mathbb{R}$ as we have seen in earlier sections. We have also seen that the subset $\mathbb{Q}$ of $\mathbb{R}$ is not complete. Thus, it is "missing" some of the points, namely, the irrational numbers.

A natural question arises: Can we find and fill up all such missing elements? Can we do it for any metric space? The answer is yes! We will see how we can construct a *"complete extension"* of any set and also, what it means to have such an extension.

We first need to understand what we mean by an extension. We have a set with a certain property (or a certain operation defined on it) or the lack of a certain desired property. We use it to define a new set which contains the original one in such a way that the property is retained (or in the case of operations, the operation remains the same when operating on elements of the original set) or the new set has the desired property. For our complete extension, we pick any space X and we wish to construct a new space Z so that $X \subseteq Z$ and Z is complete. This, however, poses a problem. The condition $X \subseteq Z$ requires that the new constructed set contains elements of the original set, which in turn requires that the construction we make must be in such a way that the original elements remain. This is a restriction we wish to overcome.

We do it by *"embedding"* X in Z. Before we get to what we mean by embedding, let's first get accustomed to some concepts that will help us build a complete picture.

Recall that the main property a metric space has is the notion of distance on a set. What if we have two different metric spaces on two different sets that behave alike (in terms of distances) on their corresponding set? Such metric spaces are said to be isometric. Formally we define:

Definition 3.6 (Isometry). *Given two metric spaces (X, d_X) and (Y, d_Y), a map $f : X \to Y$ is said to be an isometry if*

$$d_X(x_1, x_2) = d_Y(f(x_1), f(x_2)), \ \forall x_1, x_2 \in X.$$

Two metric spaces are called isometric if there is a bijective isometry from one to another.

An isometry is automatically injective since

$$f(x_1) = f(x_2) \implies d_Y(f(x_1), f(x_2)) = 0 = d_X(x_1, x_2) \implies x_1 = x_2.$$

For a fixed $a \in \mathbb{R}$, the translation function $f_a : x \mapsto x + a$ is an isometry. The identity function from $\mathbb{Q}$ to $\mathbb{R}$ is an example of an isometry that is not bijective.

Carefully examine the isometry from $\mathbb{Q}$ to $\mathbb{R}$ discussed earlier. The set $\mathbb{Q}$ is nearly identical to $\mathbb{R}$ with respect to the distance between points. It is only "missing" some points for it be exactly identical. We are now in a position to define embedding:

Definition 3.7 (Embedding). *A metric space (X, d_X) is said to be embedded in (Y, d_Y) if an isometry $f : X \to Y$ exists.*

Coming back to our original question of how to define the complete extension, we see that we now have eliminated the need for our complete set to be a superset of the original set. We can simply require the existence of an isometry from our original set to the new complete set. There is still a problem here, though! Once we have the required completion, nothing stops us from adding more and more points and thus making the set unnecessarily bigger. That is to say, nothing stops us from extending $\mathbb{Q}$ to $\mathbb{R}$ and then to $\mathbb{C}$. There is nothing wrong with such an extension, it just won't be very useful.

Thus, we add an extra criterion which will ensure that our extension remains "small enough" to be useful. We will require that the set $f(X)$ be dense in Z. This is again consistent to the fact that $\mathbb{Q}$ is dense in $\mathbb{R}$. Finally, we define the complete extension of a metric space.

Definition 3.8 (Complete extension). *A complete metric space (Y, d_Y) is said to be a complete extension of a metric space (X, d_X) if X can be embedded in Y through an isometry $f : X \to Y$ in such a way that $f(X)$ is dense in Y.*

We finally give the main result of this section.

Theorem 3.7. *Given any metric space, its complete extension exists. Moreover, such an extension is unique up to an isometry.*

By uniqueness up to an isometry we mean that if there exist any two complete extensions of a metric space, they are both isometric to each other. The proof of the above theorem, although straightforward, is rather lengthy. We therefore divide the proof into four parts (presented as subsections that follow):

For a metric space (X, d_X), we will

1. Construct a set Z by using Cauchy sequences on X and define a metric d_Z on it.

2. Prove that X can be embedded in Z.

3. Prove that Z is complete and is the desired complete extension.

4. Prove that Z is unique up to an isometry.

3.4.1 Construction of the set Z

Consider all the Cauchy sequences in X. We define an equivalence relation $\sim$ between two such Cauchy sequences as follows $(x_n) \sim (y_n) \iff \lim_{n \to \infty} d_X(x_n, y_n) = 0$.[5]

As we have seen earlier the sequence $d_n = d_X(x_n, y_n)$ is Cauchy if both (x_n) and (y_n) are Cauchy. The completeness of $\mathbb{R}$ implies that $\lim_{n \to \infty} d_n$ exists.

---

[5]We omit the subscript $n \in \mathbb{N}$ while denoting sequences to make the notations short. The reader should understand the notations automatically.

To prove that the above relation $\sim$ is an equivalence, consider the following:

1. For every sequence (x_n), $d_X(x_n, x_n) = 0$ and hence $(x_n) \sim (x_n)$, i.e., $\sim$ is reflexive.

2. The symmetry of metric d implies that $\sim$ is symmetric.

3. If $(x_n) \sim (y_n)$ and $(y_n) \sim (z_n)$, then we have $0 \leq d(x_n, z_n) \leq d(x_n, y_n) + d(y_n, z_n) \to 0$ as $n \to \infty$.

   Thus, $\lim\limits_{n \to \infty} d(x_n, z_n) = 0$ which in turn implies $(x_n) \sim (z_n)$, i.e., $\sim$ is transitive.

This equivalence $\sim$ gives rise to equivalence classes of the set of all Cauchy sequences in X. The set of all such equivalence classes is the desired Z. We denote an element of Z by $\tilde{x}$. Remember that $\tilde{x}$ is a collection of Cauchy sequences in X that are related by the equivalence $\sim$.

To proceed further, we will need a metric on Z. The natural way to create one is to define

$$d_Z(\tilde{x}, \tilde{y}) = \lim_{n \to \infty} d_X(x_n, y_n), \text{ where } (x_n) \in \tilde{x} \And (y_n) \in \tilde{y}.$$

Before we try to prove that d_Z is a metric, careful readers will note that we first have to ensure that d_Z is well-defined. This is because the choice of $(x_n) \in \tilde{x}$ and $(y_n) \in \tilde{y}$ is ambiguous. We claim that nevertheless, the limit in the definition remains the same.

For $(x_n), (x'_n) \in \tilde{x}$ and $(y_n), (y'_n) \in \tilde{y}$, we have $\lim\limits_{n \to \infty} d_X(x_n, x'_n) = 0$, $\lim_{n \to \infty} d_X(y_n, y'_n) = 0$ and that

$$d_X(x_n, y_n) \leq d_X(x_n, x'_n) + d_X(x'_n, y'_n) + d_X(y'_n, y_n),$$

which can be rewritten as

$$d_X(x_n, y_n) - d_X(x'_n, y'_n) \leq d_X(x_n, x'_n) + d_X(y_n, y'_n).$$

Starting with $d_X(x'_n, y'_n)$ and proceeding in a similar manner, we obtain

$$d_X(x'_n, y'_n) - d_X(x_n, y_n) \leq d_X(x'_n, x_n) + d_X(y_n, y'_n).$$

Combining both the above inequalities gives

$$|d_X(x_n, y_n) - d_X(x'_n, y'_n)| \leq d_X(x_n, x'_n) + d_X(y_n, y'_n).$$

The right side goes to 0 as $n \to \infty$ and subsequently so does the left. Therefore,

$$\lim_{n \to \infty} d_X(x_n, y_n) = \lim_{n \to \infty} d_X(x'_n, y'_n).$$

Since the choice of sequences was arbitrary, d_Z is well defined. We leave it as an exercise to the reader to prove that d_Z forms a metric on Z.

Exercise 3.8. Prove that the function d_Z, as defined above, is indeed a metric on Z.

3.4.2 Embedding X in Z

For any $x \in X$, consider the sequence $(x, x, x, \cdots)$. This sequence is convergent and hence Cauchy. Let $\tilde{x}$ denote the equivalence class which contains the above sequence.

Define a function $f : x \mapsto \tilde{x}$. The uniqueness of equivalence class implies that the function is well defined. Also, for $x, y \in X$, we pick $(x, x, x, \dots) \in \tilde{x}$ and $(y, y, y, \dots) \in \tilde{y}$ so that

$$d_Z\left(f\left(x\right), f\left(y\right)\right) = d_Z\left(\tilde{x}, \tilde{y}\right) = \lim_{n\to\infty} d_X\left(x, y\right) = d_X\left(x, y\right).$$

Therefore f preserves distance, i.e., it is an isometry.

We will prove the existence of a sequence in $f(X)$ that converges to x.
Let $Y = f\left(X\right)$. Now, for any arbitrary $\tilde{x} \in Z$, pick any sequence $(x_n) \in \tilde{x}$. Since (x_n) is Cauchy, $\forall \epsilon > 0$, $\exists N \in \mathbb{N}$ such that $d_X\left(x_n, x_N\right) < \epsilon/2$, $\forall n \geq N$.
Also, $f\left(x_N\right) = \tilde{x}_N \in f\left(X\right) = Y$.
To find $d_Z\left(\tilde{x}, \tilde{x}_N\right)$, we pick $(x_N, x_N, x_N, \dots) \in \tilde{x}_N$. Then,

$$d_Z\left(\tilde{x}, \tilde{x}_N\right) = \lim_{n\to\infty} d_X\left(x_n, x_N\right) \leq \frac{\epsilon}{2} < \epsilon.$$

Thus, Y is dense in Z.

3.4.3 Proving Z is complete

Let $(\tilde{z}_n)_{n \in \mathbb{N}}$ be a Cauchy sequence in Z. Since Y is dense in Z, we have $\forall n \in \mathbb{N}$, $\exists \tilde{y}_n \in Y$ such that $d_Z\left(\tilde{z}_n, \tilde{y}_n\right) < 1/n$.

$$\therefore d_Z\left(\tilde{y}_n, \tilde{y}_{m+n}\right) \leq d_Z\left(\tilde{y}_n, \tilde{z}_n\right) + d_Z\left(\tilde{z}_n, \tilde{z}_{m+n}\right) + d_Z\left(\tilde{z}_{m+n}, \tilde{y}_{m+n}\right)$$

$$< \frac{1}{n} + d_Z\left(\tilde{z}_n, \tilde{z}_{m+n}\right) + \frac{1}{m+n}, \quad \forall m \in \mathbb{N}.$$

Since $(\tilde{z}_n)$ is Cauchy, $d_Z\left(\tilde{z}_n, \tilde{z}_{m+n}\right) \to 0$ as $n \to \infty$ and therefore

$$\lim_{n\to\infty} d_Z\left(\tilde{y}_n, \tilde{y}_{m+n}\right) \leq \lim_{n\to\infty} \left(\frac{1}{n} + d_Z\left(\tilde{z}_n, \tilde{z}_{m+n}\right) + \frac{1}{m+n} \right) = 0.$$

Thus, $(\tilde{y}_n)_{n \in \mathbb{N}}$ is a Cauchy sequence. Now since f is an isometry and subsequently injective, we have $\left(f^{-1}\left(\tilde{y}_n\right)\right)_{n \in \mathbb{N}}$ is Cauchy. Or in simpler notation,

$(y_n)_{n \in \mathbb{N}}$ is Cauchy in X where $y_n = f^{-1}(\tilde{y}_n)$. Since this sequence is Cauchy, $\exists \, \tilde{y} \in Z$ which contains (y_n).

We claim that the sequence $(\tilde{z}_n)$ converges to $\tilde{y}$. To prove this, we proceed as follows:

$$d_Z(\tilde{z}_n, \tilde{y}) \leq d_Z(\tilde{z}_n, \tilde{y}_n) + d_Z(\tilde{y}_n, \tilde{y})$$
$$< \frac{1}{n} + d_Z(\tilde{y}_n, \tilde{y}).$$

To find $d_Z(\tilde{y}_n, \tilde{y})$, we have to pick a representative of each of the classes.

Since $\tilde{y}_n \in Y = f(X)$, we pick $(y_n, y_n, y_n, \ldots) \in \tilde{y}_n$.

Also, let $(y_m)_{m \in \mathbb{N}} \in \tilde{y}$ be the Cauchy sequence mentioned above. We use m as index to avoid confusion.

Now, $d_Z(\tilde{y}_n, \tilde{y}) = \lim\limits_{m \to \infty} d_X(y_n, y_m)$.

$$\therefore d_Z(\tilde{z}_n, \tilde{y}) < 1/n + \lim_{m \to \infty} d_X(y_n, y_m).$$

Finally, since $(y_m)_{m \in \mathbb{N}}$ is Cauchy, $\forall \epsilon > 0 \, \exists N_1 \in \mathbb{N}$, so that $d_X(y_n, y_m) < \frac{\epsilon}{2}$, $\forall n, m > N_1$, which implies $\lim\limits_{m \to \infty} d_X(y_n, y_m) \leq \frac{\epsilon}{2}$, $\forall n > N_1$. For n such that $n > \frac{2}{\epsilon}$ and $n > N_1$, we have

$$d_Z(\tilde{z}_n, \tilde{y}) < 1/n + \lim_{m \to \infty} d_X(y_n, y_m) < \epsilon.$$

Thus, (Z, d_Z) is a complete metric space.

3.4.4 Uniqueness of extension up to isometry

Lastly, we have to prove that the extension thus obtained is unique up to an isometry. As discussed earlier, we have to prove that if there exists another complete extension (W, d_W) of X with isometry $g : X \to W$ then Z and W are isometric.

We have isometry $f : X \to Z$ such that $f(X)$ is dense in Z. Thus for any $z \in Z$, there is a sequence $(x_n)_{n \in \mathbb{N}}$ that converges to z where $x_n \in f(X)$. Since $(x_n)_{n \in \mathbb{N}}$ is convergent, it is Cauchy as well. Now f being an isometry gives $(f^{-1}(x_n))_{n \in \mathbb{N}}$ is Cauchy in X. Finally, g being an isometry gives $((g \circ f^{-1})(x_n))_{n \in \mathbb{N}}$ is Cauchy in W. But W is complete by hypothesis and hence the sequence $((g \circ f^{-1})(x_n))_{n \in \mathbb{N}}$ converges in W to, say, w. We define a mapping $h : Z \to W$ by $h(z) = w$.

To prove that the function h is well defined, we note that the only point of ambiguity in the definition is the choice of sequence converging to z. Consider another sequence $(x'_n)_{n \in \mathbb{N}}$ that converges to z with $x'_n \in f(X)$. Thus the sequence $(x_1, x'_1, x_2, x'_2, x_3, x'_3, \cdots)$ converges to z. Following similar argument as before we obtain that the sequence $((g \circ f^{-1})(x_1), (g \circ f^{-1})(x'_1), (g \circ f^{-1})(x_2), (g \circ f^{-1})(x'_2), \cdots)$ is convergent in W. But, we already know one of its subsequences $((g \circ f^{-1})(x_n))_{n \in \mathbb{N}}$

converges to w. Thus the sequence itself converges to w. This is turn means its other subsequence $\left(\left(g \circ f^{-1}\right)(x_1'), \left(g \circ f^{-1}\right)(x_2'), \left(g \circ f^{-1}\right)(x_3'), \cdots\right)$ converges to w as well. Hence, h is well defined.

Finally for any two points $z_1, z_2 \in Z$, let $h(z_1) = w_1$ and $h(z_2) = w_2$ and let $(x_n)_{n \in \mathbb{N}}$, $(y_n)_{n \in \mathbb{N}}$ be any two sequences in $f(X)$ converging to z_1 and z_2 respectively. Therefore we have

$$d_Z(z_1, z_2) = \lim_{n \to \infty} d_Z(x_n, y_n).$$

But, f and g being isometries, we have,

$$d_Z(x_n, y_n) = d_X\left(f^{-1}(x_n), f^{-1}(y_n)\right) = d_W\left(\left(g \circ f^{-1}\right)(x_n), \left(g \circ f^{-1}\right)(y_n)\right).$$

By definition of h, we have $\left(g \circ f^{-1}\right)(x_n) \to w_1$ and $\left(g \circ f^{-1}\right)(y_n) \to w_2$ as $n \to \infty$.

$$\therefore \lim_{n \to \infty} d_W\left(\left(g \circ f^{-1}\right)(x_n), \left(g \circ f^{-1}\right)(y_n)\right) = d_W(w_1, w_2).$$

Bringing all of it together,

$$\begin{aligned}
d_Z(z_1, z_2) &= \lim_{n \to \infty} d_Z(x_n, y_n) \\
&= \lim_{n \to \infty} d_W\left(\left(g \circ f^{-1}\right)(x_n), \left(g \circ f^{-1}\right)(y_n)\right) \\
&= d_W(w_1, w_2) \\
&= d_W(h(z_1), h(z_2)).
\end{aligned}$$

Therefore, $h : Z \to W$ is an isometry.

For any $w \in W$, since $g(X)$ is dense in W, there is a sequence $(u_n)_{n \in \mathbb{N}}$ that converges to w where $u_n \in g(X)$. Proceeding in a similar manner as before, we obtain that $\left(\left(f \circ g^{-1}\right)(u_n)\right)_{n \in \mathbb{N}}$ is convergent in Z to say, z. It is then easy to see that $h(z) = w$.

Therefore h is bijective and Z and W are isometric.

3.5 Baire Category Theorem

3.5.1 Category of sets

In his dissertation, René-Louis Baire described what is now known as a Baire space. The Baire category theorem is one of the most important tools used to study complete metric spaces and is one of the reasons why such spaces are so important. But before we get ahead of ourselves, let us first understand what the category in the name of the theorem refers to:

Definition 3.9 (Rare set). *A subset $M \subseteq X$ is said to be rare if its closure has no interior points. It is also referred to as nowhere dense set.*

Definition 3.10 (Meager set). *A union of countably many rare sets is said to be of first category or a meager set. The complement of a meager set is called co-meager.*

Definition 3.11 (Non-meager set). *A subset that is not meager is said to be non-meager or of second category.*

Example 3.11. We begin with some examples from $\mathbb{R}$ as the reader is very familiar with the usual metric on it.

1. Singleton sets are rare in $\mathbb{R}$ as they are closed and contain no open interval.

2. The set $\mathbb{N}$ is rare in $\mathbb{R}$ since it is closed and no open set is contained in it.

3. The set $\mathbb{Q}$ is meager in $\mathbb{R}$ since it is a countable set and subsequently, a countable union of rare singleton sets. It is not rare since its closure is the entire $\mathbb{R}$.

4. The set $\mathbb{R}$ is non-meager (as we will see in the later part of this section). This means that $\mathbb{Q}^c$ cannot be meager.

Exercise 3.9. Prove that the union of finite number of rare subsets is rare. Notice that this is not the case with a countable union. Hence the requirement of a new nomenclature for such sets. We urge the reader to closely follow the example of $\mathbb{Q} \subseteq \mathbb{R}$. As we have already seen, $\mathbb{Q}$ is meager. However, it is not rare. In fact, it is dense in $\mathbb{R}$!

Theorem 3.8.

1. *The complement of a rare set is dense.*

2. *A subset of a meager set is meager.*

3. *The union of countably many meager sets is meager.*

Proof.

1. Let $A \subseteq X$ be a rare. Thus, it contains no open subset. Subsequently, every open set intersects A^c. Hence, A^c is dense.

2. Let, M be a meager subset of X.

   For any subset $N \subseteq M$, we have $\bar{N} \subseteq \bar{M}$. Since, $\bar{M}$ contains no open set, neither will $\bar{N}$. Hence, N is meager.

3. Let, $(M_n)_{n \in \mathbb{N}}$ be a countable collection of meager sets. Then, each M_n is the countable union of rare sets.

Since the countable union of countable collection is countable, it follows that $\bigcup\limits_{n \in \mathbb{N}} M_n$ is meager.

$\square$

The following corollary easily follows from the definition of co-meager sets.

Corollary 3.8.1.

1. *A superset of a co-meager set is co-meager.*

2. *The countable intersection of co-meager sets is co-meager.*

Now, a metric space (the definition works even for a topological space) is a *Baire space* if every non-empty open set is of second category in X.

3.5.2 Baire category theorem

Theorem 3.9. *A complete metric space is non-meager in itself.*

Proof. Suppose that (X, d) is a complete metric space that is meager in itself. That is, $X = \bigcup\limits_{k \in \mathbb{N}} M_k$ such that each M_k is rare in X.

X contains at least one open subset (itself) while $\bar{M_1}$ does not. Thus, $\bar{M_1}^c$ is a non-empty open subset.

$\therefore \exists x_1 \in B(x_1, \epsilon_1) \subseteq \bar{M_1}^c$ where we can make $\epsilon_1 < \frac{1}{2}$.

Now, $\bar{M_2}$ does not contain any open set. So, it does not contain $B\left(x_1, \frac{\epsilon_1}{2}\right)$. Thus, $\bar{M_2}^c \cap B\left(x_1, \frac{\epsilon_1}{2}\right)$ is a non-empty open set.

$\therefore \exists x_2 \in B(x_2, \epsilon_2) \subseteq \bar{M_2}^c \cap B\left(x_1, \frac{\epsilon_1}{2}\right)$.

Proceeding in this manner, we will obtain a sequence (x_n) such that

$$x_n \in B(x_n, \epsilon_n) \subseteq \bar{M_n}^c \cap B\left(x_{n-1}, \frac{\epsilon_{n-1}}{2}\right).$$

Thus, we have $M_n \cap B(x_n, \epsilon_n) = \emptyset$ and $\epsilon_n < \epsilon_{n-1}/2$ with $\epsilon_1 < \frac{1}{2}$.

For $m > n$, we have $x_m \in B\left(x_n, \frac{\epsilon_n}{2}\right)$. Thus, $d(x_m, x_n) < \frac{\epsilon_n}{2}$.

But, $\epsilon_1 < \frac{1}{2}$ and $\epsilon_n < \frac{\epsilon_{n-1}}{2}$ gives us $\epsilon_n < 2^{-n}$.

$\therefore d(x_m, x_n) < \frac{\epsilon_n}{2} < 2^{-(n+1)} \to 0$ as $n \to \infty$.

Thus, the sequence (x_n) is Cauchy in X and hence, convergent. Let, $x_n \to$

x.

For, $m > n$ we have $d(x_n, x) \leq d(x_n, x_m) + d(x_m, x) < \frac{\epsilon_n}{2} + d(x_m, x)$. $d(x_m, x) \to 0$ as $m \to \infty$ and hence, $x \in B(x_n, \epsilon_n)$, $\forall n \in \mathbb{N}$.

But, as shown earlier $M_n \cap B(x_n, \epsilon_n) = \emptyset$ and therefore $x \notin M_n$, $\forall n \in \mathbb{N}$ which is contradictory to our assumption that $X = \bigcup\limits_{n \in \mathbb{N}} M_n$. $\square$

Corollary 3.9.1. $\mathbb{Q}^c$ *is non-meager in* $\mathbb{R}$.

Proof. Baire category theorem implies that $\mathbb{R}$ is non-meager. Now, since $\mathbb{R} = \mathbb{Q} \cup \mathbb{Q}^c$ and that $\mathbb{Q}$ is meager, $\mathbb{Q}^c$ must be non-meager. $\qquad\square$

Exercise 3.10. Consider discrete metric defined on a countable set. Then the whole space can be written as a countable union of singleton sets. Does this contradict the Baire category theorem? Justify.

Although the version of the theorem we have proved is more than enough for most applications, we feel compelled to present a slightly stronger version, which deals with meager sets in a complete metric space.

Theorem 3.10. *A meager set has an empty interior in a complete metric space.*

Proof. Consider a meager set $M = \bigcup\limits_{n \in \mathbb{N}} M_n$ in a complete metric space X so that each M_n is rare in X.

We will show that every open set intersects M^c and thus M contains no open subset, i.e., it has an empty interior.

Since, M_1 is rare, $\bar{M_1}^c$ is a dense open set so that for every open subset $U \subseteq X$, $\bar{M_1}^c \cap U$ is a non-empty open set.

$\therefore \exists x_1 \in B(x_1, \epsilon_1) \subseteq \bar{M_1}^c \cap U$ where $\epsilon_1 < \frac{1}{2}$.

Now, $\bar{M_2}^c$ is dense and so $B\left(x_1, \frac{\epsilon_1}{2}\right) \cap \bar{M_2}^c$ is a non-empty open set.

$\therefore \exists x_2 \in B(x_2, \epsilon_2) \subseteq B\left(x_1, \frac{\epsilon_1}{2}\right) \cap \bar{M_2}^c$.

Proceeding in this manner, we obtain a sequence (x_n) such that $x_n \in B(x_n, \epsilon_n) \subseteq \bar{M_n}^c \cap B\left(x_{n-1}, \frac{\epsilon_{n-1}}{2}\right)$.

We have seen that the sequence (x_n) constructed in this manner is Cauchy and hence convergent in X and that if $x_n \to x$ then $x \in B(x_n, \epsilon_n)$, $\forall n \in \mathbb{N}$.

Now, $x \in B(x_1, \epsilon_1) \subseteq U$ and $x \in B(x_n, \epsilon_n) \subseteq \bar{M_n}^c$.

$$\therefore x \in U \cap \left(\bigcap_{n \in \mathbb{N}} \bar{M_n}^c\right) \implies x \in U \cap \left(\bigcup_{n \in \mathbb{N}} \bar{M_n}\right)^c,$$

i.e., $U \cap M^c$ is non-empty. Thus, M^c intersects every open subset $U \subseteq X$. It is dense in X. Consequently, M has empty interior. $\qquad\square$

To understand the theorem better, consider once again the case of $\mathbb{Q}$ with the usual metric. As we have already seen, this space is not complete. $\mathbb{Q}$ itself is a meager set in this space since it is countable and singleton sets are rare. However, it does not have an empty interior; in fact, its interior is the entire set $\mathbb{Q}$.

We yet again emphasize that a meager set having an empty interior is different from it being rare since its closure may still have a non-empty interior. As is our ritual, consider the case of $\mathbb{Q}$ in $\mathbb{R}$.

In real analysis, we tend to associate nowhere dense sets with being "small". Similarly, in Lebesgue theory of integration, we associate sets of measure zero as being "small". It is then but natural to try to find any connection between the two. We will show via an example that both these concepts are very different.

Example 3.12. First, we wish to find a meager set of measure greater than zero. Equivalently, we would rather construct a co-meager set of measure zero. We begin by noting that $\mathbb{Q}$ is countable.

Let $\mathbb{Q} = \{q_k\}_{k\in\mathbb{N}}$ and let $B\left(q_k; 1/n\right)$ denote an open ball centered at q_k of radius $\frac{1}{n}$ in $\mathbb{R}$.

We define $V_n = \bigcup_{k\in\mathbb{N}} B\left(q_k, \frac{2^{-k}}{n}\right)$ so that its measure $mV_n \leq \sum_{k=1}^{\infty} 2 . \frac{2^{-k}}{n} = 2/n$. We denote the complement of V_n by $V_n{}^c$.

V_n is open because it is a union of open balls. Therefore, $\mathbb{Q} \subseteq V_n$ is contained in the interior of V_n, which in turn implies that $V_n{}^c$ does not intersect $\mathbb{Q}$. Finally, since $V_n{}^c$ is closed and every open interval in $\mathbb{R}$ intersects $\mathbb{Q}$, $V_n{}^c$ contains no open interval and subsequently has an empty interior, i.e., it is a rare set.

Thus, V_n is a co-meager set and so is $\bigcap_{n\in\mathbb{N}} V_n$. But, $m(\bigcap_{n\in\mathbb{N}} V_n) \leq mV_n \leq \frac{2}{n}$ for every $n \in \mathbb{N}$. Thus $m(\bigcap_{n\in\mathbb{N}} V_n) = 0$.

Exercise 3.11. Prove that the Baire category theorem implies that complete metric spaces are Baire spaces.

3.5.3   Applications of Baire category theorem

Baire category theorem is a non-constructive theorem, i.e., it does not give any construction method to the "thing" it is trying to prove.

One simple application of the theorem would be the following:

Theorem 3.11. *The set $[0, 1]$ is uncountable.*

Proof. If $[0, 1]$ were countable, we would have $[0, 1] = \{x_n\}_{n\in\mathbb{N}}$. But, the singleton set $\{x_n\}$ is rare $\forall n \in \mathbb{N}$ and thus the set $[0, 1] = \bigcup_{n\in\mathbb{N}} \{x_n\}$ is a meager set which, due to Baire category theorem, contradicts the fact that $[0, 1]$ is complete. $\qquad\square$

Moving ahead, let us consider the *"raindrop function"* defined on $\mathbb{R}$ (Figure 3.3)

$$f(x) = \begin{cases} 1/q, & x = \frac{p}{q}, \text{ where } \gcd(p,q) = 1. \\ 0, & \text{otherwise.} \end{cases} \tag{3.4}$$

FIGURE 3.3: The raindrop function (also known as *Thomae's function*).

It is continuous on $\mathbb{Q}^c$ but discontinuous on $\mathbb{Q}$. The natural curiosity arises whether the reverse is possible. Specifically, is there a function that is continuous on $\mathbb{Q}$ but discontinuous on $\mathbb{Q}^c$? The answer is no, and again a result of Baire category theorem. To prove this, however, we will first need to look at some concepts from real analysis.

Definition 3.12 ($\mathcal{F}_\sigma$ and $\mathcal{G}_\delta$ sets). *A countable union of closed sets is an $\mathcal{F}_\sigma$ set. A countable intersection of open sets is a $\mathcal{G}_\delta$ set.*

It is easy to see that the complement of an $\mathcal{F}_\sigma$ set is a $\mathcal{G}_\delta$ set and vice versa. Also, the set $\mathbb{Q}$ is countable and since singleton sets are closed, it is an $\mathcal{F}_\sigma$ set. Therefore, $\mathbb{Q}^c$ is a $\mathcal{G}_\delta$ set. A natural question arises whether it is possible to write $\mathbb{Q}$ as a $\mathcal{G}_\delta$ set. The answer is no. The reason is not straightforward and comes from Baire category theorem as follows.

We already know that $\mathbb{Q}$ is a $\mathcal{F}_\sigma$ set. If $\mathbb{Q}^c$ were a $\mathcal{F}_\sigma$ set then, we would have $\mathbb{R}$ as a $\mathcal{F}_\sigma$ set.

$\therefore \mathbb{R} = \bigcup_{n \in \mathbb{N}} F_n$ where each F_n is closed and is either contained in $\mathbb{Q}$ or in $\mathbb{Q}^c$. But, both $\mathbb{Q}$ and $\mathbb{Q}^c$ have empty interior so each F_n is rare. Now, $\mathbb{R}$ is complete and thus Baire category theorem implies that at least one of F_n has a non-empty interior which is a contradiction.

The concept of $\mathcal{F}_\sigma$ and $\mathcal{G}_\delta$ sets is a very important tool in the study of continuity of a function as the following theorem shows.

Theorem 3.12. *The set of points of continuity of a function is $\mathcal{G}_\delta$ set. Subsequently, the set of points of discontinuity is $\mathcal{F}_\sigma$ set.*

Proof. A function $f : X \to Y$ is continuous at point $p \in X$ iff $\forall n \in \mathbb{N}$, $\exists U \subseteq X$ which is open with $p \in U$ such that $\forall x, y \in U$ we have

$$d_Y\left(f\left(x\right), f\left(y\right)\right) < \frac{1}{n}.$$

For a fixed $n \in \mathbb{N}$, we denote by G_n the set containing all $p \in X$ for which an open neighborhood U exists so that $d_Y\left(f\left(x\right), f\left(y\right)\right) < \frac{1}{n}$, $\forall x, y \in U$.

Note that for any other point $q \in U$, the above condition holds as well, and hence $q \in G_n$, i.e., $U \subseteq G_n$. Since G_n is the union of all such open sets, G_n is open.

It is then easy to see that the set of all points at which f is continuous is $\bigcap_{n \in \mathbb{N}} G_n$, a G_δ set as required. $\qquad\qquad\qquad\qquad\qquad\qquad\qquad\qquad$ $\square$

Since as proved earlier, $\mathbb{Q}$ cannot be a G_δ set, the following corollary easily follows:

Corollary 3.12.1. *There exists no real function which is continuous at only $\mathbb{Q}$.*

The theme of the above proof is very common in most applications of Baire category theorem. We try to construct a meager set inside a complete space and hence, claim the existence of a point outside the set or use it as a contradiction. Another interesting application is the proof for existence of a highly non-monotonic real function that is continuous.

Theorem 3.13. *There exists a continuous real-valued function that is not monotonic in any subinterval of $[0,1]$.*

Proof. We know that the space $\mathscr{C}[0,1]$ with the sup metric d_∞ is complete. Let $I \subseteq [0,1]$ be any interval and let $A(I)$ and $B(I)$ be the set of all functions in $\mathscr{C}[0,1]$ that are non-decreasing and non-increasing respectively on I.

We claim that both the sets are closed in $\mathscr{C}[0,1]$.

Consider f to be the limit of a sequence (f_n) in $A(I)$.

For $x, y \in [0,1]$ with $x > y$, let $M = \frac{f(x) + f(y)}{2}$.

If $f(x) = f(y)$ then $f(x) \geq f(y)$.

For $f(x) \neq f(y)$, both are on the opposite sides of M on the number line. Pick $\epsilon > 0$ such that

$$0 < \epsilon < \frac{|f(x) - f(y)|}{2}.$$

Then, $\exists n_0 \in \mathbb{N}$ such that

$$d_\infty(f_n, f) < \epsilon, \ \forall n > n_0.$$

Thus $|f_n(x) - f(x)| \leq d_\infty(f_n, f) < \epsilon$ and similarly $|f_n(y) - f(y)| < \epsilon$.

Therefore, both $f_n(x)$ and $f(x)$ are on the same side of M and also $f_n(y)$ and $f(y)$ are one the same side of M. But, $f(x)$ and $f(y)$ are on the opposite side of M and $f_n(x) \geq f_n(y)$ since $f_n \in A(I)$.

$$\therefore f_n(x) \geq M \geq f_n(y) \implies f(x) \geq f(y).$$

That is, $f \in A(I)$ and thus $A(I)$ is closed. Similarly, we can show that $B(I)$ is closed as well. Since the choice of the interval $I \subseteq [0,1]$ was arbitrary, this is true for each interval.

Subsequently, the set $K(I) = A(I) \cup B(I)$ is closed. It also has an empty interior as we will show next.

Let $g \in K(I)$, then g is either a non-increasing or a non-decreasing continuous function. Then due to the continuity of g, for any $x_0 \in I$ and $\forall \epsilon > 0$, $\exists \delta > 0$ such that $|g(x) - g(x_0)| < \frac{\epsilon}{3}$ whenever $|x - x_0| < \delta$.

We will add to g a small "spike" function (Figure 3.4) around x_0 so that the resulting function is no longer monotonic. For this, we define a new function ζ as follows

$$\zeta(x) = \begin{cases} \frac{\epsilon}{2\delta}(x - (x_0 - \delta)), & x \in [x_0 - \delta, x_0]. \\ \frac{\epsilon}{2\delta}((x_0 + \delta) - x), & x \in [x_0, x_0 + \delta]. \\ 0, & \text{otherwise.} \end{cases} \tag{3.5}$$

Now, note that the function $\tilde{g} = g + \zeta$ is continuous. Also, it is monotonic (strictly) increasing in $(x_0 - \delta, x_0)$ since ζ is strictly increasing. However for $x_2 \in [x_0, x_0 + \delta]$, we have

$$\tilde{g}(x_2) - \tilde{g}(x_0) = (g(x_2) - g(x_0)) + (\zeta(x_2) - \zeta(x_0)).$$

The expression in the first parenthesis, even though positive, cannot exceed $\frac{\epsilon}{3}$ whereas the expression in the second parenthesis approaches the value $-\frac{\epsilon}{2}$ as x_2 moves toward $x_0 + \delta$. Therefore at some point, the above difference becomes negative overall, and hence, $\tilde{g}$ is strictly decreasing in some sub-interval of $(x_0, x_0 + \delta)$.

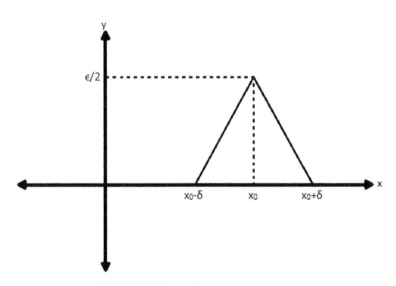

FIGURE 3.4: The spike function.

Thus the function $\tilde{g} = g + \zeta$ is continuous and non-monotonic with $d_\infty(g, \tilde{g}) < \epsilon$.

Thus, we have proven that every neighborhood of g contains a point (in $\mathscr{C}[0,1]$, which is, in fact a function) outside $K(I)$. Therefore $K(I)$ has an empty interior.

The set of intervals $\{(q - 1/m, q + 1/m) \cap [0,1] \mid q \in \mathbb{Q} \text{ and } m \in \mathbb{N}\}$ is countable and and let us denote it by $\{I_n \mid n \in \mathbb{N}\}$.

Then, the set of functions in $\mathscr{C}[0,1]$ that are monotonic in any sub-interval of $[0,1]$ is given by $\bigcup_{n \in \mathbb{N}} K(I_n)$ which is a meager subset of $\mathscr{C}[0,1]$. Since $\mathscr{C}[0,1]$ is complete, Baire category theorem dictates that $\bigcup_{n \in \mathbb{N}} K(I_n) \neq \mathscr{C}[0,1]$.

Hence there exists a continuous function which is not monotonic at any sub-interval of $[0,1]$. $\qquad\square$

Most of the applications of Baire category theorem go far beyond the scope of this book. Most notably, it can used to prove the three major theorems of functional analysis, namely the open mapping theorem, the closed graph theorem, and the uniform boundedness principle. Apart from these, it can also be used to prove that functions which have a finite derivative on at least one point in an interval are rare in the set of all continuous functions similar to how we have proved the same for monotonic functions.

Problem Set

1. We define the range of a sequence $(x_n)_{n \in \mathbb{N}}$ as the set $\{x_n \mid n \in \mathbb{N}\}$. Is the range of a subsequence always the same as that of the corresponding sequence? If not, then in which case would it be the same, if at all?

2. Is there a limit to the number of subsequences a given sequence can have? Count the number of subsequences of the following:

   (a) $1, 1, 1, 1, \ldots$

   (b) $1, 2, 1, 2, 1, 2, \ldots$ **Hint:** It is not finite

   (c) $e, \pi, e, e, e, \ldots$

3. Prove that every subsequence of a convergent sequence is convergent and it converges to the same limit. Subsequently, prove that a sequence is convergent if and only if all its subsequences are convergent to the same limit.

4. Prove that a subsequence of a Cauchy sequence is Cauchy.

5. Is it possible that the sequence of rational approximation $(a_n)_{n \in \mathbb{N}}$ of $\sqrt{2}$ is convergent to some other limit in $\mathbb{Q}$? If yes, find the limit.

   Hint: If $a_n \to a \in \mathbb{Q}$ then in $\mathbb{R}$, $a_n \to a$ (why?) and $a_n \to \sqrt{2} \implies a = \sqrt{2} \in \mathbb{Q}$.

6. Consider the set X of all bounded real functions on the closed interval $[a, b]$ with a metric d defined on it is as follows:

   For $x, y \in X$
   $$d(x, y) = \max_{t \in [a,b]} \{|x(t) - y(t)|\}.$$

   Prove that convergence in this metric implies uniform convergence. Recall that a function $f : D \subseteq \mathbb{R} \to \mathbb{R}$ is uniformly continuous if $\forall \epsilon > 0$, $\exists \delta > 0$ such that whenever $|x - y| < \delta$, we have $|f(x) - f(y)| < \epsilon$.

7. On the interval $(0, 1]$, consider the function d defined by
   $$d(x, y) = \left| \frac{1}{x} - \frac{1}{y} \right|, \quad \forall x, y \in (0, 1].$$

   Does d constitute a metric on the set? Is the metric space complete?

Biographical Notes

Baron Augustin-Louis Cauchy (21 August, 1789 to 23 May, 1857) was a French mathematician, engineer, and physicist who made pioneering contributions to several branches of mathematics, including mathematical analysis and continuum mechanics. He was one of the first to state and rigorously prove theorems of calculus, rejecting the heuristic principle of the generality of algebra of earlier authors. He almost singlehandedly founded complex analysis and the study of permutation groups in abstract algebra. Cauchy was a prolific writer; he wrote approximately eight hundred research articles and five complete textbooks on a variety of topics in the fields of mathematics and mathematical physics. Cauchy was very productive, in number of papers second only to Leonhard Euler.

René-Louis Baire (21 January, 1874 to 5 July, 1932) was a French mathematician most famous for his Baire category theorem, which helped to generalize and prove future theorems. His theory was published originally in his dissertation *Sur les fonctions de variable réelles* ("On the Functions of Real Variables") in 1899.

Chapter 4

Compact Metric Spaces

The notion of compactness is perhaps the least intuitive but plays a central role in analysis. It was Fréchet who coined the term "compact" in his 1906 doctoral thesis and gave the definition for what we now know as countable and sequential compactness. However, Alexandroff and Urysohn get credit for defining open cover compactness.

To give a vague idea, compact spaces are the ones which are not "too large"; indeed, we will see that compact spaces are always bounded (in fact, it is a generalization of finite set). So, part of saying that something is not too large means that it cannot extend to infinity. However, the open interval $(0, 1)$ in $\mathbb{R}$ is also bounded and yet, is not compact, as we shall see in the text. Therefore, "not too large" in the above sense must mean more than simply being bounded.

There are two possible definitions (as we mentioned above) that we can give for compactness: one is easier to state but gives no hint at its deeper meaning, and the other which is trickier to state but starts to get to the heart of the sense in which compact sets are "not too large".

In this chapter, we begin with the first definition of covering compactness, give examples, and prove some basic facts, and then we will give the second definition (of sequential compactness) and do the same. Finally, we shall establish characterization theorem(s) for compactness including the Lebesgue covering lemma.

4.1 Open Cover and Compact Sets

Definition 4.1 (Open cover). *Let (X, d) be a metric space and $A \subseteq X$. We say that the family of sets $\mathscr{C} = \{G_\lambda \subseteq X | \lambda \in \Lambda\}$ is an open cover of A if $\forall \lambda \in \Lambda$, the sets G_λ are open and $A \subseteq \bigcup_{\lambda \in \Lambda} G_\lambda$. Given an open cover $\mathscr{C}$ of A, we will say that the family $\mathscr{C}' = \{A_\alpha \subseteq X | \alpha \in \Delta\}$ is a subcover of $\mathscr{C}$ if $\Delta \subseteq \Lambda$, $\mathscr{C}' \subseteq \mathscr{C}$ and $\mathscr{C}'$ is an open cover for A.*

The definition is really as straightforward as it sounds. We need the sets in the family to be open, and for them to cover A. A subcover is just a subfamily of sets in the family which also covers A. Let us look at an example from $\mathbb{R}$.

Example 4.1. Consider the following sets A in the metric space $(\mathbb{R}, d)$, where $d(x, y) = |x - y|$:

1. $A = \{x \in \mathbb{R} | 0 \leq x \leq 1\}$, i.e., the closed interval $[0, 1]$. For any $\epsilon > 0$, the family of intervals $\{(\alpha - \epsilon, \alpha + \epsilon) | \alpha \in A\}$ is an open cover. There are many subcovers. One of the examples is $\{(\beta - \epsilon, \beta + \epsilon) | \beta \in \mathbb{Q} \cap [0, 1]\}$.

2. $A = \mathbb{R}$ and consider the family of intervals $\{(\alpha - 1, \alpha + 1) | \alpha \in \mathbb{Z}\}$. This family is an open cover for $\mathbb{R}$ which has no non-trivial subcovers. By this we mean that any proper subset of the family cannot cover $\mathbb{R}$.

3. $A = \left\{ \frac{1}{n} \middle| n \in \mathbb{N} \right\}$. Consider for any $\epsilon > 0$, the family $\mathscr{C} = \{G_n | n \in \mathbb{N}\}$, where $G_n = (\frac{1}{n} - \epsilon, \frac{1}{n} + \epsilon)$. The family $\mathscr{C}$ is an open cover for A. Also, by the Archimedean property of $\mathbb{N}$ in $\mathbb{R}$, $\exists n_0 \in \mathbb{N}$ such that $n_0 \epsilon > 1$. It follows from this that $\forall n \geq n_0$, $\frac{1}{n} \in G_{n_0}$. Hence, the family $\mathscr{C}$ has a finite sub-family, given by $\{G_1, G_2, \cdots, G_{n_0}\}$. Here we say that A has a finite subcover.

4. $A = (0, 1)$. If we consider the family of intervals

$$\left\{ G_n = \left(\frac{1}{2n} - \frac{1}{2n+1}, \frac{1}{2n} + \frac{1}{2n+1} \right) \middle| n \in \mathbb{N} \cup \{0\} \right\},$$

then these intervals get smaller and smaller and both the end points converge to 0.

In fact, $\bigcup_{n \in \mathbb{N} \cup \{0\}} \left(\frac{1}{2n} - \frac{1}{2n+1}, \frac{1}{2n} + \frac{1}{2n+1} \right) = \left(0, \frac{3}{2} \right) \supset (0, 1)$. Therefore, the family we have considered is an open cover for A. However, no finite subcollection of this family can cover A. This is because if we take any finite subfamily, say $\{G_{n_1}, G_{n_2}, G_{n_3}, \cdots, G_{n_k}\}$ and consider $N = \max\{n_1, n_2, \cdots, n_k\}$, then $\bigcup_{i=1}^{N} G_{n_i} \subseteq \left(\frac{1}{2^N} - \frac{1}{2^{N+1}}, \frac{3}{2} \right)$ and hence does not contain the points in A which are "sufficiently" close to 0. Note that by "sufficiently close" we mean those points which lie between 0 and $\frac{1}{2^N} - \frac{1}{2^{N+1}}$. We shall use this property further to show that the open interval $(0, 1)$ is not compact.

5. The collection of open intervals above is not an open cover for $A = [0, 1]$ because none of the intervals contain 0. We can get an open cover for A from the above family of intervals by adding another interval about 0, $(-\delta, \delta)$, where $\delta > 0$. In this case, if we choose $N \in \mathbb{N}$ sufficiently large so that $\frac{1}{2^N} - \frac{1}{2^{N+1}} < \delta$, the the set $\{G_0, G_1, \cdots, G_N\} \cup \{(-\delta, \delta)\}$ is a finite subcover for A.

Exercise 4.1. Let $X = \mathbb{Q}^c$ be the set of irrational numbers. Can you find an open cover for X?

Exercise 4.2. Let $X = \mathbb{Z}$ be the set of integers. Can you think of an open cover for X which does not admit a finite subcover?

Exercise 4.3. Give a non-trivial open cover for an arbitrary metric space.

Exercise 4.4. Let $A = \{(x,y) \in \mathbb{R}^2 | x^2 + y^2 < 4\}$, where we consider the Euclidean metric space $\mathbb{R}^2$. Can you think of an infinite cover of A which admits a finite subcover?

We shall now give the definition of compactness for a set which, at the first glance, may appear intriguing or rather intimidating! However, it is advised to the reader that attention is to be paid to the consequences of the definition, which are often easier to grasp, rather than the subtleties and technicalities of it. The definition notably has a topological flavor (i.e., it is purely defined in terms of open sets) and we will try to relate it to more familiar notions in the text.

From all the above examples, one thing which we can observe is that open covers are not interesting in themselves. Finding an open cover and then its subcover is merely a computational task. What is not so trivial, and also very intimately related to many (topological) properties of metric spaces, is that what can we say about *every* open cover for a given set? Based on this query, we have the following definition.

Definition 4.2 (Compactness). *Suppose (X, d) is a metric space and $A \subseteq X$. We will say that A is a compact set in X if every open cover of A has a finite subcover.*

Note here that by "finite subcover" we mean that the subcollection of the cover contains finitely many sets and not that the sets which cover the set have finitely many elements. More explicitly, if $\{G_\lambda | \lambda \in \Lambda\}$ is an open cover for A, then there should exist finitely many indices $\lambda_1, \lambda_2, \cdots, \lambda_n$ such that $A \subseteq \bigcup_{i=1}^{n} G_{\lambda_i}$. Then, we say that the given cover has a finite subcover given by $\{G_{\lambda_1}, G_{\lambda_2}, \cdots, G_{\lambda_n}\}$.

Let us look at the previous examples and try to find the finite subcovers whenever possible.

Example 4.2. In the very first example, we took $A = [0, 1]$ and the open cover as

$$\mathscr{C} = \{G_\alpha = (\alpha - \epsilon, \alpha + \epsilon) | \alpha \in [0, 1]\},$$

for a fixed $\epsilon > 0$. While this family has uncountably many sets, there is a lot of "overlap" in the sets. Any $x \in [0, 1]$ is "covered" by every set G_α, where $\alpha \in (x - \epsilon, x + \epsilon)$. We want to remove as much as redundancy as possible.

What we will do is, instead of taking every α in $[0, 1]$ continuously, we shall take *discrete jumps*. If we jump by an appropriate amount, we should not exclude any element from the covering. It should not take much convincing to agree that the family

$$\mathscr{C}' = \{S_0, S_\epsilon, S_{2\epsilon}, \cdots, S_{n\epsilon}\},$$

FIGURE 4.1: Pictorial representation of the finite subcover found in the example.

where, $S_{i\epsilon} = ((i-1)\,\epsilon, (i+1)\,\epsilon)$ and $n \in \mathbb{N} \cup \{0\}$ is the largest such number such that $n\epsilon < 1$ (How do we know that a largest n exists?). This is shown in Figure 4.1. It is clear that every set in $\mathscr{C}'$ is also in the family $\mathscr{C}$ and this family is finite. So, we have done our job!

Note. This example does not prove that the set $[0, 1]$ is compact. To prove so, we would have to find a finite subcover for *every* open cover of $[0, 1]$. However, in the example, we have found a finite subcover for only one of the open covers. We will, in fact, prove that this set is compact once we are equipped with proper tools to do so!

Example 4.3. We have seen that the family $\{(\alpha - 1, \alpha + 1)\,|\alpha \in \mathbb{Z}\}$ is an open cover for $\mathbb{R}$. Clearly, this family is not finite since $\mathbb{Z}$ is not finite and the family has a one-to-one correspondence with $\mathbb{Z}$. Now, suppose we remove one of the intervals $(\alpha_0 - 1, \alpha_0 + 1)$. Then, the point $\alpha_0 \in \mathbb{R}$ cannot be covered by any other interval in the family (otherwise the distance between two integers would becomes less than 1). Hence, for this cover, we cannot have a finite subcover, and therefore, we can conclude that $\mathbb{R}$ is not compact.

Example 4.4. We have considered the set $A = \left\{\frac{1}{n}\,\middle|\,n \in \mathbb{N}\right\}$ and also found an open cover $\mathscr{C} = \{G_n|n \in \mathbb{N}\}$, where $G_n = \left(\frac{1}{n} - \epsilon, \frac{1}{n} + \epsilon\right)$ which admits a finite subcover. Again, as mentioned earlier, this is not enough to show that this set is compact! We will have to do this for every open cover of A.

Now, consider the collection $\mathscr{C}' = \left\{B\left(\frac{1}{n}, r_n\right)\,\middle|\,n \in \mathbb{N}\right\}$, where $r_n = \frac{1}{n(n+1)}$. Notice that for this choice of r_n, we have

$$\frac{1}{n+1} < \frac{1}{n} - r_n < \frac{1}{n} < \frac{1}{n} + r_n < \frac{1}{n-1}.$$

Therefore, no two elements of the set A are contained in the single ball $B\left(\frac{1}{n}, r_n\right)$. Hence, if we remove any of the open balls from $\mathscr{C}'$, the collection will no longer cover A. Therefore, $\mathscr{C}'$ admits no non-trivial subcovers. Also, $\mathscr{C}'$ is infinite. Therefore, this particular open cover of A does not admit a finite subcover and thus, A is not compact.

One may ask a question: How did we know that this open cover will work to prove that A is not compact? Mathematics is not magic. Some reasoning is required for everything we do. First, we observed that if we take any two points in A, then we can have open balls around each of them so that each of the open balls contain no point of A other than the center itself. Such points are called *isolated points*. Since every point in A was isolated and A had infinitely many points, we could construct an open cover of precisely those open sets (balls) which made these points isolated in the first place. This open cover cannot have any finite subcover. Note that this method is not only applicable to this particular set A. Any set with infinitely many isolated points can be proved to be non-compact by this method.

Example 4.5. Consider the metric space ℓ^2 and the *closed unit ball* given by

$$\overline{B\left(\mathbf{0},1\right)} = \left\{\mathbf{x} \in \ell^2 | d_2\left(\mathbf{x},\mathbf{0}\right) \leq 1\right\}.$$

Now, consider the set $\{\mathbf{e}_n\} \subseteq \ell^2$, where $\mathbf{e}_n$ is that real-valued sequence where the nth term is 1 and all other terms are 0. Clearly, $d_2\left(\mathbf{e}_n,\mathbf{0}\right) = 1$ so that $\forall n \in \mathbb{N}$, $\mathbf{e}_n \in \overline{B\left(\mathbf{0},1\right)}$. Also, notice that $d_2\left(\mathbf{e}_n,\mathbf{e}_m\right) = \sqrt{2}$, whenever $n \neq m$.

Now, consider the family $\mathscr{C} = \left\{B\left(\mathbf{x},\frac{1}{2}\right) \middle| x \in \overline{B\left(\mathbf{0},1\right)}\right\}$. Suppose this has a finite subcover $\mathscr{C}' = \left\{B\left(\mathbf{x}_i,\frac{1}{2}\right) \middle| 1 \leq i \leq n\right\}$ for some $n \in \mathbb{N}$. Since $\{\mathbf{e}_n\}$ is an infinite set, at least two $\mathbf{e}_n,\mathbf{e}_m$ for $m \neq n$ must be inside one ball $B\left(\mathbf{x}_{i_0},\frac{1}{2}\right)$. Then, we have

$$d_2\left(\mathbf{e}_n,\mathbf{e}_m\right) \leq d_2\left(\mathbf{e}_n,\mathbf{x}_{i_0}\right) + d_2\left(\mathbf{e}_m,\mathbf{x}_{i_0}\right) < \frac{1}{2} + \frac{1}{2} = 1.$$

This is a contradiction! Hence, the open cover $\mathscr{C}$ cannot have any finite subcover and the closed unit ball in ℓ^2 is not compact.

Exercise 4.5. Show that the metric space (X,d), where X is an infinite set and d is the discrete metric, is not compact.

Exercise 4.6. Show that the closed unit ball in $(\mathscr{C}[0,1], d_\infty)$, where $d_\infty\left(f,g\right) = \sup\limits_{x\in[0,1]} \{|f\left(x\right) - g\left(x\right)|\}$, is not compact.

4.2 General Properties of Compact Sets

All the above examples were well and good, but none of them gave actual insights about what compact sets really are, or why they are useful. Therefore, we shall now consider some properties of compact sets specified in terms of theorems.

Theorem 4.1. *Let (X,d) be a metric space and $S \subseteq X$ be a compact set. Then,*

1. *S is bounded.*

2. *S is closed.*

3. *Every infinite subset $K \subseteq S$ has a cluster point in S.*

Proof.

1. Let $x \in X$ be fixed. Then, by Archimedean property, $\forall y \in X$, $\exists n \in \mathbb{N}$ such that $n > d(x,y)$. Therefore, the family of open balls $\{B(x,n) \,|n \in \mathbb{N}\}$ is an open cover for X and hence for S. Since S is compact, this cover has a finite subcover for S. Let $\{B(x,n_i) \,|1 \le i \le n_0\}$ be the subcover. If we take $N = \max\{n_1, n_2, \cdots, n_{n_0}\}$, then for $i = 1, 2, \cdots, n_0$, we have $B(x,n_i) \subseteq B(x,N)$. Therefore, $S \subseteq B(x,N)$ and thus, S is bounded.

2. Consider a point $p \in S^c$. For each point $y \in S$, we can construct open balls $B(y,r_y)$, where $r_y = \frac{1}{2}d(p,y)$. Now, $\forall y \in S$, $y \in B(y,r_y)$ and also being open, the family of balls $\mathscr{C} = \{B(y,r_y) \,|y \in S\}$ is a cover for S. Since S is compact, $\mathscr{C}$ has a finite subcover of open balls, say $\{B(y_i,r_{y_i}) \,|y_i \in S$ and $1 \le i \le n\}$ for some $n \in \mathbb{N}$. Also, note that from the construction of these open balls, it is evident that $p \notin B(y,r_y)$ and in particular, $\forall i \in \{1, 2, \cdots, n\}$, $p \notin B(y_i, r_{y_i})$.

   Now, let $r = \min_{1 \le i \le n} \{r_{y_i}\}$. The open ball $B(p,r)$ cannot intersect with any of the open balls $B(y_i, r_{y_i})$ and hence cannot intersect S. Therefore, $B(p,r) \subseteq S^c$ so that S^c is open and S is closed. The method of finding r is diagrammatically represented in Figure 4.2.

3. Given an infinite subset $K \subseteq S$, suppose that K does not have a cluster point in S. Then, $\forall x \in S$, $\exists r_x > 0$ such that the open ball $B(x,r_x)$ does not contain any point of K except possibly x (if $x \in K$). Now, the family $\{B(x,r_x) \,|x \in S\}$ is an open cover for S and since $K \subseteq S$, it is also a cover for K. Given that S is compact, we know that this family

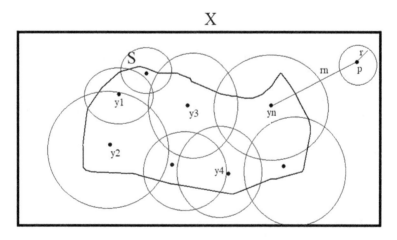

FIGURE 4.2: Compact sets are closed: Diagrammatic representation.

must have a finite subcover, say $\{B\left(x_i, r_{x_i}\right) | x_i \in S \text{ and } 1 \le i \le n\}$ for some $n \in \mathbb{N}$. Consequently, this subcover also covers K. However, each of the balls have the property that $B\left(x, r_x\right) \cap K$ has at most one element. Hence, either K is finite or this subcover cannot cover K, both of which are contradictions! Therefore, K must have a cluster point in S.

$\square$

Note that we have proved a very important property of compact sets: Every compact set is closed and bounded. What can we say about the converse? In Example 4.5, we see that the closed unit ball in ℓ^2 is not compact. However, it is both closed and bounded. Therefore, closedness together with boundedness is not a characterization of compactness. However, they are necessary conditions for a set to be compact. If we have an encounter with a set which is either not closed or not bounded, we can surely say that it is not compact.

Also, we see that any infinite subset of a compact set must have a cluster point. We call this property the *Bolzano-Weierstrass property*. The formal definition is given as follows.

Definition 4.3 (Bolzano-Weierstrass property). *A metric space (X, d) is said to have Bolzano-Weierstrass property (BWP) if every infinite subset of X has a cluster point in X.*

Theorem 4.2. *A closed subset of a compact metric space is compact.*

Proof. Let (X, d) be the given compact metric space and let $S \subseteq X$ be closed. Let $\mathscr{C} = \{G_\lambda | \lambda \in \Lambda\}$ be an open cover for S. We know that $X = S \cup S^c$ so that the family $\mathscr{C} \cup \{S^c\}$ is an open cover for X. However, X is compact and hence this family has a finite subcover. We denote this finite subcover as $\mathscr{C}'$. Note that $\mathscr{C}'$ may or may not include S^c. If it includes S^c, we make another family $\mathscr{C}''$ which has all the members of $\mathscr{C}'$ except for S^c. We do this because we know S^c will never help us cover S. Now, the family $\mathscr{C}''$ must cover S or otherwise, $\mathscr{C}'$ cannot cover X. Since $\mathscr{C}'$ is finite, $\mathscr{C}''$ is also finite and we have proved that S is compact. This is represented in Figure 4.3.

In the case when $\mathscr{C}'$ does not contain S^c, the family also acts as a cover of S, thereby proving our claim.

$\square$

Consider the family of open sets $\mathscr{F} = \left\{\left(0, \frac{1}{n}\right) \Big| n \in \mathbb{N}\right\}$. What is their intersection? This is a nested family of open sets whose right end point goes on decreasing to 0. Clearly, their intersection is empty. However, if we take any finite number of sets from this family we get a non-empty intersection. Similarly, consider the family of closed sets $\mathscr{F}' = \{[n, \infty) | n \in \mathbb{N}\}$. Again, if we take a finite number of sets from this family, their intersection is non-empty. On the other hand, if we take infinite intersection, it is empty because $\mathbb{N}$ is unbounded in $\mathbb{R}$.

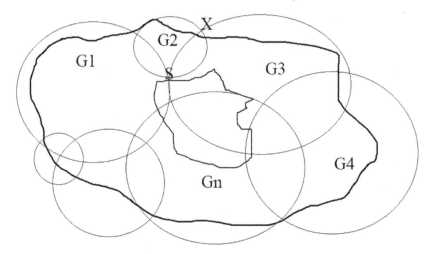

FIGURE 4.3: Pictorical representation of the fact that the finite cover for compact space is also a cover for a closed set.

Therefore, we have families of sets $\mathscr{F} = \{S_\lambda | \lambda \in \Lambda\}$ such that

$$\forall n \in \mathbb{N}, \bigcap_{i=1}^{n} S_{\lambda_i} \neq \emptyset,$$

while

$$\bigcap_{\lambda \in \Lambda} S_\lambda = \emptyset.$$

The question is: Can we have a family where the intersection still remains non-empty? The answer to this is given by compactness of the sets in the family.

Theorem 4.3. *Let (X, d) be a metric space. Suppose $\{S_\lambda | \lambda \in \Lambda\}$ is a family of compact non-empty subsets of X and that for any finite subset $\{S_{\lambda_i} | 1 \leq i \leq n\}$ of the family, we have*

$$\bigcap_{i=1}^{n} S_{\lambda_i} \neq \emptyset.$$

Then, we also have

$$\bigcap_{\lambda \in \Lambda} S_\lambda \neq \emptyset.$$

Proof. Suppose that the theorem is not true! Then, for a fixed $\lambda_0 \in \Lambda$, we will have

$$S_{\lambda_0} \cap \left(\bigcap_{\lambda \in \Lambda \setminus \{\lambda_0\}} S_\lambda \right) = \emptyset.$$

Let us denote $S = \bigcap_{\lambda \in \Lambda \setminus \{\lambda_0\}} S_\lambda$. It follows that $S_{\lambda_0} \subseteq S^c$. Also, S is an intersection of closed sets (compact sets are closed) and hence S^c is open. In fact, $S^c = \bigcup_{\lambda \in \Lambda \setminus \{\lambda_0\}} S_\lambda^c$. Therefore, the family $\{S_\lambda^c | \lambda \in \Lambda \setminus \{\lambda_0\}\}$ is an open cover for S_{λ_0}.

However, S_{λ_0} is compact, and therefore, this family must have a finite subcover, say $\{S_{\lambda_1}^c, S_{\lambda_2}^c, \cdots, S_{\lambda_n}^c\}$, for some $n \in \mathbb{N}$. Now, we have

$$S_{\lambda_0} \subseteq \bigcup_{i=1}^n S_{\lambda_i}^c = \left(\bigcap_{i=1}^n S_{\lambda_i}\right)^c.$$

Therefore, the finite intersection $S_{\lambda_0} \cap S_{\lambda_1} \cap \cdots \cap S_{\lambda_n}$ is empty, which is a contradiction to the hypothesis. Hence, the theorem must be true! □

4.3 Sufficient Conditions for Compactness

In the previous section, we saw the consequences of compactness, i.e., we saw what happens when a set is compact. However, we paid no attention to check if a given set is compact. It is evident from the examples that we have seen, proving a set to be *non-compact* is easier than proving that it is compact! This is because to prove that a set is not compact, we simply have to find one open cover which does not admit finite subcover. However, to prove that a set is compact, we need to examine every possible open cover for the set. This is tedious, requires a lot of analysis, and in many metric spaces, is challenging. In this section, we shall try to formulate certain *sufficient* conditions for compactness.

Theorem 4.4. *In the metric space $(\mathbb{R}, d)$, where $d(x, y) = |x - y|$, any closed and bounded interval $[a, b]$, with $a \leq b$ is compact.*

Proof. Suppose that this theorem is false! Let $\{G_\lambda | \lambda \in \Lambda\}$ be that open cover for $[a, b]$ which does not admit any finite subcover. Then for the point $c = \frac{a+b}{2}$, one of the intervals $[a, c]$ or $[c, b]$ does not have a finite subcover. We denote $I_0 = [a, b]$ and I_1 as that interval from the choices $[a, c]$ or $[c, b]$ which does not admit a finite subcover.

Inductively, we form a sequence of intervals I_n, for $n \in \mathbb{N}$, which do not have a finite subcover. Note that this sequence of intervals is decreasing in the sense that

$$I_0 \supseteq I_1 \supseteq I_2 \supseteq \cdots$$

Also, for each $n \in \mathbb{N}$, I_n is not covered by finitely many members of the family $\{G_\lambda | \lambda \in \Lambda\}$. And for any two points $x, y \in I_n$ we have

$$d(x - y) = |x - y| \leq \frac{b - a}{2^n}.$$

Now, we denote $I_n = [a_n, b_n]$ and the set of end points as $I^+ = \{b_1, b_2, \cdots\}$ and $I^- = \{a_1, a_2, \cdots\}$. The set I^- has an upper bound, namely b, in $\mathbb{R}$ and thus has a supremum. Similarly, the set I^+ has a lower bound, namely a, in $\mathbb{R}$ and thus has an infimum. Also, we have $\sup I^- \leq \inf I^+$ (why?). Therefore, $\exists x_0 \in \mathbb{R}$ such that $\sup I^- \leq x_0 \leq \inf I^+$ and therefore, $x_0 \in I_n$ for each $n \in \mathbb{N}$.

This is to say that $\bigcap_{n \in \mathbb{N}} I_n \neq \emptyset$. Also, the family $\{G_\lambda | \lambda \in \Lambda\}$ covers each of the I_n so that $\exists \lambda_0 \in \Lambda$ such that $x_0 \in G_{\lambda_0}$. Since G_{λ_0} is open, $\exists r > 0$ such that $B(x_0, r) \subseteq G_{\lambda_0}$.

By the Archimedean property of $\mathbb{N}$ in $\mathbb{R}$, $\exists n_0 \in \mathbb{N}$ such that $r 2^{n_0} > b - a$. Therefore, $\forall y \in I_{n_0}$, we have

$$d(y, x_0) \leq \frac{b - a}{2^{n_0}} < r.$$

Hence, $I_{n_0} \subseteq B(x_0, r) \subseteq G_{\lambda_0}$. This means that I_{n_0} is covered by a single open set from the cover of $[a, b]$, which contradicts the construction of I_{n_0}. Hence, our initial assumption that $[a, b]$ is not compact must be wrong. $\square$

We now move to a more general case of $\mathbb{R}^n$, where we in fact characterize the compactness of sets.

Theorem 4.5 (Heine-Borel). *In the Euclidean space $(\mathbb{R}^n, d_2)$, a set is compact if and only if it is closed and bounded.*

Proof. The necessary part is already proved in Theorem 4.1. For the sufficient part, we give a sketch of the proof and leave the "textbook" proof as an exercise to the reader. The strategy remains the same as in the case of $\mathbb{R}$, except that we now have to play with multiple coordinates rather than a single coordinate. Also, analogous to breaking the intervals into subintervals, in $\mathbb{R}^n$, we can carve boxes and since a closed and bounded set must fit into one of these boxes, we can use the fact that closed subset of a compact set is compact to prove that the given set is compact. $\square$

Note that the Heine-Borel theorem works also for $\mathbb{R}$. As an application, we consider the following example.

Example 4.6 (Cantor set). Consider the cantor set in $\mathbb{R}$ constructed by the intersection $C = \bigcap_{i=1}^{\infty} C_i$, where each C_i is a union of 2^n closed subintervals of $[0, 1]$. To see the complete construction, we first divide the interval $[0, 1]$ into three equal parts and then remove the middle open interval $\left(\frac{1}{3}, \frac{2}{3}\right)$. We are left with two closed intervals $\left[0, \frac{1}{3}\right]$ and $\left[\frac{2}{3}, 1\right]$ each of length $\frac{1}{3}$. We define

$C_1 = \left[0, \frac{1}{3}\right] \cup \left[\frac{2}{3}, 1\right]$. Next, we divide each of these closed intervals into three equal parts again and then remove the middle open interval. Again, we are left with $4 = 2^2$ closed intervals, the union of which we call C_2 and the length of each remaining interval is $\frac{1}{3^2}$. We continue this process (inductively) to generate C_n and thus C as defined above.

First, notice that each C_i is closed being a finite union of closed sets, and finally C is closed, being an intersection of closed sets. Also, $C \subseteq [0, 1]$ so that C is bounded. Hence, by Heine-Borel theorem, the Cantor set is compact.

Note. It is important to recognize that the proof of compactness fails for intervals in $\mathbb{R}$ which are either not closed or not bounded. We deal with these cases as follows:

1. Suppose that an interval $I \subseteq \mathbb{R}$ is closed but not bounded, i.e., it is of the form $(-\infty, b]$, $[a, \infty)$ or $\mathbb{R}$ itself. While we can subdivide this interval, we cannot guarantee that the resulting sequence of intervals constructed are such that their lengths decrease. Therefore, the result cannot hold in this case.

2. Suppose that an interval $I \subseteq \mathbb{R}$ is bounded but not closed, i.e., it is of the form (a, b), $(a, b]$ or $[a, b)$ for $a < b$. Here, we can subdivide the interval to form a sequence I_n with decreasing lengths but the existence of an x_0 which shall be in each I_n is not guaranteed. Therefore, the result again does not hold!

It is fair at this point to wonder how could a set be closed and bounded and yet not compact. We have already seen an example, ℓ^2, where the closed and bounded set, namely the closed unit ball, was not compact. Therefore, it seems that in more general sets than $\mathbb{R}^n$ with more pathological metrics that d_2 it is possible for converse to fail. We now see some examples for the same.

Example 4.7. Consider the metric space (ℓ^∞, d_∞). Consider the closed ball $B\left(\mathbf{0}, \frac{1}{2}\right)$ in ℓ^∞. It is clear that the set is both bounded and closed. Now, consider an open cover for $\overline{B}\left(\mathbf{0}, \frac{1}{2}\right)$, given by $\mathscr{C} = \{G_n | n \in \mathbb{N}\}$, where $G_n = \left\{\mathbf{x} \in \ell^\infty \,\middle|\, |x_i| < 1 \text{ for } i < n, |x_i| < \frac{1}{2} \text{ for } i \geq n\right\}$. This family is an open cover for $\overline{B}\left(\mathbf{0}, \frac{1}{2}\right)$ with no finite subcover. Therefore, $\overline{B}\left(\mathbf{0}, \frac{1}{2}\right)$ is not compact.

Example 4.8. The reader has already proved in Exercise 4.5 that an infinite set with discrete metric is not compact. However, we see that the whole space is closed and since for any two points x, y in the space, $d(x, y) \leq 1$, the space is also bounded. Hence, this is another example where closed and bounded sets are not compact. The same technique can be used to prove that any infinite subset of a discrete metric space is not compact although it is closed and bounded.

4.4 Sequential Compactness

It seems that the motivation of defining sequential compactness in Fréchet's 1906 thesis came from his desire to generalize a theorem due to Bolzano and Weirstrass (which states that every bounded infinite set of real numbers has a cluster point) to an abstract topological (metric) space.

Definition 4.4 (Sequential compactness). *A subset A of a metric space (X, d) is said to be sequentially compact if every sequence in A has a convergent subsequence in A.*

Note. We say that every sequence in A must have a convergent subsequence "in A". This means the subsequence must have its limit inside A. Otherwise, it may be possible that there are subsequences which converge but their limits are outside A, and therefore, A may not be compact.

We usually abbreviate sequential compactness as compactness but sometimes we distinguish explicitly between the two definitions: sequential compactness and open covering compactness.

Remark. Since a metric space can always be viewed as a subset of itself, it makes sense to ask whether or not a metric space itself is compact (sequentially). The point is that unlike the definitions of open and closed, compactness is not a relative notion; we do not ask whether or not a space is compact in some larger space. For instance, we have seen that closed intervals $[a, b]$ as compact subsets of $\mathbb{R}$, but the point here is that $[a, b]$ is compact when viewed as a metric space in its own, regardless of the fact that it is a subset of $\mathbb{R}$.

So, according to this definition, a space is not too large if no matter what sample points we choose, we can always find a subsequence among them which converges.

Example 4.9. In Example 4.4, we have seen that the set $A = \left\{ \frac{1}{n} \middle| n \in \mathbb{N} \right\}$ is not compact. If we consider the points of that set itself, they form a sequence which converges to 0 in $\mathbb{R}$. However, $0 \notin A$. Since any subsequence of a convergent sequence converges to the same limit, there is no subsequence which converges in A. This is a classic example of working of the definition of sequential compactness.

Exercise 4.7. Show that the set $A = \mathbb{Q} \cap [0, 1]$ is not compact.

Exercise 4.8. Show that the Hilbert cube (Example 2.7) is compact subset of ℓ^2.

The following theorem shows that the Bolzano-Weirstrass property is equivalent to sequential compactness.

Theorem 4.6. *A metric space (X, d) is sequentially compact if and only if it has the BWP.*

Proof. First, we assume that X is sequentially compact. Let $A \subseteq X$ be an infinite set. We can extract a sequence $(x_n)_{n \in \mathbb{N}}$ of distinct points from A. Since X is sequentially compact, this sequence has a convergent subsequence, say (x_{n_k}) converging to $x \in X$. We claim that x is a cluster point of A.

Let $\epsilon > 0$. Since (x_{n_k}) converges to x, $\exists n_0 \in \mathbb{N}$ such that $\forall k \geq n_0$ we have $x_{n_k} \in B(x, \epsilon)$. Since (x_n) consists of distinct elements from A, so does (x_{n_k}). Hence, there is at least one $x_{n_k} \neq x$ in the open ball $B(x, \epsilon)$. This proves that x is a cluster point of A. Since A was arbitrary, we have proved that X has BWP.

Conversely, let us assume that X has BWP and let $(x_n)_{n \in \mathbb{N}}$ be any sequence in X. Let A denote the range set of the sequence $(x_n)_{n \in \mathbb{N}}$.

Case I: If A is finite, at least one term of the sequence (x_n) is repeated infinitely many times. The subsequence consisting of this term is constant and hence convergent in X.

Case II: If A is infinite, then by BWP, A must have a cluster point in X. Let x be the cluster point. Therefore, $\forall \epsilon > 0$, $\exists y \neq x \in A$ such that $y \in B(x, \epsilon)$. In particular, $\forall k \in \mathbb{N}$, $\exists x_{n_k} \neq x \in A$ such that $x_{n_k} \in B\left(x, \frac{1}{k}\right)$. Therefore, we get a subsequence (x_{n_k}). This subsequence has the property that $\forall k \in \mathbb{N}$, $d(x_{n_k}, x) < \frac{1}{k}$. Now, if $\epsilon > 0$ is arbitrarily chosen, then we can find a $k_0 \in \mathbb{N}$ such that $k_0 \epsilon > 1$ (by Archimedean property). Hence, $\forall k \geq k_0$, we have $d(x_{n_k}, x) < \frac{1}{k} \leq \frac{1}{k_0} < \epsilon$. Hence, the subsequence is convergent in X and X is sequentially compact. This can be seen in Figure 4.4. $\square$

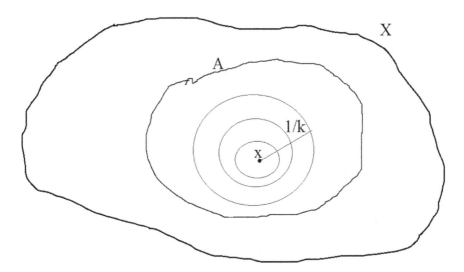

FIGURE 4.4: Capturing the (sub)sequence.

From Theorem 4.1 and Theorem 4.6, we immediately have the following corollary.

Corollary 4.6.1. *A compact metric space is sequentially compact.*

4.5 Compactness: Characterizations

In this section, we will discuss the main result of the present chapter which gives us important characterizations of compactness. To do this, we first introduce the following.

Definition 4.5 (Totally bounded). *We say that a metric space (X, d) is totally bounded if $\forall \epsilon > 0$, there exists finitely many points $x_i \in X$ for $i = 1, 2, \cdots, n$ such that*

$$X = \bigcup_{i=1}^{n} B(x_i, \epsilon).$$

Typically, the number n of points required to cover X through open balls, increases as ϵ becomes small. But no matter how small ϵ gets, for a totally bounded metric space, we shall always get finitely many points as mentioned above.

We say that a subset $A \subseteq X$ is an ϵ-net for X if given any $x \in X$, $\exists y \in A$ such that $d(x, y) < \epsilon$. Thus, we can conclude that X is totally bounded if and only if there is a finite ϵ-net for X for any $\epsilon > 0$.

Note.

1. Since a finite union of bounded sets is bounded, a totally bounded set is bounded.

2. In the metric space $(\mathbb{R}, d)$, with d given by the absolute value function, every bounded subset of $\mathbb{R}$ is totally bounded. Indeed, if $A \subseteq \mathbb{R}$ is bounded, then $A \subseteq (-\delta, \delta)$ for some $\delta > 0$. Given $\epsilon > 0$, the set A is certainly covered by the intervals $\left\{ \left(-\delta + \frac{(i-1)\epsilon}{2}, -\delta + \frac{i\epsilon}{2} \right) \Big| 1 \leq i \leq n \right\}$, where $n \in \mathbb{N}$ is such that $n\epsilon \geq 4\delta$.

3. A subset of $\mathbb{R}^n$ is bounded if and only if it is totally bounded. In fact, any bounded subset of $\mathbb{R}^n$ lies in some cube $C_n = [-N, N]^n$ for some $N \in \mathbb{N}$. But C_n can be written as a union of finitely many subcubes with arbitrarily small length and thus arbitrarily small diameter. The decomposition of a square into smaller squares of arbitrarily small lengths is shown in Figure 4.5.

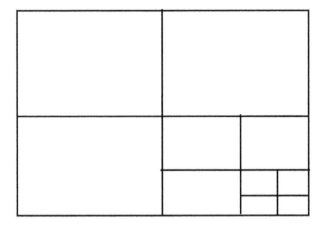

FIGURE 4.5: Decomposing a square into finitely many squares of length arbitrarily small.

Example 4.10.

1. An infinite subset of a discrete metric space is bounded but not totally bounded. This is because if x is any point in the set, then for any $\epsilon < 1$ we have $B(x, \epsilon) = \{x\}$ so that no collection of finitely many open balls can cover the infinite set.

2. Closed unit balls in ℓ^2 (Example 4.5) and $\mathscr{C}[0, 1]$ (Exercise 4.6) are bounded but not totally bounded.

Exercise 4.9. Let (X, d) be a metric space which is bounded but not totally bounded. Show that there is a positive number δ and a sequence $(x_n)_{n \in \mathbb{N}}$ in X such that $\forall n \neq m$, $d(x_n, x_m) \geq \delta$.

Exercise 4.10. Prove that every compact metric space has a finite diameter.

Exercise 4.11. Prove that a metric space is totally bounded if and only if it is bounded in every equivalent metric.

Exercise 4.12. Prove that a totally bounded metric space is separable.

Definition 4.6 (Lebesgue number). *Let (X, d) be a metric space and $\mathscr{C} = \{G_\lambda | \lambda \in \Lambda\}$ be an open cover for X. A real number $l > 0$ is called a Lebesgue number for $\mathscr{C}$ if every set $A \subseteq X$ with $\mathrm{diam}(A) < l$ is contained in G_λ for some $\lambda \in \Lambda$.*

A Lebesgue (covering) number is a certain "uniformity" across the covering. The mere fact that $\{G_\lambda | \lambda \in \Lambda\}$ covers X ensures that each point $x \in X$ belongs to at least one G_λ and since G_λ are open, each point is the center of some open ball contained in G_λ. If we now move to another point $y \in X$, we may be forced to decrease the radius of our open ball in order to squeeze it into G_λ. Under special circumstances, it may not be necessary to take radii below a certain level as we move from point to point over the entire space. That "certain level" is given by Lebesgue number.

Example 4.11. Consider an open cover $\{(-1, 2)\}$ for the set $A = [0, 1]$. Then, any real number $l > 0$ is a Lebesgue number for this cover. This is because whatever set $S \subseteq A$ we consider, whenever diam $(S) < l$, $S \subseteq (-1, 2)$ always holds.

Example 4.12. Consider the open sets $U_0 = \left(-\frac{1}{5}, \frac{1}{5}\right)$ and $U_\alpha = \left(\frac{\alpha}{2}, \frac{5}{4}\right)$ for $0 < \alpha \leq 1$. Then, the family $\{U_0\} \cup \{U_\alpha | 0 < \alpha \leq 1\}$ is an open cover for $A = [0, 1]$. Here, $l = \frac{1}{5}$ is a Lebesgue number for this cover. This is because whenever for a subset $S \subseteq A$, diam $(S) < l$ holds, its infimum exists and inf $S \geq 0$. If inf $S = 0$, then $S \subseteq \left(-\frac{1}{5}, \frac{1}{5}\right)$ and if inf $S > 0$ then $S \subseteq U_{\inf S}$.

Exercise 4.13. Let $U_0 = \left(-\frac{1}{8}, \frac{1}{8}\right)$ and $U_\alpha = \left(\frac{\alpha}{2}, 2\right)$ for $0 < \alpha \leq 1$. Find a Lebesgue number for the family $\{U_0\} \cup \{U_\alpha | 0 < \alpha \leq 1\}$ which is an open cover for $[0, 1]$.

Exercise 4.14. Is a Lebesgue number unique?

Exercise 4.15. Let $(\mathbb{R}, d)$ be the metric space with metric given by absolute value function and $A = [0, 1] \subseteq \mathbb{R}$. Find a Lebesgue number for the cover of A given by

$$\mathscr{C} = \left\{ \left(-1, \frac{1}{4}\right), \left(\frac{1}{6}, \frac{2}{3}\right), \left(\frac{1}{3}, \frac{3}{2}\right) \right\}.$$

Now to prove the characterization theorem for compact metric spaces, we need the following lemma.

Theorem 4.7 (Lebesgue covering lemma). *In a sequentially compact metric space (X, d), every open cover has a Lebesgue number. Equivalently, if a metric space has BWP, then every open cover has a Lebesgue number.*

Proof. Suppose that the theorem is not true. Let $\{G_\lambda | \lambda \in \Lambda\}$ be an open cover for X for which there is no Lebesgue number. This is to say that $\forall l > 0$, $\exists x_l \in X$ such that the open ball $B\left(x_l, \frac{l}{4}\right)$, for which diam $\left(B\left(x_l, \frac{l}{4}\right)\right) \leq \frac{l}{2} < l$ is not contained in any of the open sets G_λ. In particular, $\forall n \in \mathbb{N}$, $\exists x_n \in X$ such that $\forall \lambda \in \Lambda$, $B\left(x_n, \frac{l}{4n}\right) \not\subseteq G_\lambda$. Since X is sequentially compact, $(x_n)_{n \in \mathbb{N}}$ has a convergent subsequence and consequently, has a cluster point. Let that cluster point be named c. Since the family mentioned covers X, $\exists \lambda_0 \in \Lambda$ such that $c \in G_{\lambda_0}$. Also, G_{λ_0} being open, $\exists \delta > 0$ such that $B(c, \delta) \subseteq G_{\lambda_0}$. Also, being a cluster point, there is at least one point x_N of the sequence

$(x_n)_{n\in\mathbb{N}}$ such that $x_N \neq c$ and $x_N \in B\left(c, \frac{\delta}{2}\right)$ and $\frac{1}{N} < 2\delta$. Now, consider any $y \in B\left(x_N, \frac{1}{4N}\right)$. Then, we have

$$d(y,c) \leq d(y,x_N) + d(x_N,c) < \frac{1}{4N} + \frac{\delta}{2} < \frac{\delta}{2} + \frac{\delta}{2} = \delta.$$

Hence, $B\left(x_N, \frac{1}{4N}\right) \subseteq B(c,\delta) \subseteq G_{\lambda_0}$, which is a contradiction to the construction of the sequence $(x_n)_{n\in\mathbb{N}}$.

Therefore the theorem is correct and every open cover has a Lebesgue number. □

We now characterize all the concepts we have dealt with so far.

Theorem 4.8 (Characterization of compactness). *Let A be a subset of a metric space (X,d). Then, the following statements are equivalent:*

1. *A is compact.*

2. *A is sequentially compact.*

3. *A is totally bounded and complete.*

Proof. First let us clear the way by proving the easier implications $(1) \Rightarrow (2)$ and $(2) \Rightarrow (3)$.

Assume the conditions of (1), i.e., A is compact. Let $(a_n)_{n\in\mathbb{N}}$ be a sequence in A and suppose that this sequence does not have a convergent subsequence in A. This means either the sequence has no cluster points at all or if it has one, it is not in A. Therefore, $\forall x \in A, \exists r_x > 0$ such that $B(x,r_x)$ has at most finitely many terms of the sequence $(a_n)_{n\in\mathbb{N}}$. Also, the family $\{B(x,r_x)\,|\,x \in A\}$ is an open cover for A. Since A is compact, this has a finite subcover, say $\{B(x_k,r_{x_k})\,|\,1 \leq k \leq n_0\}$ for some $n_0 \in \mathbb{N}$.

Now, we note that since $(a_n)_{n\in\mathbb{N}}$ has no cluster point, its range cannot be finite. Also, each $B(x_k,r_{x_k})$ contains at most finitely many terms of the sequence $(a_n)_{n\in\mathbb{N}}$ so that $\bigcup_{i=1}^{n_0} B(x_i,r_{x_i})$ contains finitely many terms of the sequence. This means that either the range of the sequence is finite or the remaining terms are outside of A, both of which are contradictions!

This proves $(1) \Rightarrow (2)$.

Now, assume the conditions of (2), i.e., A is sequentially compact. Let $(a_n)_{n\in\mathbb{N}}$ be a Cauchy sequence in A. Since A is sequentially compact, the sequence $(a_n)_{n\in\mathbb{N}}$ has a convergent subsequence and hence, being Cauchy, it is convergent and in fact to the same limit as its subsequence. Therefore, A is complete.

Now, suppose that A is not totally bounded. Then, $\exists\epsilon > 0$ such that A cannot be covered by finitely many open balls. Let $a_1 \in A$ be arbitrarily chosen. Now, choose $a_2 \in A$ with the property that $a_2 \notin B(a_1,\epsilon)$. By induction, we can construct the sequence $(a_n)_{n\in\mathbb{N}}$, where $a_n \notin B(a_i,\epsilon)$, for $i = 1,2,\cdots,n-1$. Notice that the method of construction of the sequence guarantees that for any two distinct points of the sequence $(a_n)_{n \in \mathbb{N}}$, the distance is at least ϵ. Hence, this sequence cannot have a convergent subsequence, contradicting the hypothesis of (2). Therefore, A must be totally bounded and (2) $\Rightarrow$ (3) is proved!

To complete the proof of the theorem, it is enough to show (3) $\Rightarrow$ (1). This gives us the illusion that there is only one step remaining to complete the proof! Unfortunately, a direct attack on this implication seems to be hopelessly difficult. Maybe we should try establishing the implications (3) $\Rightarrow$ (2) and (2) $\Rightarrow$ (1).

The implication (3) $\Rightarrow$ (2) is not so easy; however, it is "manageable".

So, let us assume the hypothesis of (3), i.e., A is totally bounded and complete. Let $(a_n)_{n \in \mathbb{N}}$ be a sequence in A for which we have to find a cluster point. Once this is achieved, the corresponding convergent subsequence is also obtained. In order to "catch" a cluster point, we have to construct a nested sequence of "traps" $(F_k)_{k \in \mathbb{N}}$, diminishing in size. These traps will narrow down to the cluster point of the sequence.

More precisely, we define, inductively, subsets F_k for $k \in \mathbb{N}$ such that

(i) $(F_k)_{k \in \mathbb{N}}$ is decreasing in the sense $F_1 \supseteq F_2 \supseteq F_3 \supseteq \cdots$

(ii) Each F_k contains infinitely many terms of the sequence $(a_n)_{n \in \mathbb{N}}$.

(iii) The size of F_k shrinks to 0; more precisely, there is a sequence of open balls $B(x_k, r_k)$ with the property that $\lim_{k \to \infty} r_k = 0$ such that $F_k \subseteq B(x_k, r_k)$.

Set $F_1 = A$. Assume that the sets $F_1, F_2, \cdots, F_{k-1}$ have been constructed. We construct F_k in the following manner. Cover A by finitely many balls of radius $\frac{1}{k}$, say $B\left(y_1, \frac{1}{k}\right), B\left(y_2, \frac{1}{k}\right), \cdots, B\left(y_N, \frac{1}{k}\right)$. This is possible because A is totally bounded. Since F_{k-1} contains infinitely many terms of the sequence $(a_n)_{n \in \mathbb{N}}$, one of the sets $F_{k-1} \cap B\left(y_i, \frac{1}{k}\right)$, for $i = 1, 2, \cdots, N$, contains infinitely many terms of the sequence, say $F_{k-1} \cap B\left(y_j, \frac{1}{k}\right)$. Set $F_k = F_{k-1} \cap B\left(y_j, \frac{1}{k}\right)$. It is clear that the sequence $(F_k)_{k \in \mathbb{N}}$ has all the properties required.

Now, in each F_k, choose a point b_k. We claim that the sequence $(b_k)_{k \in \mathbb{N}}$ is Cauchy. In fact, by property (iii), $F_k \subseteq B(x_k, r_k)$ for each $k \in \mathbb{N}$ and $\lim_{k \to \infty} r_k = 0$. Therefore, after a certain stage $k_0 \in \mathbb{N}$ any two terms of the sequence $(b_k)_{k \in \mathbb{N}}$ have the property that $d(b_k, b_m) < 2r_k$ and hence the sequence is Cauchy.

Since A is complete, this Cauchy sequence is convergent, to say $c \in A$. Let $\delta > 0$. We will show that $B(c, \delta)$ contains infinitely many terms of $(a_n)_{n \in \mathbb{N}}$, thereby completing the proof. To prove this claim, it is enough to show that $B(c, \delta)$ contains F_k for some k. Choose a large enough $k_0 \in \mathbb{N}$ so that $d(b_{k_0}, c) < \frac{\delta}{3}$ and $r_{k_0} < \frac{\delta}{3}$. Since $b_{k_0} \in F_{k_0} \subseteq B(x_{k_0}, r_{k_0})$, we have $d(x_{k_0}, b_{k_0}) < \frac{\delta}{3}$. Hence,

$$d(x_{k_0}, c) \leq d(x_{k_0}, b_{k_0}) + d(b_{k_0}, c) < \frac{\delta}{3} + \frac{\delta}{3} = 2\frac{\delta}{3}.$$

Now, the ball $B(x_{k_0}, r_{k_0})$ has a radius less than $\frac{\delta}{3}$ and is within $2\frac{\delta}{3}$ distance of the center of the ball $B(c, \delta)$, which is a ball of radius $\delta = \frac{\delta}{3} + 2\frac{\delta}{3}$. Hence, we have $B(x_{k_0}, r_{k_0}) \subseteq B(c, \delta)$ and consequently, $F_{k_0} \subseteq B(c, \delta)$. This completes the proof $(3) \Rightarrow (2)$!

It remains to prove that $(2) \Rightarrow (1)$. This, perhaps is the most difficult part of the proof. Once we overcome this difficulty, the proof will be completed.

We assume the condition in (2), i.e., A is sequentially compact. Let $\mathscr{C} = \{G_\lambda | \lambda \in \Lambda\}$ be an open cover for A. By Lebesgue covering lemma, there is a Lebesgue number $l > 0$ for this cover. What should we do next? This is a question worth pondering. The answer to this question (given below) is rather surprising!

Earlier, we have proved that $(2) \Rightarrow (3)$. Therefore, we have condition (3) in our hands, i.e., A is totally bounded and complete. Since A is totally bounded, A can be covered by finitely many open balls, say $B(x_i, \frac{l}{4})$, for $i = 1, 2, \cdots, n_0$, for some $n_0 \in \mathbb{N}$. Now, $\text{diam}(B(x_i, \frac{l}{4})) \leq \frac{l}{2} < l$. Since l is a Lebesgue number for the covering $\mathscr{C}$, each of the open balls $B(x_i, \frac{l}{4})$ is contained in exactly one of the G_λ. Hence, the finite subcollection $\{G_{\lambda_i} | 1 \leq i \leq n_0\}$ of $\mathscr{C}$ covers A. Hence, A is compact. $\qquad\square$

Problem Set

1. Does every metric space have an open cover?

2. Find an open cover for the metric space (A, d), where $A = \{\cos x | x \in \mathbb{R}\}$ and d is the restriction of the usual metric on $\mathbb{R}$ to A.

3. For the following spaces and subsets, determine which of the sets are compact and sequentially compact:

   (a) Any metric space and any finite subset.

   (b) $\mathbb{R}$ with the usual metric and

   $\{x \in (-1, 1) | \text{ the only digits appearing in the decimal expansion of } x \text{ are } 0 \text{ and } 1\}$.

   (c) $\mathbb{C}$ with the usual metric and $\{e^{nr\pi\iota} | n \in \mathbb{Z}\}$, where r is a fixed real number.

   (d) $\mathbb{R}^2$ with the Euclidean metric and $\{(x, y) \in \mathbb{R}^2 | x + y = 1\}$.

4. Prove that the spaces $\mathbb{R}^n$, ℓ^1, ℓ^2, c_0 and ℓ^∞ are not compact. Here c_0 denotes that subset of $\mathbb{R}^{\mathbb{N}}$ which contains all sequences that converge to 0.

5. Let $X_1, X_2, \cdots, X_n$ be a finite collection of compact subsets of a metric space. Prove that $X_1 \cup X_2 \cup \cdots \cup X_n$ is compact. Show, by an example, that this cannot be generalized to infinite union.

6. Prove that a compactness is preserved in equivalent metrics.

7. Prove that the product of compact spaces is compact.

8. Is there any relation between Lebesgue number and ϵ-nets?

Biographical Notes

Pavel Sergeyevich Alexandrov (sometimes romanized Paul Alexandroff) (7 May, 1896 to 16 November, 1982) was a Soviet mathematician. He wrote about three hundred papers, making important contributions to set theory and topology. In topology, the Alexandroff compactification and the Alexandrov topology are named after him.

Pavel Samuilovich Urysohn (3 February, 1898 to 17 August, 1924) was a Soviet mathematician of Jewish origin who is best known for his contributions in dimension theory, and for developing Urysohn's metrization theorem and Urysohn's lemma, both of which are fundamental results in topology. His name is also commemorated in the terms Urysohn universal space, Frechét-Urysohn space, Menger-Urysohn dimension, and Urysohn integral equation. He and Pavel Alexandrov formulated the modern definition of compactness in 1923.

Bernard Bolzano, born Bernardus Placidus Johann Nepomuk Gonzal Bolzano; (5 October, 1781 to 18 December, 1848) was a Bohemian mathematician, logician, philosopher, theologian, and Catholic priest of Italian extraction, also known for his antimilitarist views. Bolzano made several original contributions to mathematics. His overall philosophical stance was that, contrary to much of the prevailing mathematics of the era, it was better not to introduce intuitive ideas such as time and motion into mathematics. Bolzano gave the first purely analytic proof of the fundamental theorem of algebra, which had originally been proven by Gauss from geometrical considerations. He also gave the first purely analytic proof of the intermediate value theorem (also known as Bolzano's theorem). Today he is mostly remembered for the Bolzano-Weierstrass theorem, which Karl Weierstrass developed independently and published years after Bolzano's first proof and which was initially called the Weierstrass theorem until Bolzano's earlier work was rediscovered.

Karl Theodor Wilhelm Weierstrass (31 October, 1815 to 19 February, 1897) was a German mathematician often cited as the "father of modern analysis". Despite leaving university without a degree, he studied mathematics and trained as a teacher, eventually teaching mathematics, physics, botany, and gymnastics. Weierstrass formalized the definition of the continuity of a function, proved the intermediate value theorem and the Bolzano-Weierstrass theorem, and used the latter to study the properties of continuous functions on closed and bounded intervals.

Heinrich Eduard Heine (16 March, 1821 to October, 1881) was a German mathematician. Heine became known for results on special functions and in real analysis. In particular, he authored an important treatise on spherical harmonics and Legendre functions (*Handbuch der Kugelfunctionen*). He also investigated basic hypergeometric series. He introduced the Mehler-Heine formula.

Felix Edouard Justin Emile Borel (7 January, 1871 to 3 February, 1956) was a French mathematician and politician. As a mathematician, he was known for his founding work in the areas of measure theory and probability. Along with René-Louis Baire and Henri Lebesgue, Emile Borel was among the pioneers of measure theory and its application to probability theory. The concept of a Borel set is named in his honor. One of his books on probability introduced the amusing thought experiment that entered popular culture under the name infinite monkey theorem or the like. He also published a series of papers (1921–1927) that first defined games of strategy.

Chapter 5

Connected Spaces

In the previous chapter, we studied a property of metric spaces, which we called compactness. Indeed, it was the generalization of finite sets. In this chapter, we shall study another such property, namely, connectedness. Intuitively speaking, we say that something is "connected" if it cannot be broken into two chunks far away from one another. Although this intuition appears to be vague, in the sense of the terms used, this will be our motivation in defining connectedness of a metric space. Once this is achieved, as usual, we shall start looking at some consequences and characterizations of connected spaces, along with finding connected spaces among those we know by experience. Finally, before ending the chapter we shall introduce some stronger concepts such as total connectedness and disconnectedness. Although it seems that there is nothing interesting in connectedness of a metric space (especially due to such short length of this chapter), we advise the reader to go through continuity (Chapter 6) as well, since many interesting results arise as a result of continuous functions. Also it becomes a lot easier to find (or prove) connected sets in a metric space once we have studied the concept of continuity. This chapter, though, scratches the surface of the concept of connected spaces, thereby giving a kick-start to its study.

5.1 Connectedness

As mentioned earlier, we want to say that a metric space is connected if it cannot be broken into "two chunks" far apart. However, we need to convert all the intuition into proper mathematical form. Clearly, by "breaking" our space, we mean that we would want to have the whole space as a union of two of its subsets. Also, to convert "far apart" into mathematics, we would either like to take their intersection (which should come out to be empty) or the distance between them (which should turn out to be positive). However, a major obstacle is, how should we define these "chunks"? From Chapter 2, we have been emphasizing that open sets play a major role in the study of metric spaces and their properties. Therefore, what would be more logical than to define these chunks in terms of open sets? Based on this reasoning, we now give the formal definition of connected space.

Definition 5.1 (Connected space). *A metric space (X, d) is said to be connected if it cannot be written as a union of two non-empty disjoint open sets. In other words, if $X = A \cup B$, where A and B are disjoint and open, then either $A = \emptyset$ or $B = \emptyset$.*

Immediately, by taking the negation of the statement, we see that a metric space is disconnected if it can be written as a disjoint union of non-empty open sets. That is to say, X is disconnected if there are two non-empty open sets A and B with $A \cap B = \emptyset$ and $A \cup B = X$. In such a case, the pair (A, B) is called a *disconnection* or a *separation* of X.

Let us look at a few examples of metric spaces and inspect whether they are connected or not.

Example 5.1. Recall that in a discrete space, every set is open. Thus, if X is the discrete space with at least two elements, then for any fixed $x \in X$, $\{x\}$ and $X \setminus \{x\}$ are non-empty and open. Also, they have the properties, $\{x\} \cap (X \setminus \{x\}) = \emptyset$ and $\{x\} \cup (X \setminus \{x\}) = X$. Thus, we can say that the discrete space with at least two elements is disconnected.

Example 5.2. Consider the set $\mathbb{R}^* = \mathbb{R} \setminus \{0\}$, equipped with the usual metric on $\mathbb{R}$. Since this is a subspace of $(\mathbb{R}, |\cdot|)$, the open sets in $(\mathbb{R}^*, |\cdot|)$ are precisely those obtained by intersecting open sets of $\mathbb{R}$ with $\mathbb{R}^*$. In particular, $(-\infty, 0) = (-\infty, 0) \cap \mathbb{R}^*$ and $(0, \infty) = (0, \infty) \cap \mathbb{R}^*$ are open in $(\mathbb{R}^*, |\cdot|)$. Also, $(-\infty, 0) \cap (0, \infty) = \emptyset$ and $(-\infty, 0) \cup (0, \infty) = \mathbb{R}^*$. Hence, we can conclude that $\mathbb{R}^*$ is also disconnected.

In Example 5.2, we effectively saw that if a point from whole of $\mathbb{R}$ is removed, then the resulting metric (sub)space is disconnected. However, the reader must note that it does not mean $\mathbb{R}$ will become connected if we just add this point back! To claim that $\mathbb{R}$ is connected, we will have to show that there is no disconnection possible for $\mathbb{R}$. Indeed, we shall look at this a bit later.

Exercise 5.1. Let X be finite with a metric d. Is, in general, X connected or disconnected? When can we surely say that it is connected?

Let us look with some depth at examples above. In the discrete case, we saw that the sets in disconnection were not only open but also closed. In Example 5.2, let us see if this holds. Consider the sets $A = (-\infty, 0)$ and $B = (0, \infty)$. Clearly, $A^c = B$ and $B^c = A$, in $\mathbb{R}^*$ and both (complements) are open. Hence, A and B are also closed. Therefore, we have observed an important property of disconnection! Whenever we have a disconnection in terms of two open sets, these two sets are also closed. Also, since both are non-empty and disjoint, they must be proper subsets of the whole space. We make this observation a theorem.

Theorem 5.1. *For a metric space (X, d), the following are equivalent:*

1. *X is connected.*

2. *X cannot be written as a disjoint union of non-empty closed subsets.*

3. *The only subsets of X which are both open and closed are X and $\emptyset$.*

Proof. As usual, we shall prove the statement in a cyclic manner. However, we shall do so taking the contrapositive.

Let $X = A \cup B$, where $A \cap B = \emptyset$, A and B are closed. Therefore, A^c and B^c must be open. Clearly, $A^c \cup B^c = (A \cap B)^c = (\emptyset)^c = X$. And, $A^c \cap B^c = (A \cup B)^c = X^c = \emptyset$. Thus, A^c and B^c form a disconnection of X so that X is disconnected. Therefore, we have proved $(1) \Rightarrow (2)$.

Similarly, suppose that there is a proper non-empty subset A of X, which is both open and closed. Therefore, A^c is also both open and closed. We know that $X = A \cup A^c$ so that X is written as a disjoint union of non-empty closed subsets ($A^c \neq \emptyset$, since A is a proper subset of X). Hence, we have proved $(2) \Rightarrow (3)$.

Finally, let X be disconnected so that $X = A \cup B$, where $A \cap B = \emptyset$ and A, B are open. Notice that $A^c = X \setminus A = (A \cup B) \setminus A = B \setminus A = B$, since $A \cap B = \emptyset$. Therefore, A is also closed. Since $B \neq \emptyset$, $A^c \neq \emptyset$ and thus, A is proper subset of X which is both open and closed in X, thereby proving $(3) \Rightarrow (1)$. $\qquad\square$

Let us make some analysis on the connectedness of $\mathbb{R}$, which we know by experience.

Example 5.3. Consider the metric space $(\mathbb{R}, |\cdot|)$. We want to see if this is connected. If not, then $\mathbb{R}$ will have a proper non-empty subset which is both open and closed. Let this subset be A. Notice that since A is a proper subset, there is some $x \in A^c$. Also, either $A \cap (-\infty, x] \neq \emptyset$ or $A \cap [x, \infty) \neq \emptyset$. Without loss of generality, let $A \cap (-\infty, x] \neq \emptyset$. Then, this set is bounded above (by x) and therefore has a least upper bound, say l. Since A is closed, $l \in A$. Also, l is a limit point of the interval $(l, x]$, and we also have $(l, x] \subseteq A^c$. Thus, l is a limit point of A^c. Since A is both open and closed, so is A^c and therefore, $l \in A^c$, contradicting the fact that a set and its complement can have nothing in common. Therefore, there cannot be any such set! Thus, we have proved that $\mathbb{R}$ is connected.

Let us now look at the subsets of connected spaces.

5.1.1   Connected subsets

A natural question to ask at this point is: What can we say about connectedness of subsets of a metric space? To answer this question, let us first understand that the subspace topology, induced from the parent metric, effectively does not change the structure of the subset under consideration. Thus, for any subset, if we just look at its subspace topology, we can answer if it is connected or not. Therefore, we make the following definition.

Definition 5.2 (Connected subset). *A subset $S \subseteq X$, where (X, d) is a metric space, is connected if the metric (sub)space (S, d) is connected.*

Looking at the definition, we can ask if connectedness can be inherited. That is, if X is connected, then will all its subsets be connected? Conversely, we can also ask if there are any connected subsets in a disconnected space.

Looking at Examples 5.2 and 5.3, we can immediately say that a subset of connected space need not be connected. Therefore, connectedness is not an inheritable property. Also, from Exercise 5.1, the reader must have noticed that singleton sets are connected, so that disconnected spaces may have connected subsets.

Given a set X, we immediately have at least two subsets, namely $\emptyset$ and X itself. Previously, we have talked about connectedness of X. Now, we shall address the question on the connectedness of $\emptyset$. We claim that it is connected. To see that this claim is true, suppose that someone wants to prove us wrong. Then, that person would have to come up with two non-empty disjoint subsets of $\emptyset$, whose union is $\emptyset$. Since is this not possible, our claim must be true! (We remind the reader that such statements are vacuously true, i.e., they are true because the premise itself does not hold.)

In the exercise that follows, the reader is supposed to inspect subsets of $\mathbb{R}$ for connectedness.

Exercise 5.2. Check if the sets $\mathbb{N}$ and $\mathbb{Z}$ are connected in $\mathbb{R}$. **Hint:** First see what are all the open sets inherited in the subspace topology.

Example 5.4. Let us see if $\mathbb{Q}$ is connected. Let $z \in \mathbb{Q}^c$. Consider two sets $A = \mathbb{Q} \cap (-\infty, z]$ and $B = \mathbb{Q} \cap [z, \infty)$. We claim that $A \cup B = \mathbb{Q}$. To see this, first observe that $A \cup B \subseteq \mathbb{Q}$, by their definition. Let $x \in \mathbb{Q}$. Since the law of trichotomy[1] holds in $\mathbb{R}$, and we know that $x \neq z$, either $x < z$ or $x > z$. Therefore, either $x \in A$ or $x \in B$. This proves that $\mathbb{Q} = A \cup B$.

Now, $A \cap B = (\mathbb{Q} \cap (-\infty, z]) \cap (\mathbb{Q} \cap [z, \infty)) = \mathbb{Q} \cap ((-\infty, z] \cap [z, \infty)) = \mathbb{Q} \cap \{z\} = \emptyset$. Also, since $(-\infty, z]$ and $[z, \infty)$ are closed in $\mathbb{R}$, A and B are also closed in $\mathbb{Q}$ (in the subspace topology). Therefore, by Theorem 5.1, we can conclude that $\mathbb{Q}$ is disconnected.

Exercise 5.3. Is $\mathbb{Q}^c$ connected in $\mathbb{R}$?

From the above example and exercises, it seems that the subsets of $\mathbb{R}$ are failing to be connected because either they do not have any accumulation points or their accumulation points are outside of the sets. Let us look at one

---

[1] The law of trichotomy suggests that given any two elements in $\mathbb{R}$, exactly one of the following three things hold: $x = y$, $x < y$ or $x > y$. In fact, this can be true for any totally ordered set.

example of a set, for which the accumulation points are inside the set and then comment upon its connectedness.

Example 5.5. Consider the set $A = \left\{ \frac{1}{n} \middle| n \in \mathbb{N} \right\} \cup \{0\}$. First, we shall find the open sets in $(A, |\cdot|)$. To do so, we shall simply see how open balls look.

First, consider a point $\frac{1}{n} \in A \setminus \{1\}$. Then, we know that

$$\frac{1}{n+1} < \frac{1}{n} < \frac{1}{n-1}.$$

Also,

$$\frac{1}{n} - \frac{1}{n+1} = \frac{1}{n(n+1)} < \frac{1}{n(n-1)} = \frac{1}{n-1} - \frac{1}{n}.$$

Therefore, consider the open ball

$$B\left(\frac{1}{n}, \frac{1}{n(n+1)}\right) = \left\{ x \in A \middle| \left| x - \frac{1}{n} \right| < \frac{1}{n(n+1)} \right\}.$$

Clearly, $B\left(\frac{1}{n}, \frac{1}{n(n+1)}\right) = \left\{\frac{1}{n}\right\}$ (verify!). Similarly, for the singleton set $\{1\}$, we have $B\left(1, \frac{1}{2}\right) = \{1\}$ (verify!). Therefore, it seems that the induced metric is equivalent to the discrete metric (since all singletons are open). However, we have not looked at one singleton set, namely $\{0\}$. If we are able to prove that this is open, then indeed it would have the open sets the same as those in discrete spaces and thus A would immediately become disconnected.

Consider any open ball $B(0, \epsilon)$. Then, by the Archimedean property, there is some $n \in \mathbb{N}$ such that $n\epsilon > 1$, or $\frac{1}{n} < \epsilon$ so that $\frac{1}{n} \in B(0, \epsilon)$. In particular, $\{0\}$ is not open! So, is A connected?

Intuitively, it does not seem so! This is because we can actually "see" elements of A far apart from one another. Let us now try to prove our intuition.

Consider the open ball $B(0, 1)$. Clearly, $B(0, 1) = A \setminus \{1\}$ (verify!). Since open balls are open, this set is open. Also, $A = (A \setminus \{1\}) \cup \{1\}$, where $\{1\}$ is open. Hence, there is a disconnection of A and A is not connected.

This process can be seen in Figure 5.1.

Indeed, we have not found many connected sets in $\mathbb{R}$. A natural question arises: Are there any connected subsets of $\mathbb{R}$? We address this question in the next result.

Theorem 5.2. *A subset of $\mathbb{R}$ is connected if and only if it is an interval.*

Proof. Suppose that a subset J of $\mathbb{R}$ is not an interval. Then, there are elements $x, y \in J$ and $z \in J^c$ such that $x < z < y$. Now, consider the two sets $A = J \cap (-\infty, z)$ and $B = J \cap (z, \infty)$. Clearly, A and B are open and disjoint in J. Also, $A \cup B \subseteq J$. Now, let $a \in J$. Since $z \notin J$, $a \neq z$. Also, by the law of trichotomy in $\mathbb{R}$, either $a < z$ or $a > z$, i.e., $a \in A$ or $a \in B$. Therefore, we have $J = A \cup B$ and hence is not connected. Therefore, we have proved that

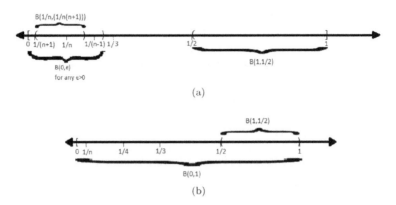

FIGURE 5.1: The working of Example 5.5: (a) Constructing open balls around each point in A and (b) finding a disconnection of A.

if a subset of $\mathbb{R}$ is not an interval, then it is disconnected. Equivalently, if a subset of $\mathbb{R}$ is connected, then it is an interval.

Conversely, let a subset J of $\mathbb{R}$ be an interval, and let, if possible, J be disconnected. Then, there is a separation A and B of J, where A and B are non-empty, closed and disjoint. Consider $x \in A$ and $y \in B$. Without loss of generality, let $x < y$. Since J is an interval, $[x, y] \subseteq J$ and consider the sets $A_1 = A \cap [x, y]$ and $B_1 = B \cap [x, y]$. Since A_1 is bounded above, it has a least upper bound, say l. Since A is closed and $[x, y]$ is closed, so is A_1 and therefore, $l \in A_1$. Clearly, for any $z > l$, $z \notin A_1$. Since J is an interval, all z with the property $l < z < y$ must be in J. Also, $z \notin A_1$ as mentioned above so that $z \in B_1$. Now, let $\epsilon > 0$ and consider $l + \epsilon$. Clearly, $l < \frac{l + \min\{y, l+\epsilon\}}{2} < y$ so that l is the greatest lower bound for B_1. Again, B_1 is closed so that $l \in B_1$, which is a contradiction to $A_1 \cap B_1 = \emptyset$. Therefore, an interval is connected.   □

The above results give all the connected sets in $\mathbb{R}$. Indeed, $\mathbb{R}$ itself is an interval and is connected!

Talking about connected subsets, the thought of operating (unions, intersections) connected subsets comes to the mind. Therefore, let us see if the unions and intersections of connected sets remain connected. Looking at Example 5.2, we can say that $\mathbb{R}^* = (-\infty, 0) \cup (0, \infty)$ is not connected, but $(-\infty, 0)$ and $(0, \infty)$ are connected (being intervals). Therefore, in general, the union of connected sets need not be connected.

We will look at some results about the unions and intersections of connected sets once we have dealt with continuity and have seen what we call the "magic wand" of connectedness.

Exercise 5.4. Is the intersection of two connected sets connected? **Hint:** Take two non-concentric circles which intersect at at least two points.

Let us see a result concerning a connected subset of a disconnected space.

Theorem 5.3. *If $X = A \cup B$, where $A \cap B = \emptyset$ and A, B are closed, then for any connected subset E of X, $E \subseteq A$ or $E \subseteq B$.*

Proof. Let if possible, $E \cap A \neq \emptyset$ and $E \cap B \neq \emptyset$. Let $P = E \cap A$ and $Q = E \cap B$. Since A and B are closed in X, P and Q are closed in E with the property that $P \cap Q = \emptyset$. Now,

$$E = E \cap X$$
$$= E \cap (A \cup B)$$
$$= (E \cap A) \cup (E \cap B)$$
$$= P \cup Q.$$

Therefore, P and Q forms a separation of E, which is a contradiction to the fact that E is connected. Therefore, the theorem must be true! $\square$

In all the above examples and results, we saw that whenever there was a disconnection, there were two points x and y in two separate sets. Now, we ask the question that if we consider two distinct points in a connected metric space, what can be said about the sets which contain both of them?

Theorem 5.4. *A metric space (X, d) is connected if and only if for any pair of distinct points, there is a connected set containing the two points.*

Proof. First, if X is connected and $x \neq y$ are two points in X, clearly, $x, y \in X$. Here X is the required connected set, thereby proving the first part of the proof.

Conversely, let for any $x \neq y$ there is a connected set S containing the two points. Let, if possible, X be disconnected. Then there is a separation A and B of X. Consider the points $x \in A$ and $y \in B \cap (X \setminus \{x\})$. Clearly, $x \neq y$ and by hypothesis, there is a connected set S such that $\{x, y\} \subseteq S$. Now, since S is connected, by Theorem 5.3, $S \subseteq A$ or $S \subseteq B$, which is a contradiction! Hence, X must be connected. $\square$

Theorem 5.5. *Let (X, d) be a metric space and let S be a connected set in X. Then, any subset A with the property $S \subseteq A \subseteq \bar{S}$ is also connected.*

Proof. Let, if possible, A be disconnected and have the property as given in the statement of the theorem. Then, there is a disconnection of A, say, P and Q. Here, $A = P \cup Q$, P and Q are open in (A, d), non-empty, and disjoint. Therefore, there are open sets P_1, Q_1 of X such that $P = P_1 \cap A$ and $Q = Q_1 \cap A$. Thus, we may write $A = (P_1 \cap A) \cup (Q_1 \cap A)$. Now, $S \subseteq A$ implies $S \cap A = S$. Therefore, $S = S \cap ((P_1 \cap A) \cup (Q_1 \cap A)) = (S \cap (P_1 \cap A)) \cup (S \cap (Q_1 \cap A)) = (P_1 \cap S) \cup (Q_1 \cap S)$. Since P_1 and Q_1 are open in X, $P_1 \cap S$ and $Q_1 \cap S$ are open in (S, d). Thus, there will be a disconnection of S if these sets are non-empty and disjoint. The proof that these are disjoint follows from the fact that P and Q are disjoint. Therefore, we shall only prove that these are non-empty.

To see that these are non-empty, first we observe that since $P_1 \cap A \neq \emptyset$, $P_1 \cap \bar{S} \neq \emptyset$. Let $x \in P_1 \cap \bar{S}$. Therefore, x is a limit point of S and P_1 is an open neighborhood of x. Thus, $P_1 \cap S \neq \emptyset$. Similarly, $Q_1 \cap S \neq \emptyset$. This finally implies that S is disconnected, which is a contradiction! Hence, A must be connected. $\qquad\qquad\square$

Corollary 5.5.1. *Closure of a connected set is connected.*

Let us ask the converse question for Corollary 5.5.1, i.e., if the closure of a set is connected, can we say that the set itself is connected? To answer this question, consider Example 5.2. Clearly, $\mathbb{R}^*$ is disconnected in $\mathbb{R}$ and $\overline{\mathbb{R}^*} = \mathbb{R}$, which is connected. Therefore, the converse of Corollary 5.5.1 is, in general, not true.

Exercise 5.5. Prove that if in a metric space (X, d), an open set G is disconnected, then its disconnection is open in X.

5.2 Components

In the previous section, we saw a few examples of connected and disconnected sets. We make a small observation here: In Example 5.1, we saw that any set containing two or more points is disconnected in the discrete metric. On the other hand, the singleton sets in discrete space were connected, by Exercise 5.1. Similarly, in Example 5.2, we saw that the two intervals $(-\infty, 0)$ and $(0, \infty)$ were connected but any set properly containing them was disconnected. Therefore, in a sense, these connected sets were largest. Although, we know that using largest is wrong! This is because while we order sets with $\subseteq$, we do not have a total order and therefore we do not know if these sets are really larger than every other connected set. However, we do know that these are maximal (see Chapter 1, for reference). Based on this observation, we now define what is called connected components of a metric space.

Definition 5.3 (Connected components). *In a metric space (X, d), a set $A \subseteq X$ is said to be a connected component of X if it is a maximally connected set in X. In other words if $A \subseteq B$ and B is connected, then $A = B$.*

Note. Henceforth, we shall not use the phrase "connected components" whenever it can be avoided. Rather we will use only "components", and the reader must understand that they are connected.

It is clear that the components of the discrete space are singleton sets and that of $\mathbb{R}^*$ are $(-\infty, 0)$ and $(0, \infty)$.

Exercise 5.6. What are the components of a connected metric space?

It seems that it is not interesting to find components in a connected metric space. Therefore, we shall look at a few disconnected spaces and see what could its components be.

Example 5.6. From Example 5.5, we take the set A and try to find its components. First, we observe that $A \setminus \{0\}$ has a topology equivalent to the discrete one, so that each of $\left\{\frac{1}{n}\right\}$ is a component. Indeed, if we take any set $S \subseteq A \setminus \{0\}$, which is not singleton, then it can be written as a union of two open sets so that it is disconnected, while singletons are connected (from Exercise 5.1).

Therefore, let us look at any set S which properly contains $\{0\}$. In that case, there is some $n \in \mathbb{N}$ such that $0, \frac{1}{n} \in S$. Therefore, consider the open ball $B\left(0, \frac{1}{n}\right)$ which is open in S and the set $\left\{\frac{1}{k} \middle| k \leq n\right\} \cap S$, which is also open in S (why?). We leave it to the reader to verify that these two sets actually form the disconnection for S so that any set properly containing $\{0\}$ is not connected. Therefore, $\{0\}$ is also a component.

Remark. For a first time reader, it is natural to think that only discrete space has the property that singletons are the only components. However, the above example shows that it is not so. We advise the reader to take some time and go through the proof thoroughly!

Exercise 5.7. What are the components of $\mathbb{Q}$? **Hint:** Use density of rationals and irrationals in $\mathbb{R}$.

Let us look at a few properties of components and how we can write the whole space in terms of its components.

Theorem 5.6. *Let (X, d) be a metric space and $\mathscr{C}$ be a collection of all components of X. Then,*

1. *For $A, B \in \mathscr{C}$, either $A = B$ or $A \cap B = \emptyset$.*

2. *Each component of X is closed.*

3. $X = \bigcup_{A \in \mathscr{C}} A.$

Proof. To prove the theorem, we will assume a result, which we will prove in Chapter 6 (Corollary 6.25.1). We state the result here: "A union of connected sets with non-empty intersection is connected."

1. Let $A, B \in \mathscr{C}$. If $A \cap B = \emptyset$, then we are through! Therefore, assume that $A \cap B \neq \emptyset$. Since A and B are components, they are connected and by the statement mentioned above, we have $A \cup B$ is connected. However, we have $A \subseteq A \cup B$ and $B \subseteq A \cup B$ so that by the definition of components, $A = A \cup B = B$.

2. Let A be a component of X. We know that $A \subseteq \bar{A}$ and by Corollary 5.5.1, we also know that $\bar{A}$ is connected. Therefore, by the definition of components, $A = \bar{A}$ so that A is closed.

3. It is clear that $\bigcup\limits_{A \in \mathscr{C}} A \subseteq X$. Therefore, let $x \in X$. We shall show that it must lie in some component of X. Consider the set

$$A_x = \{S \subseteq X | S \text{ is connected and } x \in S\}.$$

Then, $A_x \neq \emptyset$ since $\{x\} \in A_x$. Consider

$$A = \bigcup_{S \in A_x} S.$$

We will show that A is a component of X. Let B be a connected set such that $A \subseteq B$. Then, we know that $x \in A \subseteq B$ so that $B \in A_x$. Thus, $A = B$. Also, from the result mentioned above, A is connected since $x \in \bigcap\limits_{S \in A_x} S$. Therefore, A is indeed a component and the result is proved!

$\square$

A natural question to ask is: Are components open? We leave it as an exercise to the reader to answer this question, by looking at the examples we have discussed for components.

Exercise 5.8. Are the components of any metric space necessarily open?

5.3   Totally Disconnected Spaces

We have studied connectedness of metric spaces to some depth (if not much), and it has been made clear what components do. To quickly summarize, a disconnected space is "shattered" into its components. While they are "far apart", they still make the whole space, just not in a single piece. Therefore, it is very natural to look at a highly shattered space. How will this space look? Although we cannot experiment on metric spaces, smash them, and see what remains when they are completely shattered, intuition can help us. Based on our intuition, we shall define our highly shattered spaces and call them totally disconnected.

Definition 5.4 (Totally disconnected spaces). *A metric space (X, d) is totally disconnected if the only components are singleton sets.*

From the examples studied above, it is clear that any discrete space and the set A from Example 5.5 are totally disconnected. We shall see one more example for totally disconnected spaces.

Example 5.7 (Cantor's set). Recall the construction of Cantor's set C from Example 4.6 (Chatper 4). Let S be a non-empty connected subset of C. Then, $S \in C_n$ for each n. Also, each C_n is a union of pairwise disjoint closed intervals so that C_n itself is disconnected. Then, by Theorem 5.3, we can conclude that S lies in exactly one of these subintervals of length $\frac{1}{3^n}$. Since this is true for all n, we must have diam $(S) \leq \frac{1}{3^n}$ for each $n \in \mathbb{N}$ so that diam $(S) = 0$, i.e., S is a singleton set. Since the only connected sets in C are singletons, they are also the components. Hence, the Cantor's set C is totally disconnected.

Problem Set

1. What can we say about a non-empty connected subset of $\mathbb{R}$ which contains only rational points?

2. Give an example of nested sets A_n, i.e., $A_n \subseteq A_{n-1}$ for each $n \in \mathbb{N}$, such that each A_n is connected but $\bigcap_{n \in \mathbb{N}} A_n$ is not connected.

Chapter 6

Continuity

Around the secondary school level, when we were being taught the concept of functions for the first time, a major thing we used to do is to graph them. When we graphed them, consciously or unconsciously, we made certain decisions about the function being "good" or "not good". And as it has been observed, as students, we (used to) like the functions which we could graph *"without lifting our pencil"*.

Later in our studies, maybe in higher secondary education, most of us came across the mathematical concept of "not lifting the pencil". Rigorously, that was called continuity. A very vague and not-so-rigorous definition was given to us at that time by means of limits.

Those who have taken a course in analysis might have already encountered the rigorous definition of continuity in terms of epsilon and delta.

As we have been doing so far, we will take into account what we know from our experience in the real numbers, and since we know that $\mathbb{R}$ is a metric space, try to generalize it for any metric space. In this chapter, we shall see what we mean by continuity of functions on arbitrary metric spaces. However, first we review what we know from our experience of real numbers.

6.1 Continuity of Real Valued Functions

As mentioned earlier, we were first introduced to continuity, by means of limits. Let us recall that if $f : D \to \mathbb{R}$ is a (real-valued) function, where $D \subseteq \mathbb{R}$ is its domain, then we say that f is continuous *at a point* $x_0 \in D$ if the following holds:

1. $f(x_0)$ is well-defined.

2. $\lim\limits_{x \to x_0} f(x)$ exists.

3. $\lim\limits_{x \to x_0} f(x) = f(x_0)$.

Later, we were introduced to the rigorous definition (in a course of analysis), which is as follows.

Definition 6.1 (Continuity). *A function* $f : D \subseteq \mathbb{R} \to \mathbb{R}$ *is said to be continuous at a point* $x_0 \in D$, *if* $\forall \epsilon > 0$, $\exists \delta > 0$ *such that* $\forall x \in D$ *with* $|x - x_0| < \delta$, *we have* $|f(x) - f(x_0)| < \epsilon$.

Before moving toward understanding the definition, let us note that continuity is a *local* property, i.e., it is defined at a particular point in the domain. If a function is continuous at every point in its domain, we say that it is *continuous throughout the domain*. Therefore, whenever we are asked about the continuity of a function, we must ask back: At what point in the domain?

Now, let us look at what this means geometrically. Since for real valued functions, we can *graph* the function on a plane, we try to understand continuity through it. First, we give a small analogy for the situation. Suppose that f is a machine which gives the product $f(x)$ when we supply it with x. As the creators of this machine, we know that if we put exactly x_0, we shall get the exact product $f(x_0)$. However, the customer (or the manufacturer who is using this machine) wants to get *as close as possible* to the exact product. Therefore, they give us the error tolerance (ϵ) and say that the product should not deviate more than this value from the actual (exact) product. In return, we (creators) give the customer a condition (δ) and tell them that if they supply within the bounds of δ, their error will be managed.

Moving on to the graph of function f, suppose that we want the error to be ϵ about the point $f(x_0)$. Suppose we are able to *find* a δ so that whenever the input is within an error range of δ from x_0, the output is also within an error range of ϵ from $f(x_0)$. Since $\epsilon > 0$ is arbitrary, this means that we can get as close to $f(x_0)$ as possible. This will be our main motivation to define continuity for an arbitrary metric space. A pictorial representation of this concept is given in Figure 6.1.

Also, we can get the definition of discontinuity by just negating the definition of continuity, i.e., if we are able to find an $\epsilon > 0$ such that no matter how close we get our inputs to x_0, the error is at least ϵ for some input, then we can say that the function is discontinuous at x_0. Mathematically, this can be written as

$$\exists \epsilon > 0 \text{ such that } \forall \delta > 0, \exists x \in D \text{ with } |x - x_0| < \delta \text{ and } |f(x) - f(x_0)| \geq \epsilon.$$

We believe that the reader is thorough with continuity in $\mathbb{R}$ and therefore we do not give its examples. Rather, we now move on to define continuous functions on arbitrary metric spaces.

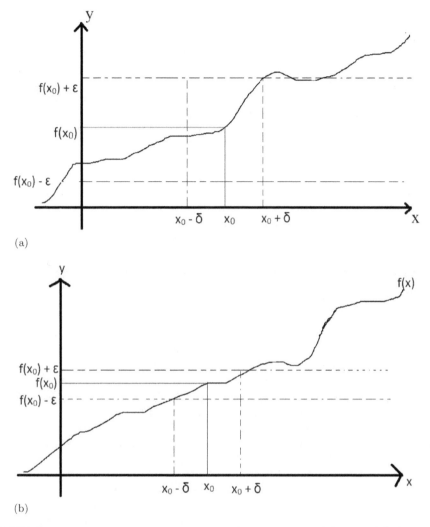

FIGURE 6.1: Pictorial representation of dependence of δ on ϵ. (a) Choice of δ with "large" ϵ. (b) Choice of δ with "small" ϵ.

6.2 Continuous Functions in Arbitrary Metric Spaces

In the previous section, we looked at the meaning of continuity in $\mathbb{R}$ and we wanted to take that as a motivation for generalizing the definition to arbitrary metric spaces. First, let us ask ourselves: What are the things that are required? We must be given an error range ϵ for elements in the codomain for which we have to find an error range δ for elements in domain. Also, those error ranges cannot be directly applied to the elements of domain and codomain. Recall that $|\cdot|$ defines a metric on $\mathbb{R}$. Therefore, these error ranges are for the distances between the elements of the domain and codomain. With all the motivation we obtained from the description in the previous section, we now define continuity as follows:

Definition 6.2 (Continuity). *A function $f : X \to Y$ defined from a metric space (X, d_X) to another metric space (Y, d_Y) is said to be continuous at a point $x_0 \in X$, if $\forall \epsilon > 0$, $\exists \delta > 0$ such that $\forall x \in X$ with $d_X(x, x_0) < \delta$, we have $d_Y(f(x), f(x_0)) < \epsilon$.*

Remark. Notice that both the domain and the codomain must be metric spaces to talk about continuity. If any of them does not have a metric, then we cannot comment on the continuity of a function, since we do not know (at least for now) the meaning of "closeness" without a metric.

Now, let us look at some continuous functions that we can have on metric spaces.

Example 6.1 (Constant function). Consider metric spaces (X, d_X) and (Y, d_Y). Fix $y_0 \in Y$ and define the constant function $f : X \to Y$ as $f(x) = y_0$, $\forall x \in X$. We know that in $\mathbb{R}$, constant functions are continuous. Can we have the same in this case? To answer this question, fix $x_0 \in X$ and let $\epsilon > 0$ be given. We need to *find* a $\delta > 0$ such that whenever we have $x \in X$ with $d_X(x, x_0) < \delta$, we also have $d_Y(f(x), f(x_0)) < \epsilon$. Note that f is a constant function and therefore, $\forall x \in X$, $f(x) = f(x_0) = y_0$. Hence, $d_Y(f(x), f(x_0)) = 0 < \epsilon$ for any $x \in X$. Therefore, in particular, if we choose $\delta = 1$, we have continuity of f at $x_0 \in X$. Also, the choice of x_0 was arbitrary, which makes the function f continuous throughout X.

Remark. In the above example, any $\delta > 0$ works. However, for a textbook proof, we always *need* to mention what δ are we choosing. In this case, we have chosen $\delta = 1$. The reader may choose $\delta = \frac{1}{2}$ or $\delta = 2$ or any other positive real number.

Example 6.2 (Identity function). Consider a metric space (X, d) and the identity function $id : X \to X$ as $id(x) = x$, $\forall x \in X$. Fix an $x_0 \in X$ and let $\epsilon > 0$ be given. Again, we need to find a $\delta > 0$ which will help us prove the continuity. Observe that $d(id(x), id(x_0)) = d(x, x_0)$. Therefore, if we choose $\delta = \epsilon$, we are through! Hence, identity function on a metric space is continuous.

Example 6.3. Recall that in a metric space (X, d), the distance of a point $x \in X$ from a set $A \subseteq X$, is given by

$$d(x, A) = \inf \{d(x, a) \mid a \in A\}.$$

If we fix the subset A, then the above expression is a function of x alone. We denote it as $d_A : X \to \mathbb{R}$. Now, we want to check its continuity. Here, we shall use the usual metric ($|\cdot|$) on $\mathbb{R}$. First, we shall see certain inequalities that follow.

Let $x, y \in X$ and $a \in A$ be arbitrarily chosen. Then, by triangle inequality, we have

$$d(a, x) \le d(a, y) + d(x, y),$$

or, what is same,

$$d(a, y) \ge d(a, x) - d(x, y).$$

However, $d(a, x) \ge d_A(x)$, so that we have the inequality,

$$d(a, y) \ge d_A(x) - d(x, y).$$

Hence, $d_A(x) - d(x, y)$ is a lower bound to the set $\{d(a, y) | a \in A\}$, so that

$$d_A(y) \ge d_A(x) - d(x, y)$$

or, what is the same,

$$d_A(x) - d_A(y) \le d(x, y).$$

Interchanging the roles of x and y, we also arrive at

$$d_A(y) - d_A(x) \le d(x, y).$$

This leads us to

$$|d_A(x) - d_A(y)| \le d(x, y)$$

so that if $\epsilon > 0$ is given, we can choose $\delta = \epsilon$ and whenever $d(x, y) < \delta$, we have

$$|d_A(x) - d_A(y)| \le d(x, y) < \delta = \epsilon.$$

Hence, the function d_A is continuous on X. In fact, we can comment something more on this function, but we shall keep it until later when we would be equipped with the appropriate tools.

Remark. Example 6.3 shows that there are *abundant* non-constant continuous functions from a metric space to $\mathbb{R}$. In fact, given any two points $x, y \in X$, we can construct a function $f : X \to \mathbb{R}$ such that $f(x) = 0$ and $f(y) = 1$. Hence, we now have a method to *separate* two points in a metric space through continuous functions. But, this is not it! We can also separate two disjoint closed sets in a similar manner. We shall see this later in this text.

Exercise 6.1. Let $x \in \mathbb{R}$. What is the value of $d_{\mathbb{Q}}(x)$?

We have seen in Chapter 3 the concept of an isometry between metric spaces. Let us quickly recall the definition. A function $f : (X, d_X) \to (Y, d_Y)$ is an isometry if for all $x, y \in X$, we have $d_Y(f(x), f(y)) = d_X(x, y)$. Thus, if we want to control the distance between $f(x)$ and $f(y)$, all we need to do is control it in x and y by the same amount. Thus, it tells us that any isometry is continuous.

Writing this with mathematical rigor, we see that given an $\epsilon > 0$, we can choose $\delta = \epsilon$ so that whenever $d_X(x, y) < \delta$, we immediately have $d_Y(f(x), f(y)) < \epsilon$. In other words, f is continuous. We make this observation as a result.

Theorem 6.1. *Any isometry between two metric spaces is continuous.*

Before moving ahead, we would like to comment upon two important functions in $\mathbb{R}^n$, namely rotations and translations. First we will define them and then will comment upon their continuity.

Definition 6.3 (Translation). *A translation $T_a : \mathbb{R}^n \to \mathbb{R}^n$ is defined as*

$$T_a(x_1, x_2, \cdots, x_n) = (x_1 + a_1, x_2 + a_2, \cdots, x_n + a_n)$$

where $a = (a_1, a_2, \cdots, a_n) \in \mathbb{R}^n$ is a fixed vector.

To define rotation in $\mathbb{R}^n$ for $n \geq 3$, a bit of knowledge from inner product spaces is required. Therefore, we shall be satisfied by defining rotations in $\mathbb{R}^2$, where we have the knowledge of X- and Y-axes and their property of being perpendicular.

Definition 6.4 (Rotation). *A rotation in $\mathbb{R}^2$ by an angle $\theta \in [0, 2\pi)$ is defined as $R_\theta : \mathbb{R}^2 \to \mathbb{R}^2$, where*

$$R_\theta(x, y) = (x \cos \theta + y \sin \theta, -x \sin \theta + y \cos \theta).$$

The geometric representation of translation and rotation is shown in Figure 6.2.

Now, let us look at a major property of translations and rotations in the Euclidean metric. Notice that for translations, we can use any other metric as well (this will be clear in the explanation that follows). However, for rotations we will not be able to do so (since we will be wanting to use the identities of sine and cosine which include square terms). Also, we know that on $\mathbb{R}^n$,

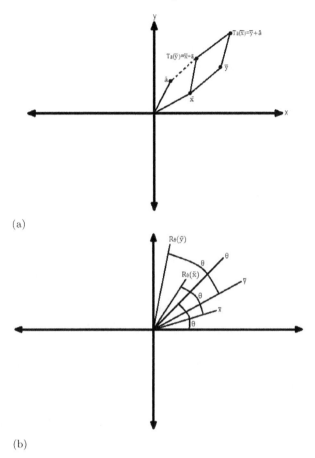

(a)

(b)

FIGURE 6.2: Geometric representation of translations and rotations in $\mathbb{R}^2$. (a) Translation in $\mathbb{R}^2$. (b) Rotation in $\mathbb{R}^2$.

any two p-metrics are equivalent. Therefore, the properties (of continuity) of translation and rotations will remain same in all other metrics as well.

Consider two points $\mathbf{x} = (x_1, x_2, \cdots, x_n), \mathbf{y} = (y_1, y_2, \cdots, y_n) \in \mathbb{R}^n$, where $\mathbb{R}^n$ is equipped with the Euclidean metric d_2. Therefore, we have

$$d_2\left(T_{\mathbf{a}}\left(\mathbf{x}\right), T_{\mathbf{a}}\left(\mathbf{y}\right)\right) = \sqrt{\sum_{i=1}^{n} |(x_i + a_i) - (y_i + a_i)|^2}$$

$$= \sqrt{\sum_{i=1}^{n} |x_i - y_i|^2}$$

$$= d_2\left(\mathbf{x}, \mathbf{y}\right).$$

Similarly, we have for two points $\mathbf{x} = (x_1, x_2)$, $\mathbf{y} = (y_1, y_2) \in \mathbb{R}^2$,

$$d_2\left(R_\theta\left(\mathbf{x}\right), R_\theta\left(\mathbf{y}\right)\right)$$
$$= \left(\left|(x_1 \cos\theta + x_2 \sin\theta) - (y_1 \cos\theta + y_2 \sin\theta)\right|^2 + \right.$$
$$\left. \left|(-x_1 \sin\theta + x_2 \cos\theta) - (-y_1 \sin\theta + y_2 \cos\theta)\right|^2\right)^{\frac{1}{2}}$$
$$= \left(\left|x_1 - y_1\right|^2 + \left|x_2 - y_2\right|^2\right)^{\frac{1}{2}}$$
$$= d_2\left(\mathbf{x}, \mathbf{y}\right).$$

With this we have proved that translations and rotations are isometries and hence by Theorem 6.1, we can conclude that translations and rotations are continuous mappings.

Although this definition of continuity helps us to prove the continuity of functions, it can sometimes be tedious and really difficult to deal with. One such example is the square function in real numbers. While it is possible to prove the continuity through the ϵ-δ definition, it becomes tedious and time consuming. Therefore, we now move to some other characterizations of continuity, which can also be considered as equivalent definitions.

6.2.1 Equivalent definitions of continuity and other characterizations

We know that to define continuity, we have used the notion of "getting close to the exact value". In sequences, we have seen that convergent sequences also "get close enough" to some point in the metric space. Also, every metric space has some "special sets" which we call open and these are, in fact, a result of defining a metric. Therefore, the equivalent definitions of continuity will be given in terms of these concepts.

Theorem 6.2 (Equivalent definitions of continuity). *Let (X, d_X) and (Y, d_Y) be metric spaces. Let $f : X \to Y$ be defined. Then, the following are equivalent.*

1. *For every sequence (x_n) in X such that $x_n \to x$ for some $x \in X$, we also have $f(x_n) \to f(x)$ in Y.*

2. *f is continuous at $x \in X$.*

3. *Given an open set $V \subseteq Y$ containing $f(x)$, there is an open set $U \subseteq X$ containing x such that $f(U) \subseteq V$.*

Proof. We will prove the theorem in a cyclic manner: $(1) \Rightarrow (2) \Rightarrow (3) \Rightarrow (1)$.

To prove $(1) \Rightarrow (2)$, we shall use the method of contradiction. Therefore, assume that (1) is true, but f is not continuous at x. Therefore, $\exists \epsilon > 0$ such that $\forall \delta > 0$, $\exists x' \in X$ with $d_X(x', x) < \delta$ and $d_Y(f(x'), f(x)) \geq \epsilon$. In particular, $\forall n \in \mathbb{N}$, $\exists x_n \in X$ such that $d_X(x_n, x) < \frac{1}{n}$ but $d_Y(f(x_n), f(x)) \geq \epsilon$.

Hence, (x_n) is a sequence in X which converges to x; however, $f(x_n)$ does not converge. This is a contradiction and hence $(1) \Rightarrow (2)$ holds.

Now, we shall prove $(2) \Rightarrow (3)$. Let $V \subseteq Y$ be the given open set which contains $f(x)$. Therefore, $\exists \epsilon > 0$ such that $B_Y(f(x), \epsilon) \subseteq Y$. Since f is continuous at x, $\exists \delta > 0$ such that $\forall x' \in X$ with $d_X(x', x) < \delta$, we have $d_Y(f(x'), f(x)) < \epsilon$. So, we consider the open set $B_X(x, \delta) \subseteq X$. Clearly, $f(B_X(x, \delta)) \subseteq V$ and we are through!

Finally, to prove $(3) \Rightarrow (1)$, consider a sequence (x_n) in X which converges to x. Also, for a given $\epsilon > 0$, consider $B_Y(f(x), \epsilon)$. Since this open ball is open in Y, $\exists U \subseteq X$ an open set containing x such that $f(U) \subseteq B_Y(f(x), \epsilon)$. Also, U being open, $\exists \delta > 0$ such that $B_X(x, \delta) \subseteq U$. Since $x_n \to x$, $\exists n_0 \in \mathbb{N}$ such that $\forall n \geq n_0$, $x_n \in B_X(x, \delta)$. Therefore, $f(x_n) \in f(B_X(x, \delta)) \subseteq f(U) \subseteq B_Y(f(x), \epsilon)$, $\forall n \geq n_0$. Hence, $f(x_n) \to f(x)$. $\qquad \square$

As mentioned earlier, these equivalent definitions help us in checking if the given function is continuous or not with more ease than that provided by our initial definition. We will see that most of the time, the sequence definition will help us checking if a function is continuous. However, we will also see that the open set definition can also help if we have *abundant* open sets. We will come to this later. However, the reader must note that when we move to arbitrary topological spaces, which are in fact motivated by metric spaces, we do not have the first two definitions of continuity. Therefore, sooner or later, we will have to use the open set definition. Indeed, in this text, we shall use the sequence definition in most places. This is because, wherever possible, using algebra of limits (of sequences) is easier than finding ϵ-δ or open sets. Let us begin with an example.

Example 6.4. Consider $M(2, \mathbb{R})$, the set of all 2×2 matrices with real entries. We equip this set with the metric $d : M(2, \mathbb{R}) \times M(2, \mathbb{R}) \to \mathbb{R}$, defined as

$$d(A, B) = \sqrt{|a_{11} - b_{11}|^2 + |a_{12} - b_{12}|^2 + |a_{21} - b_{21}|^2 + |a_{22} - b_{22}|^2}$$

where $A = \begin{bmatrix} a_{11} & a_{12} \\ a_{21} & a_{22} \end{bmatrix}$ and $B = \begin{bmatrix} b_{11} & b_{12} \\ b_{21} & b_{22} \end{bmatrix}$. We look at the function

$$\det : M(2, \mathbb{R}) \to \mathbb{R}.$$

To check the continuity, we shall use the sequence definition. First, observe that for any sequence of matrices (A_n) in $M(2, \mathbb{R})$, where $A_n = \begin{bmatrix} a_n & b_n \\ c_n & d_n \end{bmatrix}$, $A_n \to A$ for some matrix $A = \begin{bmatrix} a & b \\ c & d \end{bmatrix}$ if and only if $a_n \to a$, $b_n \to b$, $c_n \to c$, and $d_n \to d$. Hence, if $A_n \to A$, then we have

$$\det(A_n) = a_n d_n - b_n c_n,$$

which, by algebra of limits converges to $ad - bc = \det(A)$. Hence, det is a continuous function. In fact, for any $n \in \mathbb{N}$, $\det : M(n, \mathbb{R}) \to \mathbb{R}$ is a continuous function. However, we do not include the proof here. If the reader wants to prove this result, a small hint is that it goes on the same lines as done for $M(2, \mathbb{R})$, but also includes some results from permutations.

Do you appreciate the ease with which we could prove the continuity of determinant as a function? If we were to use the ϵ-δ definition, the same proof would have been tedious and too difficult to handle. The exercises that follow will enable you to take a grasp on other definitions of continuity, especially the sequence definition.

Exercise 6.2. Check the continuity of the following functions:

1. $f : \mathbb{R}^n \times \mathbb{R}^n \to \mathbb{R}^n$, defined as $f(\mathbf{x}, \mathbf{y}) = \mathbf{x} + \mathbf{y}$. (Vector addition)

2. $f : \mathbb{R} \times \mathbb{R}^n \to \mathbb{R}^n$, defined as $f(\alpha, \mathbf{x}) = \alpha \cdot \mathbf{x}$. (Scalar multiplication)

3. $f : \mathbb{R}^n \times \mathbb{R}^n \to \mathbb{R}$, defined as $f(\mathbf{x}, \mathbf{y}) = \sum_{i=1}^{n} x_i y_i$, where $\mathbf{x} = (x_i)_{i=1}^{n}$, $\mathbf{y} = (y_i)_{i=1}^{n} \in \mathbb{R}^n$. (Usual inner product)

Exercise 6.3. Check if the following functions are continuous:

1. $f : M(n, \mathbb{R}) \to M(n, \mathbb{R})$, defined as $f(A) = A^T$, where A^T denotes the transpose of A.

2. $f : M(n, \mathbb{R}) \to M(n, \mathbb{R})$, defined as $f(A) = A^2$. What can you say about A^k, for $k \in \mathbb{N}$?

Exercise 6.4. Let X, Y be metric spaces and $f, g : X \to Y$ be continuous functions. Prove that the set $\{x \in X | f(x) = g(x)\}$ is closed in X. What can you say about the set $\{x \in X | f(x) \neq g(x)\}$?

Example 6.5. Recall that every subset of a discrete metric space is open. Let (X, d) be a discrete metric space and (Y, d_Y) be any other metric space. Let $f : X \to Y$ be any function. For an arbitrary $x \in X$, consider an open set $V \subseteq Y$ such that $f(x) \in V$. Since every set in X is open, $\{x\} \subseteq X$ is also open and satisfies the property that $f(\{x\}) = \{f(x)\} \subseteq V$. Hence, f is continuous. Since the choice of f was arbitrary, we have proved that every function from a discrete space to any other space is continuous.

Example 6.6. Consider a function $f : X \to Y$, where Y is the discrete space. What can we say about the continuity of f? To see this, let us look at the identity function $id : \mathbb{R} \to \mathbb{R}$, where the domain is equipped with the usual metric $(|\cdot|)$ and the codomain is equipped with the discrete metric. Let $x \in \mathbb{R}$. In the discrete metric, $\{x\}$ is an open set which contains x. Now, if $U \subseteq \mathbb{R}$ is open with respect to the usual metric, and contains x then it also contains elements other than x. In particular, $x + \frac{r}{2} \in U$ for some $r > 0$. However, $id(x + \frac{r}{2}) = x + \frac{r}{2} \notin \{x\}$. Hence for any open set U of $(\mathbb{R}, |\cdot|)$, $id(U) \nsubseteq \{x\}$. Therefore, id is not continuous.

Remark. Examples 6.5 and 6.6 are of importance. This is because they tell us the true value of open sets in a metric space. When we have a different collection of open sets, the same function can behave differently. Even the identity function, which seems to be continuous, is discontinuous in some cases! We therefore advise the reader to take a pause and go through the above two examples and assimilate them.

Also, in Example 6.5, we saw a *fantastic* property of discrete spaces. This was possible because the discrete space has a lot of open sets. In fact, it can be proved that if a metric space has more open sets than the other, then any function from the first (which has more open sets) to the second will be continuous. However, we do not cover it in this text. It will be more clear and interesting once the reader is introduced to general topology.

Let us look at another example of a continuous function.

Example 6.7. Consider the metric spaces $(\mathbb{R}^n, d_2)$ and $(\mathbb{R}, |\cdot|)$. We define a function $\Pi : \mathbb{R}^n \to \mathbb{R}$, as

$$\Pi(x_1, x_2, \cdots, x_n) = x_1.$$

This is called the projection onto the first coordinate. Consider a sequence $\left(\left(x_1^{(m)}, x_2^{(m)}, \cdots, x_n^{(m)} \right) \right)_{m \in \mathbb{N}}$ in $\mathbb{R}^n$, which converges to $(x_1, x_2, \cdots, x_n)$. Recall that this is true if and only if the coordinates converge individually, i.e., $x_i^{(m)} \to x_i$ for $i = 1, 2, \cdots, n$. Therefore, the projection function is continuous.

Remark. Often, the projection function defined in Example 6.7 is called the *projection on the first coordinate* and is denoted by Π_1. In general, we can define the *projection on ith coordinate* and can denote it as Π_i. Apart from this, there are also projections onto some subspaces,[1] by taking into consideration only a few coordinates and discarding the others. All such projections are continuous. The proof goes on the same lines as Example 6.7 and therefore we do not include it here.

We have seen projections of Euclidean space $\mathbb{R}^n$. However, we can abstract from our experience in geometry and define projections from product spaces to the coordinate spaces. We give the formal definition below.

Definition 6.5 (Projection). *Let (X, d_X) and (Y, d_Y) be metric spaces and let $(X \times Y, d)$ be the product space. A function $\Pi_X : X \times Y \to X$ defined by*

$$\Pi_X(x, y) = x$$

is called the projection map onto the coordinate space X. Similarly, we can define the projection map Π_Y onto the coordinate space Y.

---

[1] The readers who have not taken a linear algebra course can consider a subspace as a subset of $\mathbb{R}^n$, in which some coordinates are not considered. However, we do encourage them to go through vector spaces and subspaces since those concepts occur almost everywhere in mathematics.

Let us talk about continuity of the projection map. We know that a sequence in the product space is convergent if and only if the corresponding coordinate sequences are convergent, i.e., $(x_n, y_n) \to (x, y)$ if and only if $x_n \to x$ and $y_n \to y$. Therefore, it is clear that $\Pi_X (x_n, y_n) \to \Pi_X (x, y)$. Therefore, the projections are indeed continuous.

Exercise 6.5. Generalize the concept of projection on an n-product space. Is the projection continuous?

Let us now look at some other characterizations of continuity. These characterizations are again in terms of sets and their properties and their proof will use the open set definition of continuity.

Theorem 6.3. *A function $f : X \to Y$, where X and Y are metric spaces, is continuous if and only if the inverse image of open sets in Y is open in X.*

Proof. Let f be continuous and $V \subseteq Y$ be open. We need to prove that $f^{-1}(V)$ is open in X. Let $x \in f^{-1}(V)$. Then, $f(x) \in V$. Since V is open, $\exists r > 0$ such that $B_Y(f(x), r) \subseteq V$. Since f is continuous, there is an open set U such that $x \in U$ and $f(U) \subseteq B_Y(f(x), r)$ and hence, $\exists r' > 0$ such that $B_X(x, r') \subseteq U$. We also have $U \subseteq f^{-1}(f(U)) \subseteq f^{-1}(B_Y(f(x), r)) \subseteq f^{-1}(V)$. Thus, we have $B_X(x, r') \subseteq f^{-1}(V)$ so that $f^{-1}(V)$ is open.

Conversely, consider that inverse image of open sets is open. Let $V \subseteq Y$ be an open set containing $f(x)$. Then, $f^{-1}(V)$ is open in X and contains x. Also, $f(f^{-1}(V)) \subseteq V$. Therefore, f is continuous. □

An amazing application of the above theorem is the following example.

Example 6.8. Consider the metric space $(\mathbb{R}^2, d_2)$ and the set $(a, b) \times (c, d)$, often termed as a *rectangle*. We have seen that this set is open (see Figure 2.14, Chapter 2). We wish to prove this through continuous functions. The question is: Given only this set, where should we start finding our desired continuous function? Observe that any point in this set is an ordered pair (x, y), where $a < x < b$ and $c < y < d$. Therefore, if we consider the two individual projections, we can get to the individual open intervals (a, b) and (c, d), respectively.

So, consider $\Pi_1 : \mathbb{R}^2 \to \mathbb{R}$ and $\Pi_2 : \mathbb{R}^2 \to \mathbb{R}$, given by $\forall (x, y) \in \mathbb{R}^2$, $\Pi_1(x, y) = x$ and $\Pi_2(x, y) = y$. Geometrically, these projections yield the X- and the Y-axis respectively. We also know from Example 6.7 that these two functions are continuous. Therefore, we now must start to look at open sets in the codomain and their inverse images, which might lead to our set $(a, b) \times (c, d)$.

However, also notice that if we consider only one of the projections, we always get an *infinite strip*. This can be seen in Figures 6.3 and 6.4. Therefore, what we need to have is a combination of inverse images of the projections. Therefore, consider the sets $\Pi_1^{-1}(a, b)$ and $\Pi_2^{-1}(c, d)$, which are open in $\mathbb{R}^2$. Clearly, $(a, b) \times (c, d) = \Pi_1^{-1}(a, b) \cap \Pi_2^{-1}(c, d)$ (Verify!). Since finite intersection of open sets is open, our set is open in $\mathbb{R}^2$. This is shown in Figure 6.5.

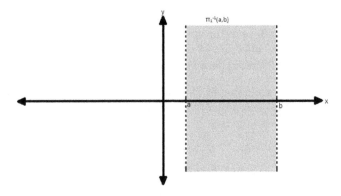

FIGURE 6.3: Inverse image of (a, b) under Π_1.

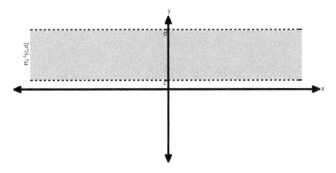

FIGURE 6.4: Inverse image of (c, d) under Π_2.

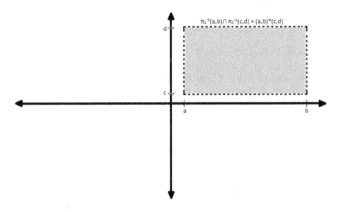

FIGURE 6.5: $(a, b) \times (c, d)$ as an intersection of the two inverse images.

Remark. In the above example, any equivalent metric can be used. Therefore, the set $(a, b) \times (c, d)$ is open in any of the metric space $(\mathbb{R}^n, d_p)$ for $p \geq 1$, which have the same collection of open sets. It is also open in discrete space since every set in a discrete space is open.

Exercise 6.6. Show that the set of all invertible matrices in $M(n, \mathbb{R})$ is open. This set is often represented by $GL(n, \mathbb{R})$, and stands for *general linear group of invertible* $n \times n$ *real matrices*.

Exercise 6.7. Let $f : \mathbb{R} \to \mathbb{R}$ be a function such that $\forall a \in \mathbb{R}$, the sets $f^{-1}(a, \infty)$ and $f^{-1}(-\infty, a)$ are open. Prove that f is continuous.

Based on similar lines as in Theorem 6.3, we can also prove the following result. We leave it as an exercise to the reader.

Theorem 6.4. *A function* $f : X \to Y$, *where* X *and* Y *are metric spaces, is continuous if and only if inverse image of closed sets in* Y *is closed in* X.

Exercise 6.8. Prove Theorem 6.4.

We now give an application of this result.

Example 6.9. Consider the metric space $M(n, \mathbb{R})$ and the set of all nilpotent matrices. From linear algebra, we know that a matrix A is nilpotent[2] if and only if all the eigenvalues of A are zero, which is equivalent to saying $A^n = 0$, from Cayley-Hamilton theorem.[3] Also, the function $f : M(n, \mathbb{R}) \to M(n, \mathbb{R})$, given by $f(A) = A^n$, is continuous. And, the set $\{0\}$, is closed in $M(n, \mathbb{R})$. Hence, the set of all nilpotent matrices, given by $f^{-1}(\{0\})$, is closed in $M(n, \mathbb{R})$.

We now give other characterizations of continuous functions.

Theorem 6.5. *A function* $f : X \to Y$, *where* X *and* Y *are metric spaces, is continuous if and only if for every subset* $A \subseteq X$, *we have* $f(\bar{A}) \subseteq \overline{f(A)}$.

Proof. Let f be continuous. We know that $\overline{f(A)}$ is closed in Y, so that $f^{-1}\left(\overline{f(A)}\right)$ is closed in X. Also, $f(A) \subseteq \overline{f(A)}$. This gives

$$A \subseteq f^{-1}(f(A)) \subseteq f^{-1}\left(\overline{f(A)}\right).$$

---

[2]The exact definition of a nilpotent matrix is as follows: A square matrix A is nilpotent if $A^k = 0$, where 0 denotes the matrix with all entries zero, for some positive integer k.

[3]Cayley-Hamilton theorem states that every square matrix (with real/complex entries) satisfies its own characteristic equation. The characteristic equation for a square matrix A is given by $\det(A - \lambda I) = 0$.

Hence, $\bar{A} \subseteq f^{-1}\left(\overline{f(A)}\right) = f^{-1}\left(\overline{f(A)}\right)$. This further leads to $f\left(\bar{A}\right) \subseteq \overline{f(A)}$.

Conversely, let for every subset $A \subseteq X$, $f\left(\bar{A}\right) \subseteq \overline{f(A)}$ hold. Let F be any closed set in Y. Then, we have

$$f\left(\overline{f^{-1}(F)}\right) \subseteq \overline{f(f^{-1}(F))} \subseteq \bar{F} = F.$$

This leads us to $\overline{f^{-1}(F)} \subseteq f^{-1}(F)$. Therefore, $f^{-1}(F)$ is closed in X and f is continuous. □

Exercise 6.9. Let X and Y be metric spaces. Prove that a function $f : X \to Y$ is continuous if and only if for every subset $B \subseteq Y$, we have $\overline{f^{-1}(B)} \subseteq f^{-1}\left(\bar{B}\right)$.

Exercise 6.10. Let D be a dense set in a metric space X, and Y be another metric space. Let $f : X \to Y$ be continuous and surjective. Prove that $f(D)$ is dense in Y.

A natural question to ask at this stage is: Can we have a characterization of continuity in terms of interior of a set? We have seen characterizations in terms of open and closed sets and how it leads to a characterization in terms of image of closure of a set. Can we have similar results? The following example answers this question negatively.

Example 6.10. Consider $f : \mathbb{R} \to \mathbb{R}$, where $\mathbb{R}$ is equipped with the usual metric, defined as $f(x) = \sin x$. Clearly, f is continuous. Consider $A = (0, 3\pi)$. Then, $\text{Int}(A) = (0, 3\pi)$ and $\text{int}(f(A)) = (-1, 1)$. Hence, $f(\text{Int}(A)) = [-1, 1] \nsubseteq (-1, 1) = \text{Int}(f(A))$.

Let us look at another example where the characterization of continuity in terms of reverse inclusion also cannot hold.

Example 6.11. Consider the *Dirichlet function* $f : \mathbb{R} \to \mathbb{R}$, defined by

$$f(x) = \begin{cases} 1, & x \in \mathbb{Q}. \\ 0, & x \notin \mathbb{Q}. \end{cases}$$

We consider that $\mathbb{R}$ is equipped with the usual metric, $|\cdot|$. For any set $A \subseteq \mathbb{R}$, we have $f(A) \subseteq \{0, 1\}$. Hence, $\text{Int}(f(A)) \subseteq f(\text{Int}(A))$ always holds, while f is clearly discontinuous. To prove the discontinuity, let $x_0 \in \mathbb{R}$. If $x_0 \in \mathbb{Q}$, consider a sequence (x_n) of irrationals such that $x_n \to x_0$. Then, $(f(x_n))$ is the constant sequence that takes the value 0 everywhere and hence does not converge to $f(x_0) = 1$. Similarly, if x_0 is irrational, a sequence of rationals does the job! Therefore, the characterization of continuity cannot be done in terms of images of interior of a set.

However, we can characterize continuity in terms of inverse images of interiors of sets, as the following theorem states.

Theorem 6.6. *Let $f : X \to Y$ be a function, where X and Y are metric spaces. f is continuous if and only if for every subset $B \subseteq Y$, we have $f^{-1}(Int(B)) \subseteq Int(f^{-1}(B))$.*

Proof. Let f be continuous and $B \subseteq Y$ be given. We know that $\text{Int}(B)$ is open in Y and therefore, $f^{-1}(\text{Int}(B))$ is open in X. Also, $\text{Int}(B) \subseteq B$ and therefore, $f^{-1}(\text{Int}(B)) \subseteq f^{-1}(B)$. Finally, we have

$$\text{Int}\left(f^{-1}(\text{Int}(B))\right) = f^{-1}(\text{Int}(B)) \subseteq \text{Int}\left(f^{-1}(B)\right).$$

Conversely, let for every subset $B \subseteq Y$, $f^{-1}(\text{Int}(B)) \subseteq \text{Int}(f^{-1}(B))$ hold. Let G be any open set. Then, we have $f^{-1}(\text{Int}(G)) = f^{-1}(G) \subseteq \text{Int}(f^{-1}(G))$. Hence, $f^{-1}(G)$ is open in X and f is continuous. $\square$

Exercise 6.11. Consider a function $f : X \to Y \times Z$, where X, Y, Z are metric spaces and $Y \times Z$ has the product metric. Suppose $f(x) = (f_1(x), f_2(x))$, where $f_1 : X \to Y$ and $f_2 : X \to Z$ are well-defined. Then, prove that f is continuous if and only if both f_1 and f_2 are continuous.

6.2.2 Results on continuity

In this section, we shall see some consequences of continuity. We start with the easiest, i.e., when the codomain of functions is $\mathbb{R}$, equipped with the usual metric. In the proof that shall follow, we shall again use the sequence definition of continuity, and the reader will appreciate how it makes the job easy!

Theorem 6.7. *Let $f, g : X \to \mathbb{R}$ be continuous functions. Then, $f + g$, fg and $\alpha \cdot f$ are continuous for every $\alpha \in \mathbb{R}$. Also, if $g(x) \neq 0$ for any $x \in X$, then $\frac{f}{g}$ is also continuous. Here, $\cdot$ denotes the scalar multiplication and is defined as*

$$(\alpha \cdot f)(x) = \alpha f(x).$$

Proof. We shall prove only the first part of the proof and the remaining parts will be left as an exercise to the reader.

Consider a sequence (x_n) in X such that $x_n \to x$. Since f and g are continuous, we know that $f(x_n) \to f(x)$ and $g(x_n) \to g(x)$. Hence, $(f + g)(x_n) = f(x_n) + g(x_n) \to f(x) + g(x) = (f + g)(x)$, by the algebra of limits (of sequences) in $\mathbb{R}$. Therefore, $f + g$ is also continuous. $\square$

Exercise 6.12. Prove the remaining parts of Theorem 6.7.

Remark. Theorem 6.7 is often called *algebra of continuous functions* and is very useful in checking continuity of functions with *difficult* expressions. From Theorem 6.7, we can further prove that the set of all continuous functions from a metric space to $\mathbb{R}$, themselves from a vector space. However, it is not only restricted to the codomain being $\mathbb{R}$. If the codomain is any vector space V, equipped with a metric (where the algebra of sequences hold), the set of continuous functions, often denoted by $\mathscr{C}(X, V)$, forms a vector space.

Theorem 6.8. *Composition of continuous functions is continuous.*

Proof. Let X, Y, Z be metric spaces and $f : X \to Y$, $g : Y \to Z$ be given continuous functions. We want to prove that $g \circ f : X \to Z$ is continuous. Let (x_n) be a sequence in X converging to x. Since f is continuous, the sequence $(f(x_n))$ converges to $f(x)$ in Y. Consequently, by the continuity of g, the sequence $(g(f(x_n)))$ converges to $g(f(x))$ in Z. This concludes the proof! □

Theorem 6.9. *Let X, Y be metric spaces and $D \subseteq X$ be a dense set. Let $f, g : X \to Y$ be continuous functions such that $\forall x \in D$, $f(x) = g(x)$. Then, $f = g$.*

Proof. Suppose that the theorem is false! Then, $\exists x_0 \in X$ such that $f(x_0) \neq g(x_0)$. Using the Haussdroff property of Y, $\exists U, V \subseteq Y$, open such that $f(x_0) \in U$, $g(x_0) \in V$ and $U \cap V = \emptyset$. Since f is continuous, there is an open set $W_1 \subseteq X$ such that $x_0 \in W_1$ and $f(W_1) \subseteq U$. Also, g is continuous so that $\exists W_2 \subseteq X$, open such that $x_0 \in W_2$ and $g(W_2) \subseteq V$. Let $W = W_1 \cap W_2$. Then, $x_0 \in W$, and W is open in X. Since D is dense in X, $\exists x \in D \cap W$. Here $f(x) = g(x)$ and also, $f(x) \in U \cap V$, which is a contradiction! Hence, the theorem must be true. □

We now move to an important theorem due to Urysohn. This theorem tells us how to *separate* two disjoint closed subsets in a metric space through continuous real valued functions.

Theorem 6.10 (Urysohn's lemma). *Let A, B be two non-empty closed disjoint sets in a metric space X. Then, there is a continuous function $f : X \to \mathbb{R}$ such that $\forall x \in X$, $0 \leq f(x) \leq 1$ and $\forall x \in A$, $f(x) = 0$ and $\forall x \in B$, $f(x) = 1$.*

Proof. Since we want a continuous function which satisfies certain properties on certain sets, we are tempted to use d_A and d_B. Clearly, if we use only d_A, we shall get $f(x) = 0$ for any $x \in A$. However, f need not be 1 on B. Therefore, somehow, we want $\frac{d_A(x)}{d_A(x)}$, whenever $x \in B$. This can be achieved easily by defining the function as

$$f(x) = \frac{d_A(x)}{d_A(x) + d_B(x)}.$$

Now, we need to check only two things: f is continuous and it takes values between 0 and 1. First, observe that if $d_A(x) \neq 0$, then we can have

$$f(x) = \frac{1}{1 + \dfrac{d_B(x)}{d_A(x)}}.$$

So that

$$0 < f(x) \leq 1, \ \forall x \in X.$$

Also, for $x \in A$, $f(x) = 0$. Hence, $0 \leq f(x) \leq 1, \ \forall x \in X$.

Now, by the algebra of continuous functions, f is continuous and we are through! $\hfill\square$

Exercise 6.13. Let A and B be two non-empty disjoint closed subsets of a metric space X. Prove that there exist two open sets U and V such that $A \subseteq U$, $B \subseteq V$ and $U \cap V = \emptyset$.

Remark. The property shown in Exercise 6.13 is important in general topological spaces. Such spaces are called *normal*. It gives a way to *separate* closed sets through open sets. However, Urysohn's lemma gives a stronger condition. It separates closed sets through continuous functions.

Now, suppose that we are provided with continuous functions on open sets of a metric space. Can we have a function continuous on the whole metric space? A similar question can be asked for closed sets. The following theorem answers these questions positively.

Theorem 6.11 (Gluing lemma). *Let X, Y be metric spaces.*

1. *Let $\{U_i | i \in I\}$ be an indexed family of open sets of X such that $\bigcup\limits_{i \in I} U_i = X$. Also, let $f_i : U_i \to Y$ be given continuous functions for each $i \in I$ with the property that $f_i(x) = f_j(x)$, for all $x \in U_i \cap U_j$, $\forall i, j \in I$. Then the function $f : X \to Y$, defined as $f(x) = f_i(x)$, whenever $x \in U_i$ for some $i \in I$, is well-defined and continuous on X.*

2. *Let $\{F_i | i \in I_n\}$ be a finite family of closed sets of X, where $I_n = \{1, 2, \cdots, n\}$ is the finite index set, such that $\bigcup\limits_{i \in I_n} F_i = X$. Also, let $f_i : F_i \to Y$ be given continuous functions such that $f_i(x) = f_j(x)$, whenever $x \in F_i \cap F_j$, $\forall i, j \in I_n$. Then, the function $f : X \to Y$ defined as $f(x) = f_i(x)$, whenever $x \in F_i$ for some $i \in I_n$ is well-defined and continuous on X.*

Proof.

1. Observe that under all the conditions given, f is well-defined. We do not include the "textbook" proof for well-defined-ness of f here. Let $V \subseteq Y$ be a given open set. We want to prove that $f^{-1}(V)$ is open in X. First, we claim that $f^{-1}(V) \cap U_i = f_i^{-1}(V) \cap U_i$ for every $i \in I$.

   To prove this, observe that $x \in f^{-1}(V) \cap U_i$ iff $x \in U_i$ and $f(x) \in V$, which is equivalent to $f_i(x) \in V$. Hence, $f^{-1}(V) \cap U_i = f_i^{-1}(V) \cap U_i$.

Now,

$$f^{-1}(V) = f^{-1}(V) \cap X$$

$$= f^{-1}(V) \cap \left(\bigcup_{i \in I} U_i \right)$$

$$= \bigcup_{i \in I} \left(f^{-1}(V) \cap U_i \right)$$

$$= \bigcup_{i \in I} \left(f_i^{-1}(V) \cap U_i \right).$$

Since, $\forall i \in I$, f_i is continuous, $f^{-1}(V)$ is a union of open sets and hence open. Therefore, f is continuous.

2. Left as an exercise to the reader.

□

Remark. At first, gluing lemma might seem trivial to the reader. However, it must be kept in mind that however trivial the lemma looks, it has its own beauty. We shall shortly see how gluing lemma is helpful to us in many cases. In fact, from gluing lemma, we can see that for a disconnected metric space X, we can have a non-constant continuous function $f : X \to \{0, 1\}$. Indeed, we shall comment on this later in this text.

Exercise 6.14. Prove part (2) of Theorem 6.11.

Exercise 6.15. In Exercise 6.14, the reader must have noticed that the collection of closed sets must be finite for the proof to work. Intuitively, it seems that if we have an arbitrary collection of closed sets, then gluing lemma may not hold. Find a counterexample for the same. In particular, find a collection of closed sets $\{F_i | i \in I\}$, where I is not a finite index set, and continuous functions $f_i : F_i \to Y$ satisfying the conditions of gluing lemma such that the function $f : X \to Y$ defined by $f(x) = f_i(x)$, whenever $x \in F_i$ for some $i \in I$ is not continuous on X.

The following exercises are a few applications of gluing lemma.

Exercise 6.16. Let $f, g : [0, 1] \to X$ be continuous, where X is a metric space. Further assume that $f(1) = g(0)$. Define $h : [0, 1] \to X$ as

$$h(x) = \begin{cases} f(2x), & 0 \le x \le \dfrac{1}{2}. \\ g(2x - 1), & \dfrac{1}{2} \le x \le 1. \end{cases}$$

Comment on the continuity of h.

Exercise 6.17. Consider a function $f : \mathbb{R}^n \to \mathbb{R}^n$, given by

$$f(\mathbf{x}) = \begin{cases} \mathbf{x}, & \text{if } ||\mathbf{x}|| \leq 1. \\ \dfrac{\mathbf{x}}{||\mathbf{x}||}, & \text{if } ||\mathbf{x}|| > 1. \end{cases}$$

where $||\cdot||$ denotes the *length* of a vector in $\mathbb{R}^n$ and is given by

$$||(x_1, x_2, \cdots, x_n)|| = \sqrt{x_1^2 + x_2^2 + \cdots + x_n^2}.$$

Is f continuous?

We now give two amazing applications of gluing lemma. The reader will appreciate the theorem after assimilating these applications. The first application is from complex analysis.

We denote by $\mathbb{C}$, the set of all complex numbers and $\mathbb{C}^*$, the set of extended complex numbers, i.e., $\mathbb{C} \cup \{\infty\}$. We know from high school mathematics that if $z \in \mathbb{C}$, then a number $\theta \in \mathbb{R}$ such that $z = |z|\, e^{\iota\theta}$ is called the argument of z. Also, any two arguments of the same complex number differ by integer multiple of 2π. It is well known that there is no continuous function $\theta : \mathbb{C}^* \to \mathbb{R}$ such that $z = |z|\, e^{\iota\theta(z)}$ for $z \in \mathbb{C}^*$. However, if we just exclude the non-positive real axis from $\mathbb{C}^*$, then we can have a continuous function. The following theorem gives the details for the same.

Theorem 6.12. *Let* $X = \mathbb{C} \setminus \{z \in \mathbb{C} | z \in \mathbb{R} \text{ and } z \leq 0\}$. *Then, there is a continuous function* $\alpha : X \to (-\pi, \pi)$ *such that* $z = |z|\, e^{\iota\alpha(z)}$.

Proof. Consider the following open half planes: $H_1 = \{z \in \mathbb{C} | \text{Re}(z) > 0\}$, $H_2 = \{z \in \mathbb{C} | \text{Im}(z) > 0\}$, and $H_3 = \{z \in \mathbb{C} | \text{Im}(z) < 0\}$. Notice that $H_1 \cup H_2 \cup H_3 = X$.

Let $z \in H_1$. Then, $\text{Re}(z) = |z| \cos\theta$ for some $\theta \in \left(-\frac{\pi}{2}, \frac{\pi}{2}\right)$. Therefore, define $\alpha_1 : H_1 \to \left(-\frac{\pi}{2}, \frac{\pi}{2}\right)$ as $\alpha_1(z) = \sin^{-1}\left(\frac{\text{Im}(z)}{|z|}\right)$. Similarly, we can define $\alpha_2 : H_2 \to (0, \pi)$ and $\alpha_3 : H_3 \to (-\pi, 0)$ as $\alpha_2(z) = \cos^{-1}\left(\frac{\text{Re}(z)}{|z|}\right)$ and $\alpha_3(z) = \cos^{-1}\left(\frac{\text{Re}(z)}{|z|}\right)$. All the functions, α_1, α_2, and α_3 are continuous and agree on their common domain and hence, by a simple application of gluing lemma, we can have a continuous function $\alpha : X \to (-\pi, \pi)$ with the desired property. $\square$

We conclude this section by giving another application of gluing lemma, which is called the Tietze extension theorem. It tells us that if we are provided with a continuous function on a closed subset, then we can *extend* it to the whole space without disturbing the continuity.

Theorem 6.13 (Tietze extension theorem). *Let A be a closed subset of a metric space X. Given a continuous function $g : A \to [1, 2]$, there is a continuous function $f : X \to [1, 2]$ which is an extension of g, i.e., $\forall a \in A$, $f(a) = g(a)$.*

Proof. We define $f : X \to [1, 2]$ as

$$f(x) = \begin{cases} g(x), & \text{when } x \in A. \\ \dfrac{\inf\{g(a)\,d(x,a)\,|a \in A\}}{d_A(x)}, & \text{when } x \notin A. \end{cases}$$

Note that since $\forall a \in A$, we have $1 \leq g(a) \leq 2$, $1 \leq f(x) \leq 2$ also holds $\forall x \in X$. Also, on A, $f = g$. Hence, f is an extension of g. All that remains to prove is that f is continuous.

To do so, we wish to incorporate the gluing lemma. We know that $X = A \cup (X \setminus A)$. However, for the gluing lemma to work, both sets should be either open or closed. Since A is already closed, we close the second set by taking its closure. We have, $X = A \cup \left(\overline{X \setminus A}\right)$. It is clear that f is continuous on A, since g is continuous. We now wish to prove that f is continuous on $\overline{X \setminus A}$.

Note that when we take a point in $\overline{X \setminus A}$, it can either be in the interior or on the boundary. Those points on the boundary also lie inside A, since A is closed. On the other hand, those in the interior lie completely outside A. Therefore, we shall consider them as two different cases.

Case I: Let $x \in X \setminus A$. Since A is closed, $d_A(x) > 0$. Also, d_A is a continuous function. Therefore, we only need to prove that $h : X \setminus A \to [1, 2]$ defined as $h(u) = \inf\{g(a)\,d(u,a)\,|a \in A\}$ is continuous (Figure 6.6).

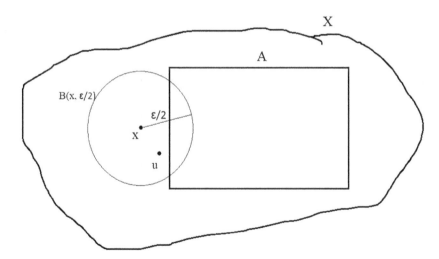

FIGURE 6.6: Pictorial representation of Case I of the Tietze extension theorem.

Let $\epsilon > 0$ be given. Consider $u \in (X \setminus A) \cap B\left(x, \frac{\epsilon}{2}\right)$. Then, for any $a \in A$, we have

$$d(x, a) \le d(x, u) + d(a, u) < \frac{\epsilon}{2} + d(a, u).$$

Multiplying by $g(a)$ on both sides, we get

$$g(a) d(x, a) < g(a) \frac{\epsilon}{2} + g(a) d(a, u) \le \epsilon + g(a) d(a, u).$$

On taking the infimum as $a \in A$ varies, we get

$$h(x) \le \epsilon + h(u).$$

On interchanging the roles of x and u, we also get

$$h(u) \le \epsilon + h(x),$$

which further leads us to

$$|h(x) - h(u)| \le \epsilon.$$

Since $\epsilon > 0$ was arbitrary, we can conclude that h is continuous on $X \setminus A$ and so is f.

Case II: Now, let $x \in \partial A$. Notice that $\partial A = \partial(X \setminus A)$ and in this case $x \in A$. Let $\epsilon > 0$ be given. By the continuity of g on A, $\exists \delta > 0$ such that $|g(a) - g(x)| < \epsilon$ for all $a \in A \cap B(x, \delta)$ (Figure 6.7).

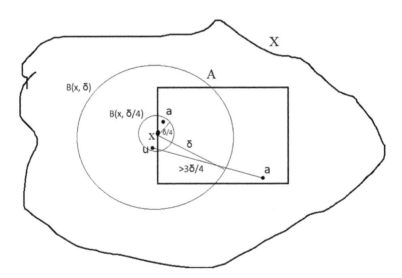

FIGURE 6.7: Pictorial representation of Case II of the Tietze extension theorem.

In particular, we have $|f(u) - f(x)| < \epsilon$ for all $u \in A \cap B\left(x, \frac{\delta}{4}\right)$.

The same thing should now holds for $u \in (X \setminus A) \cap B\left(x, \frac{\delta}{4}\right)$. Notice that if $a \in A \setminus B(x, \delta)$, then

$$d(a, u) \geq d(x, a) - d(x, u) > \delta - \frac{\delta}{4} = 3\frac{\delta}{4}.$$

Since $g(a) \geq 1$, we get

$$\inf\left\{g(a) d(u, a) \,|a \in A \setminus B(x, \delta)\right\} \geq \frac{3\delta}{4}.$$

On the other hand, since $x \in A \cap B\left(x, \frac{\delta}{4}\right)$, we have

$$g(x) d(x, u) \leq 2d(x, u) < \frac{\delta}{2} < \frac{3\delta}{4}.$$

Hence, for $u \in (X \setminus A) \cap B\left(x, \frac{\delta}{4}\right)$, we have

$$\inf\left\{g(a) d(a, u) \,|a \in A\right\} = \inf\left\{g(a) d(a, u) \,|a \in A \cap B(x, \delta)\right\}.$$

A similar argument shows that

$$\inf\left\{d(u, a) \,|a \in A \cap B(x, \delta)\right\} = \inf\left\{d(u, a) \,|a \in A\right\} = d_A(u).$$

And, from the continuity of g, we have

$$g(x) - \epsilon < g(a) < g(x) + \epsilon.$$

So, that we have

$$(g(x) - \epsilon) d_A(u) \leq \inf\left\{g(a) d(u, a) \,|a \in A \cap B(x, \delta)\right\} \leq (g(x) + \epsilon) d_A(u).$$

Finally, we get

$$|f(u) - g(x)| \leq \epsilon.$$

Since on A, we have $f = g$, continuity of f follows also on ∂A. $\qquad\square$

6.3 Uniform Continuity

In the previous section, we saw what continuity of functions meant and made a major remark. We said that continuity of a function is a *local property*, i.e., it is defined for each point in the domain of a function and can vary with varying points. However, it is natural to ask: Can we have a global property similar to continuity? First, note that when we want to make a local property into a global property, we are actually strengthening the conditions. So, if at all such a property exists, we expect the functions exhibiting this property to be necessarily continuous. Apart from this, the functions must satisfy certain other conditions, which we will see in some time.

To *upgrade* continuity, let us look at the definition and find some places which can be upgraded. According to the very first definition of continuity given in this text, given a function $f : X \to Y$, where X, Y are metric spaces, for each given $\epsilon > 0$, we need to find a $\delta > 0$ such that some conditions are satisfied. However, before all this, we are provided with a point $x_0 \in X$ and depending on whether or not all conditions of the definition are satisfied, we comment upon the continuity of f at point x_0. Hence, the choice of δ not only depends on ϵ, but also (in general) depends on x_0. To see this, let us look at $f : (0, 1) \to \mathbb{R}$, given by $f(x) = \frac{1}{x}$. As we approach 0, $f(x)$ becomes unbounded. Therefore, to remain within the ϵ distance in the codomain, we need to make our δ distance smaller and smaller in the domain.

Therefore, to upgrade continuity, we want another condition that the choice of δ should be independent of x_0. Although, it will (in general) always depend on ϵ. Motivated by this, we now give the formal definition of what is called *uniform continuity*.

Definition 6.6 (Uniform continuity). *Let $f : X \to Y$ be a function, where X, Y are metric spaces. Then, f is said to be uniformly continuous if $\forall \epsilon > 0$, $\exists \delta > 0$ such that $\forall x_1, x_2 \in X$ with $d_X(x_1, x_2) < \delta$, we have $d_Y(f(x_1), f(x_2)) < \epsilon$.*

From the above description, it is clear that $\frac{1}{x} : (0, 1) \to \mathbb{R}$ is not uniformly continuous. We will now prove it.

Example 6.12. Consider $\epsilon = 1$ and let $\delta > 0$ be arbitrary. By Archimedean property, $\exists n_0 \in \mathbb{N}$ such that $n_0 > \frac{1}{\delta}$. Consider two points $\frac{1}{n_0}, \frac{1}{2n_0} \in (0, 1)$. We have $\left| \frac{1}{n_0} - \frac{1}{2n_0} \right| = \frac{1}{2n_0} < \delta$. However, $|n_0 - 2n_0| = n_0 \geq 1$. Hence, $\frac{1}{x}$ is not uniformly continuous.

What can we say about the same function on $[1, \infty)$?

Example 6.13. We shall estimate $|f(x) - f(y)|$ for $x, y \in [1, \infty)$. From this estimate, we can choose our δ, if possible. If not possible, we get to know how to construct the counterexample by choosing the correct ϵ.

Observe that

$$|f(x) - f(y)| = \left| \frac{y - x}{xy} \right|$$
$$\leq |x - y|,$$

since $x \geq 1$ and $y \geq 1$. Therefore, given $\epsilon > 0$, if we choose $\delta = \epsilon$, we are through! Hence, $\frac{1}{x}$ is uniformly continuous on $[1, \infty)$.

Remark. It is to be noted that $\frac{1}{x}$ is not uniformly continuous over any interval with boundary 0 and is uniformly continuous over any closed interval where it is well-defined.

One of the major concerns of this *upgradation* is that the original properties must remain intact. This means that every uniformly continuous function must be necessarily continuous. The proof of this fact is left as an exercise to the reader.

Exercise 6.18. Prove that every uniformly continuous function is continuous on its domain.

Exercise 6.19. Prove that composition of uniformly continuous functions is uniformly continuous.

Let us look at another example from $\mathbb{R}$.

Example 6.14. Let $f : \mathbb{R} \to \mathbb{R}$ be defined as $f(x) = x^2$. We want to check if this function is uniformly continuous. Indeed, it is continuous. If $|x|$ is large enough, $f(|x|)$ becomes much larger. Therefore, for small increments in x, we can have large differences in the value of the function. Therefore, our guess is that x^2 is not uniformly continuous.

To see this, consider $\epsilon = 1$ and let $\delta > 0$ be arbitrary. Then, by the Archimedean property, $\exists n \in \mathbb{N}$ such that $n\delta > 1$. Now, $\left|n + \frac{1}{n} - n\right| = \frac{1}{n} < \delta$. However,

$$\left| f\left(n + \frac{1}{n}\right) - f(n) \right| = \left| 2 + \frac{1}{n^2} \right| \geq 2 > 1.$$

Hence, x^2 is not uniformly continuous on $\mathbb{R}$. What happens if we take a closed and bounded interval?

Consider $[-M, M]$. Then, $\forall x, y \in [-M, M]$, we have $|x + y| \leq 2M$. Hence,

$$\left| x^2 - y^2 \right| = |x + y|\,|x - y|$$

$$\leq 2M\,|x - y|.$$

Therefore, for a given $\epsilon > 0$, if we choose $\delta = \frac{\epsilon}{2M}$, then we are through! In fact, such a proof works for any bounded interval and not only closed ones.

The two above examples might lead us to think that whenever function becomes unbounded in its domain, it is not uniformly continuous. However, this is not true as the next exercise will show.

Exercise 6.20. Prove that the identity function on any metric space (X, d) to itself is uniformly continuous. In particular $id : \mathbb{R} \to \mathbb{R}$ is uniformly continuous, although it is unbounded.

Exercise 6.21. Let $f : \mathbb{R} \to \mathbb{R}$ be a function such that $f'(x)$ exists for all $x \in \mathbb{R}$ and is bounded on $\mathbb{R}$. Comment upon the uniform continuity of f.

Exercise 6.22. Show that any function from a discrete space to any other metric space is uniformly continuous.

Before moving to some consequences of uniform continuity, let us look at another example of a uniformly continuous function.

Example 6.15. Let (x_n) be a sequence in a metric space (X, d). Define $f : X \to \mathbb{R}$ as

$$f(x) = \inf \{d(x, x_n) \,|\, n \in \mathbb{N}\}.$$

We want to estimate $|f(x_1) - f(x_2)|$ for arbitrary $x_1, x_2 \in X$. Notice that if we consider $A = \{x_n \,|\, n \in \mathbb{N}\}$, the range set of the sequence, then $f(x) = d_A(x)$. From Example 6.3, we know that if we choose $\delta = \epsilon$, we can not only prove the continuity, but also uniform continuity.

Remark. As promised at the end of Example 6.3, we now formally state that the function d_A is uniformly continuous for any non-empty set $A \subseteq X$ of a metric space.

Let us now look at some consequences of uniform continuity.

Let $f : X \to Y$ be a uniformly continuous function and $A \subseteq X$ be a bounded set. We want to know the fate of $f(A)$, i.e., we want to ask: Is $f(A)$ bounded? Although it may seem that the answer is "yes", the next example shows that this is not true.

Example 6.16. Consider $f : \mathbb{N} \to \mathbb{R}$, defined as $f(n) = n$. Here, $\mathbb{N}$ is equipped with the discrete metric and $\mathbb{R}$ is equipped with the usual metric. Then, $\mathbb{N}$ is bounded (in discrete metric); however, $f(\mathbb{N}) = \mathbb{N}$ is not bounded in $\mathbb{R}$. Also, from Exercise 6.22, we know that f is uniformly continuous. Hence, we conclude that uniformly continuous functions need not take bounded sets to bounded sets.

However, uniformly continuous functions do preserve Cauchyness of sequences. The details of this fact are given in the following result.

Theorem 6.14. *A uniformly continuous function takes Cauchy sequences to Cauchy sequences.*

Proof. Let $f : X \to Y$ be a uniformly continuous function, where X, Y are metric spaces. Let (x_n) be a Cauchy sequence in X. By the uniform continuity, we know that

$$\forall \epsilon > 0, \exists \delta > 0 \text{ such that } \forall x_n, x_m \in \{x_n \,|\, n \in \mathbb{N}\} \text{ with } d_X(x_n, x_m) < \delta$$

$$\text{we have } d_Y(f(x_n), f(x_m)) < \epsilon.$$

Since (x_n) is Cauchy, $\exists n_0 \in \mathbb{N}$ such that $d_X(x_n, x_m) < \delta$ for all $n, m \geq n_0$. Hence, for the same n_0, we also have $d_Y(f(x_n), f(x_m)) < \epsilon$ for $m, n \geq n_0$. Therefore, $(f(x_n))_{n \in \mathbb{N}}$ is Cauchy in Y. □

Exercise 6.23. Show that a function $f : (X, d_X) \to (Y, d_Y)$ is uniformly continuous if and only if it satisfies the following condition: if (s_n) and (t_n) are two sequences in X such that $d_X(s_n, t_n) \to 0$ as $n \to \infty$, then $d_Y(f(s_n), f(t_n)) \to 0$ as $n \to \infty$.

As the next step, we now ask the question if uniformly continuous functions can be extended. If we are given a uniformly continuous function $f : A \to Y$, where $A \subseteq X$, X, Y are metric spaces, we want to know whether there exist a uniformly continuous function $g : X \to Y$ such that $g = f$ on A. Indeed, there are certain conditions that need to be satisfied by the set A and the metric space Y.

In particular, we saw in Theorem 6.14 that uniformly continuous functions preserve Cauchy sequences. This is a very "nice" property and we would like to make use of it. However, Cauchy sequences are no use to us if the metric space is not complete. Therefore, in order to have some concrete result, we would want Y to be a complete metric space. In addition, since we are using sequences, the closedness of A is not sufficient as in Theorem 6.13. We would now want the set A to be dense.

With all these conditions satisfied, we can define $g : X \to Y$ as

$$g(x) = \lim_{n \to \infty} f(a_n)$$

where (a_n) is a sequence in A which converges to x. Note that this is possible due to density of A in X. Then, we are left to check that g is well-defined and uniformly continuous. We first give a concrete statement of this result and then provide all the details in the proof.

Theorem 6.15. *Let (X, d_X) be a metric space and (Y, d_Y) be a complete metric space. Let $A \subseteq X$ be dense in X and $f : A \to Y$ be uniformly continuous. Then, there is a unique function $g : X \to Y$ which is uniformly continuous and moreover, $\forall a \in A$, $f(a) = g(a)$, i.e., g extends f.*

Proof. Let $x \in X$. Since A is dense in X, there is a sequence (a_n) in A such that $a_n \to x$. Consequently, (a_n) is Cauchy. Since f is uniformly continuous, the sequence $(f(a_n))$ is Cauchy in Y and by the completeness of Y, it is convergent. To be able to define g as in the above description, it must be well defined, i.e., if there are two sequences (a_n) and (b_n) converging to the same x, then $\lim_{n \to \infty} f(a_n) = \lim_{n \to \infty} f(b_n)$ must hold.

Let $y_1 = \lim_{n \to \infty} f(a_n)$ and $y_2 = \lim_{n \to \infty} f(b_n)$ and let $\epsilon > 0$ be arbitrary. By the uniform continuity of f, $\exists \delta > 0$ such that $\forall c_1, c_2 \in A$ with $d_X(c_1, c_2) < \delta$, we have $d_Y(f(c_1), f(c_2)) < \frac{\epsilon}{3}$.

Now, by the definition of convergence of sequences, $\exists n_0, n_1 \in \mathbb{N}$ such that $\forall n \geq n_0$ we have $d_X(a_n, x) < \frac{\delta}{2}$ and $\forall n \geq n_1$, we have $d_X(b_n, x) < \frac{\delta}{2}$. Therefore, for $n \geq n_2 = \max\{n_0, n_1\}$, we have

$$d_X(a_n, b_n) \leq d_X(a_n, x) + d_X(b_n, x) < \delta.$$

Hence, by the uniform continuity of f we have,

$$d_Y(f(a_n), f(b_n)) < \frac{\epsilon}{3}.$$

Also, by the convergence of sequences $(f(a_n))$ and $(f(b_n))$ in Y, $\exists n_3, n_4 \in \mathbb{N}$ such that $\forall n \geq n_3$, we have $d_Y(y_1, f(a_n)) < \frac{\epsilon}{3}$ and $\forall n \geq n_4$, we have $d_Y(y_2, f(b_n)) < \frac{\epsilon}{3}$. Hence, $\forall n \geq n_5 = \max\{n_2, n_3, n_4\}$, we have

$$d_Y(y_1, y_2) \leq d_Y(y_1, f(a_n)) + d_Y(f(a_n), f(b_n)) + d_Y(y_2, f(b_n)) < \epsilon.$$

Therefore, we can conclude that $y_1 = y_2$. Now, we define $g : X \to Y$ as $g(x) = \lim\limits_{n \to \infty} f(a_n)$, where (a_n) is a sequence in A such that $a_n \to x$. Clearly, g is well-defined from the above description. First, we check if g extends f.

Let $x \in A$. Then, the constant sequence (a_n), where $a_n = x$, $\forall n \in \mathbb{N}$, converges to x. Also,

$$g(x) = \lim\limits_{n \to \infty} f(a_n) = \lim\limits_{n \to \infty} f(x) = f(x).$$

Therefore, g is an extension of f. Now, we look at the uniform continuity of g.

Let $\epsilon > 0$ be given. Then, by the uniform continuity of f, $\exists \delta > 0$ such that $\forall c_1, c_2 \in A$ with $d_X(c_1, c_2) < \delta$, we have $d_Y(f(c_1), f(c_2)) < \frac{\epsilon}{3}$.

Now, let $x_1, x_2 \in X$ with $d_X(x_1, x_2) < \frac{\delta}{3}$. Since A is dense in X, there are sequences (a_n) and (b_n) which converge to x_1 and x_2 respectively. By the definition of convergence, $\exists n_0, n_1 \in \mathbb{N}$ such that $\forall n \geq n_0$, we have $d_X(a_n, x_1) < \frac{\delta}{3}$ and $\forall n \geq n_1$, we have $d_X(b_n, x_2) < \frac{\delta}{3}$. Hence, for $n \geq n_2 = \max\{n_0, n_1\}$, we have,

$$d_X(a_n, b_n) \leq d_X(a_n, x_1) + d_X(x_1, x_2) + d_X(b_n, x_2) < \delta.$$

Consequently, $d_Y(f(a_n), f(b_n)) < \frac{\epsilon}{3}$.

By the definition of g, $f(a_n) \to g(x_1)$ and $f(b_n) \to g(x_2)$. Hence, $\exists n_3, n_4 \in \mathbb{N}$ such that $\forall n \geq n_3$ we have $d_Y(f(a_n), g(x_1)) < \frac{\epsilon}{3}$ and for $n \geq n_4$, we have $d_Y(f(b_n), g(x_2)) < \frac{\epsilon}{3}$. Hence, for $n \geq n_5 = \max\{n_2, n_3, n_4\}$ we have,

$$d_Y\left(g\left(x_1\right),g\left(x_2\right)\right) \le d_Y\left(g\left(x_1\right),f\left(a_n\right)\right) + d_Y\left(f\left(a_n\right),f\left(b_n\right)\right)$$

$$+\, d_Y\left(f\left(b_n\right),g\left(x_2\right)\right)$$

$$< \frac{\epsilon}{3} + \frac{\epsilon}{3} + \frac{\epsilon}{3} = \epsilon.$$

This proves the uniform continuity of g. Note that the uniqueness follows from Exercise 6.18 and Theorem 6.9. □

We now give an application of Theorem 6.15. Before we formally state the result and give the details of the proof, let us recall how we learned exponentiation, starting from primary school. Initially, when we were taught multiplication, we were told that a *square* of a number is the number multiplied by itself, *cube* is the number multiplied with itself three times, and so on. Soon, we abstracted the meaning of a^n, for a real number a and $n \in \mathbb{N}$. Then, we were introduced to radicals and started looking for *square roots*, *cube roots*, and so on. It is only in the analysis course that one is exposed to the nth root of a non-negative real number in a rigorous fashion. We assume that the reader is thorough with the existence of nth root. Once we had nth roots, we started looking at integer powers of those roots and we represented them by $a^{\frac{m}{n}}$, where $m \in \mathbb{Z}$ and $n \in \mathbb{N}$. Thus, so far we rigorously defined a^r for $r \in \mathbb{Q}$. But, we have been using the exponentiation a^x for any real number x. How do we know that $a^{\sqrt{2}}$ makes sense? In the following result, which is an application of Theorem 6.15, we shall answer this question. In particular we shall show that a^x can be defined for any $x \in \mathbb{R}$ and $a \ge 0$.

Theorem 6.16. *Let $a \in \mathbb{R}^+$ and $M \in \mathbb{N}$. Then, the function $f : \mathbb{Q} \cap [-M, M] \to \mathbb{R}$, defined as $f\left(r\right) = a^r$ is uniformly continuous. Hence, this extends to a function on $[-M, M]$, which we denote by a^x.*

Proof. All we need to prove is that f is uniformly continuous. The extension will follow from the density of $\mathbb{Q}$ and Theorem 6.15. We shall consider the case $a \ge 1$. The case $0 < a < 1$ reduces to our case by considering $b = \frac{1}{a}$.

Let $x, x + h \in \mathbb{Q} \cap [-M, M]$. Then, by the increasing nature of a^r on $\mathbb{Q}$, we have

$$\left|a^{x+h} - a^x\right| = a^x \left|a^h - 1\right| \le a^M \left|a^h - 1\right|.$$

Now, we know that as $h \to 0$, $a^h \to 1$. Hence, the right hand side of the inequality can be controlled as much as we please. Hence, f is uniformly continuous on $\mathbb{Q} \cap [-M, M]$. □

Remark. Since $M \in \mathbb{N}$ used in the above theorem was arbitrary, a^x is defined on all of $\mathbb{R}$.

We end this section by giving an *upgradation* of uniform continuity.

Definition 6.7 (Lipschitz continuity). *A function $f : X \to Y$, where X, Y are metric spaces, is said to be Lipschitz continuous if $\exists L > 0$ such that $\forall x_1, x_2 \in X$, we have*

$$d_Y \left(f\left(x_1\right), f\left(x_2\right) \right) \leq L d_X \left(x_1, x_2\right).$$

The constant L is called Lipschitz constant.

Exercise 6.24. Prove that every Lipschitz continuous function is uniformly continuous.

Remark. Lipschitz continuity is stronger in both continuity and uniform continuity and is useful in the theory of ordinary differential equations. In particular, it is used to prove the existence and uniqueness of a solution to an initial value problem.

6.4   Continuous Functions on Compact Spaces

In this section we shall study how continuous functions behave when their domain is compact. We shall also study the fate of the compact domain in terms of its continuous image. We start with the following result.

Theorem 6.17. *A continuous function f from a compact metric space (X, d_X) to any metric space (Y, d_Y) is bounded, i.e., the range $f(X)$ is a bounded subset of Y.*

Proof. We fix a $y \in Y$. Notice that for any $y' \in Y$, $\exists n \in \mathbb{N}$ such that $n > d(y, y')$ (by Archimedean property). Therefore, we have $\bigcup_{n \in \mathbb{N}} B_Y(y, n) = Y$. Also, each of the open balls is an open set and since f is continuous, each of $f^{-1}(B_Y(y, n))$ is open in X, and $\bigcup_{n \in \mathbb{N}} f^{-1}(B_Y(y, n)) = f^{-1}\left(\bigcup_{n \in \mathbb{N}} B_Y(y, n) \right) = f^{-1}(Y) = X$. Hence, $\{f^{-1}(B_Y(y, n)) \,|\, n \in \mathbb{N}\}$ is an open cover for X.

Since X is compact, this has a finite subcover, say, $\{f^{-1}(B_Y(y, n)) \,|\, n \in I_N\}$ for some $N \in \mathbb{N}$, where $I_N = \{1, 2, \cdots, N\}$. Notice that $\forall n_1 \leq n_2$, $B_Y(y, n_1) \subseteq B_Y(y, n_2)$. Hence, $f^{-1}(B_Y(y, N)) = X$ so that $f\left(f^{-1}(B_Y(y, N)) \right) = f(X) \subseteq B_Y(y, N)$. Therefore, f is bounded. □

Now, let us consider a subset $S \subseteq \mathbb{R}^n$, for some $n \in \mathbb{N}$. Also, suppose that every continuous real valued function on S is bounded. What can we say about the compactness of S? Let us, for instance, assume that S is not compact. Then, by the Hiene-Borel theorem, either S is not bounded or not closed. Suppose that S is not bounded. Then, for a fixed $p \in S$, $\forall r > 0$, $S \nsubseteq B(p, r)$. Hence, the function $f : S \to \mathbb{R}$ defined as $f(x) = d(x, p)$ is not bounded. On the other hand, if S is not closed, then there is an accumulation point x_0 of S which is not in S. In this case, the function $f : S \to \mathbb{R}$ defined by $f(x) = \frac{1}{d(x, x_0)}$ is unbounded.

Based on this, we have proved the following result, which is indeed a characterization of compactness in terms of continuous functions.

Theorem 6.18. *A subset $X \subseteq \mathbb{R}^n$ is compact if and only if every continuous real valued function on X is bounded.*

Exercise 6.25. Suppose that (X, d) is a non-compact metric space. Give an example of a function $f : X \to \mathbb{R}$ such that f is continuous but not bounded.

Now, we give an important consequence of continuity on compact sets, called the extreme value theorem. We have been using this result for long, especially when dealing with $\mathscr{C}[a, b]$.

Theorem 6.19 (Extreme value theorem). *A real valued continuous function on a compact metric space attains its bounds.*

Proof. Let X be a compact metric space and $f : X \to \mathbb{R}$ is continuous. Then, we know that $f(X)$ is bounded, i.e., $\exists m, M \in \mathbb{R}$ such that $m \le f(x) \le M$, $\forall x \in X$. All we need to prove is that $\exists x_0, x_1 \in X$ such that $f(x_0) = m = \inf\{f(x) \,|\, x \in X\}$ and $f(x_1) = M = \sup\{f(x) \,|\, x \in X\}$.

Suppose that f does not attain its upper bound, i.e., there is no $x \in X$ such that $f(x) = M$. Consider the collection of open sets U_n, where each set is given as

$$U_n = \left\{ x \in X \,\middle|\, f(x) < M - \frac{1}{n} \right\}.$$

The reader is advised to verify that each U_n is indeed open.

Now, by the Archimdean property, $\forall x \in X$, $\exists n \in \mathbb{N}$ such that $n > \frac{1}{M - f(x)}$. Therefore, $\bigcup_{n \in \mathbb{N}} U_n = X$, or what is the same, $\{U_n \,|\, n \in \mathbb{N}\}$ is an open cover for X. Also, notice that $U_n \subseteq U_{n+1}$. Since X is compact, we conclude that $X = U_N$ for some $N \in \mathbb{N}$ and hence $\forall x \in X$, $f(x) < M - \frac{1}{N}$. This leads us to $\sup\{f(x) \,|\, x \in X\} \le M - \frac{1}{N} < M$, which is a contradiction. Similar arguments can be made for the infimum. $\square$

Now, we give one application of the above result, a theorem due to Dini. But before that, we brush up some concepts that one might have encountered in the analysis course. Recall that we can define a sequence of functions in a similar manner as we defined a sequence of real numbers. We have done this for real valued functions on $\mathbb{R}$. Then follows point-wise and uniform convergence of sequence of functions. If X, Y are metric spaces, then we can do the same for a sequence of functions $f_n : X \to Y$. The definitions of point-wise and uniform convergence then remain similar to the real valued case. We encourage the reader to *write* these definitions with mathematical rigor.

Theorem 6.20 (Dini's theorem). *Let X be a compact metric space and $f_n :$ $X \to \mathbb{R}$ be continuous for each $n \in \mathbb{N}$. Moreover, let the sequence of functions (f_n) be monotone, i.e., $\forall x \in X$, let the sequence of real numbers $(f_n(x))$ be monotone. If (f_n) is a point-wise convergent sequence such that its point-wise limit is continuous, then it converges uniformly to its point-wise limit.*

Proof. For the proof, we assume that (f_n) is increasing. The decreasing case is analogous. Let $f_n \to f$ point-wise. Define $g_n : X \to \mathbb{R}$ as $g_n(x) = f(x) - f_n(x)$. Then, g_n is continuous for each $n \in \mathbb{N}$ and decreases to 0. For a given $\epsilon > 0$, consider the sets

$$U_k = \{x \in X | g_k(x) < \epsilon\}.$$

Then, $U_k = g_k^{-1}(-\infty, \epsilon)$, each of which are open due to continuity of g_k. Also, we observe that $U_k \subseteq U_{k+1}$. Since g_n decreases to 0, $\forall x \in X$, $\exists k \in \mathbb{N}$ such that $|g_k(x)| = g_k(x) < \epsilon$. Hence, $\bigcup_{k \in \mathbb{N}} U_k = X$, i.e., $\{U_k | k \in \mathbb{N}\}$ is an open cover for X.

Since X is compact, it has a finite subcover and in fact, $X = U_N$ for some $N \in \mathbb{N}$. Notice that this N is independent of x. Therefore, $\forall n \geq N$ and $\forall x \in X$, we have

$$|f(x) - f_n(x)| = |g_n(x)| < \epsilon.$$

Hence, $f_n \to f$ uniformly. $\qquad\square$

We now look at the fate of compact sets under continuous functions.

Theorem 6.21. *Continuous image of a compact space is compact.*

Proof. Let $f : X \to Y$ be a continuous function and X be a compact metric space. Let $\{U_i | i \in I\}$ be an open cover for $f(X)$. Then, by the continuity of f, the collection $\{f^{-1}(U_i) | i \in I\}$ is an open cover in X. Since X is compact, it has a finite subcover, say $\{f^{-1}(U_i) | i \in I_n\}$ for some $n \in \mathbb{N}$. Then, we have

$$\bigcup_{i=1}^{n} f^{-1}(U_i) = X.$$

So that,

$$f(X) = f\left(\bigcup_{i=1}^{n} f^{-1}(U_i)\right) = \bigcup_{i=1}^{n} f\left(f^{-1}(U_i)\right) \subseteq \bigcup_{i=1}^{n} U_i.$$

Hence, $f(X)$ has a finite subcover and therefore is compact. $\qquad\square$

Remark. The theorem also works for compact subsets of a metric space. Therefore, continuous image of a compact (sub)set is compact.

When we are dealing with compact spaces, we can say something more about continuous functions.

Theorem 6.22. *A continuous function from a compact metric space to any other metric space is uniformly continuous.*

Proof. Let $f : X \to Y$ be a continuous function, where X is a compact space. For each $x \in X$ and a given $\epsilon > 0$, by the continuity of f, $\exists \delta_x$ such that whenever $d_X (x, y) < \delta_x$, we have $d_Y (f (x), f (y)) < \frac{\epsilon}{2}$. Consider the open cover $\left\{ B_X \left(x, \frac{\delta_x}{2} \right) \middle| x \in X \right\}$ for X. Since X is compact, it admits a finite subcover, say $\left\{ B \left(x_i, \frac{\delta_i}{2} \right) \middle| i \in I_n \right\}$ for some $n \in \mathbb{N}$, where $\delta_i = \delta_{x_i}$.

Choose $\delta = \min_{i \in I_n} \{ \delta_i \}$. Then, for $x, y \in X$ with $d_X (x, y) < \frac{\delta}{2}$, we also have $x \in B \left(x_i, \frac{\delta_i}{2} \right)$ so that

$$d (x_i, y) \leq d (x, x_i) + d (x, y) < \frac{\delta_i}{2} + \frac{\delta}{2} \leq \delta_i.$$

Hence, $y \in B_X (x_i, \delta_i)$. Finally, we have

$$d_Y (f (x), f (y)) \leq d_Y (f (x), f (x_i)) + d_Y (f (x_i), f (y)) < \epsilon,$$

so that f is uniformly continuous on X. $\qquad\square$

Exercise 6.26. Let $f : (0, 1) \to \mathbb{R}$ be continuous, monotone, and bounded. Prove that it is uniformly continuous.

Recall that in Example 2.21, we used some special type of continuous functions, which were linear in subintervals. We now give a few formal definitions in that context.

Definition 6.8 (Linear function). *A function $f : \mathbb{R} \to \mathbb{R}$ is said to be linear if it is of the form $f (x) = \alpha x + \beta$ for some $\alpha, \beta \in \mathbb{R}$.*

Definition 6.9 (Piecewise linear function). *A function $f : \mathbb{R} \to \mathbb{R}$ is said to be piecewise linear on some interval $[a, b]$ if there is a partition $a = x_0 < x_1 < x_2 < \cdots < x_n = b$ such that f is linear on each of the subinterval $[x_i, x_{i+1}]$ for $i = 0, 1, \cdots, n - 1$.*

Remark. The definition of piecewise linear function can be extended to any interval (or in fact, any set) by making suitable modifications in finding the partition.

In Example 2.21, we saw that the set of piecewise linear function satisfying certain special properties at rational points in the domain is dense in $(\mathscr{C} [0, 1], d_\infty)$. We now see that it is true even for the set of all piecewise linear functions.

Example 6.17 (Density of piecewise linear functions). Consider the metric space $(\mathscr{C}[0,1], d_\infty)$ and the set D of all piecewise linear functions. We want to prove that D is dense in $\mathscr{C}[0,1]$. To do so, we would need to prove that $\forall f \in \mathscr{C}[0,1]$ and given $\epsilon > 0$, $\exists g \in D$ such that $d_\infty(f,g) < \epsilon$.

Since f is continuous on $[0,1]$, a compact interval, it is uniformly continuous. Hence, $\exists \delta > 0$ such that $\forall x, y \in [0,1]$ with $|x - y| < \delta$, we have $|f(x) - f(y)| < \frac{\epsilon}{2}$. By Archimedean property, $\exists n_0 \in \mathbb{N}$ such that $n_0 \delta > 1$.

We now partition the interval $[0,1]$ as $0 = x_0 < x_1 < x_2 \cdots < x_{n_0} = 1$, where $x_i = \frac{i}{n_0}$. Clearly, $|x_{i+1} - x_i| = \frac{1}{n_0} < \delta$. Define $g : [0,1] \to \mathbb{R}$ as

$$g(x) = n_0(x - x_i)(f(x_{i+1}) - f(x_i)) + f(x_i), \text{ for } x \in [x_i, x_{i+1}].$$

Note that g is piecewise linear, continuous, and agrees with f on the end points of each subinterval $[x_i, x_{i+1}]$. Hence, we have

$$|f(x_i) - g(x)| \le |f(x_i) - f(x_{i+1})| < \frac{\epsilon}{2}.$$

Finally, we have

$$|f(x) - g(x)| \le |f(x) - f(x_i)| + |f(x_i) - g(x)| < \epsilon.$$

Hence, $d_\infty(f,g) = \sup_{x \in [0,1]} \{|f(x) - g(x)|\} \le \epsilon$ and D is dense in $(\mathscr{C}[0,1], d_\infty)$. The working of this theorem can be seen in Figure 6.8.

Remark. The proof works for any $[a,b]$. We took the particular case $[0,1]$ due to its ease in handling.

We end this section by giving a beautiful property of bijective continuous functions on compact sets.

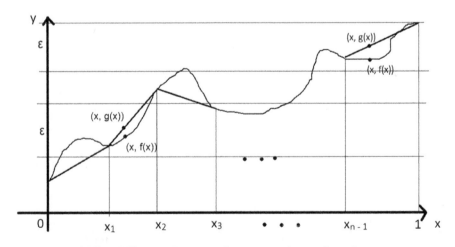

FIGURE 6.8: Density of piecewise linear functions.

Consider $f : X \to Y$, a bijective and continuous function defined on a compact space X. Let $F \subseteq X$ be a closed set. Then, we know that F is compact and hence $f(F)$ is compact in Y and hence closed (why?). Also, $f^{-1} : Y \to X$ is well-defined. Also, for any set $S \subseteq X$, its inverse image under f^{-1} is given by $f(S)$. Therefore, inverse image of closed sets under f^{-1} is closed and hence f^{-1} is continuous. Thus, we have proved the following.

Theorem 6.23. *A bijective continuous function from a compact metric space has continuous inverse.*

Remark. Indeed, such functions are special in the study of metric spaces and we shall deal with them later in this chapter. It is also easy to see that a continuous function on a compact space takes closed sets to closed sets.

Exercise 6.27. Give an example to show that Theorem 6.23 may not hold if X is not compact.

6.5 Continuous Functions on Connected Spaces

In this section, we shall study the behavior of continuous functions on connected sets and the fate of connected sets under continuous functions. However, we start with an amazing property of connected sets, which we like to call the *magic wand* for connectedness.

Theorem 6.24. *A metric space X is connected if and only if the only continuous functions from X to the discrete space $\{0, 1\}$ are constant.*

Proof. Suppose that there is a non-constant continuous function $f : X \to \{0, 1\}$. Define a set $A = \{x \in X | f(x) = 1\}$. Then, $f^{-1}(\{1\}) = A$ is both open and closed in X. Also, since f is non-constant, $\exists x \in X \setminus A$ so that A is a proper subset which is both open and closed. Hence, X is disconnected.

Taking the contrapositive, if X is connected, the only continuous functions are the constant ones. This proves the necessary part.

Now, to prove the sufficient part, let $A \subseteq X$ be both open and closed. We look at the characteristic function $\chi_A : X \to \{0, 1\}$, defined as

$$\chi_A(x) = \begin{cases} 1, & \text{if } x \in A. \\ 0, & \text{if } x \notin A. \end{cases}$$

Since A is both open and closed, $\chi_A^{-1}(\emptyset) = \emptyset$, $\chi_A^{-1}(\{0\}) = X \setminus A$, $\chi_A^{-1}(\{1\}) = A$ and $\chi_A^{-1}(\{0, 1\}) = X$ are all both open and closed. In particular, χ_A is continuous and hence constant. Therefore, either $\chi_A^{-1}(\{0\}) = X \setminus A = \emptyset$, i.e., $A = X$ or $\chi_A^{-1}(\{1\}) = A = \emptyset$. Hence, the only sets in X which are both open and closed are $\emptyset$ and X itself. Therefore, X is connected. $\qed$

Now, let us see the importance, rather the usefulness of our *magic wand*. The next result(s) tell us about the fate of unions of connected sets. Although, this could have been analyzed using the open set definitions of connectedness, we purposely put this here to show the reader that Theorem 6.24 is very helpful and makes proving/disproving connectedness easy. Indeed, this is the reason we like to call it the magic wand of connectedness. Before looking at the result, we would like to define a small technicality that we will be using further.

Definition 6.10 (Chained collection of sets). *A family $\mathscr{F}$ of sets is called a chained collection if for any $A, B \in \mathscr{F}$, there is an n-tuple of sets $(U_1, U_2, \cdots, U_n)$, where each $U_i \in \mathscr{F}$ for $i = 1, 2, \cdots, n$, and $U_1 = A$, $U_n = B$, such that $U_{i-1} \cap U_i \neq \emptyset$. Such a tuple is then called a chain from A to B.*

Theorem 6.25. *Let (X, d) be a metric space and let $\mathscr{F}$ be a chained collection of connected subsets of X. Then, $\bigcup_{S \in \mathscr{F}} S$ is also connected.*

Proof. Let $F = \bigcup_{S \in \mathscr{F}} S$. Let $f : F \to \{0, 1\}$ be a continuous function. We know that for each $S \in \mathscr{F}$, the restriction of f on S, given by $f|_S : S \to \{0, 1\}$ is constant. Let $x, y \in F$ be arbitrary. And, let $A, B \in \mathscr{F}$ such that $x \in A$ and $y \in B$. Since $\mathscr{F}$ is a chained collection, there is a chain $(U_1, U_2, \cdots, U_n)$ such that $U_1 = A$, $U_n = B$ and for $i = 2, \cdots, n$, we have $U_{i-1} \cap U_i \neq \emptyset$. Let $u_i \in U_{i-1} \cap U_i$. In particular, $u_i \in U_{i-1}$ and $u_i \in U_i$.

Now, this leads us to the fact that $x, u_2 \in A$, $u_2, u_3 \in U_2$, $u_3, u_4 \in U_3$, $\cdots$, $u_{n-1}, y \in B$. Since the restriction of f on each of these sets is constant, we have

$$f(x) = f(u_2) = f(u_3) = \cdots = f(y).$$

In particular, we have proved that for any $x, y \in F$, $f(x) = f(y)$, so that f is constant. By Theorem 6.24, we can conclude that F is indeed connected. Figure 6.9 shows the working of this theorem. $\square$

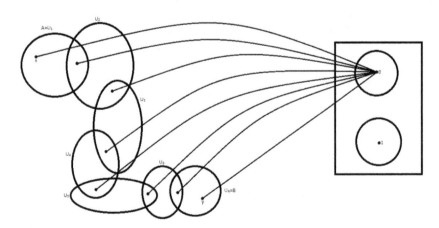

FIGURE 6.9: Diagrammatic representation of working of Theorem 6.25.

Immediately, by just looking at the definition of chained collection and properties of intersection, we have the following result.

Corollary 6.25.1. *If $\mathscr{F}$ is a collection of connected sets in a metric space, with non-empty intersection, then the union of this family is also connected.*

Exercise 6.28. Prove Corollary 6.25.1.

Exercise 6.29. Prove Theorem 5.4 using Theorem 6.25 and/or Corollary 6.25.1.

Now let us see what happens to an image of connected spaces under continuous functions. The reader will appreciate the ease with which we can prove connectedness using the above criterion in the following proof.

Theorem 6.26. *Continuous image of connected space is connected.*

Proof. Let $f : X \to Y$ be a continuous function, where X is a connected space. Let $g : f(X) \to \{0,1\}$ be continuous. Then, $g \circ f : X \to \{0,1\}$ is continuous and since X is connected, $g \circ f$ must be constant. Further, g must be constant so that $f(X)$ is connected. $\qquad\square$

Let us look at an immediate consequence of this result.

Theorem 6.27. *Any line segment in $\mathbb{R}^2$ is connected.*

Proof. Let AB be a line segment in $\mathbb{R}^2$. Define A to be origin, i.e., $A = (0,0)$ and B to be a point on the X-axis. Notice that if this is not so, then we can always translate and rotate this line segment to get it on the X-axis. Similarly, any line segment on the X-axis can be translated and rotated to get AB as required. Therefore, if we analyze the line segment on the X-axis with one end at origin, we would essentially be studying all line segments in $\mathbb{R}^2$.

Let the line segment on the X-axis have the end points $A = (0,0)$ and $B = (b,0)$, where $b > 0$. Consider the function $f : [0,b] \to \mathbb{R}^2$, defined as

$$f(x) = (x,0).$$

Then, we shall prove that f is continuous. To show this, let (x_n) be a sequence in $[0,b]$ such that $x_n \to x_0$, for some $x_0 \in [0,b]$. Then,

$$f(x_n) = (x_n,0) \to (x_0,0) = f(x_0).$$

Thus, f is continuous and hence, by Theorem 6.26, $f([0,b])$ is connected in $\mathbb{R}^2$. However, we have not yet proved that $f([0,b])$ is the line segment AB. To prove it, we shall show that f is a bijection from $[0,b]$ to the set $AB = \{(x,0) \in \mathbb{R}^2 | 0 \le x \le b\}$.

Indeed, if $f(x_1) = f(x_2)$, then $(x_1,0) = (x_2,0)$ implies $x_1 = x_2$. Thus, f is injective. And for any point $(x,0) \in AB$, since $0 \le x \le b$, we have x as the inverse image, so that f is also surjective. Therefore, we can conclude that $f([0, b])$ is the line segment AB.

This completes the proof of the fact that any line segment AB in $\mathbb{R}^2$ is connected. □

In fact, it can be easily proved that every line in $\mathbb{R}^2$ is connected, using the strategy above.

Note. We can indeed generalize this proof to $\mathbb{R}^n$ by doing the exact same analysis. However the rotation required will be complicated and will require the knowledge of inner product spaces. We urge the reader to look up to rotations of line segments in $\mathbb{R}^n$, which will be readily available in an intermediate level linear algebra book.

Exercise 6.30. Prove that $\mathbb{R}^2$ is connected. Can you generalize this to $\mathbb{R}^n$?

Exercise 6.31. Prove that the circle $\{(x, y) \in \mathbb{R}^2 | x^2 + y^2 = 1\}$ in $\mathbb{R}^2$ is connected.

Exercise 6.32. Prove that the set of all invertible matrices given by $GL(n, \mathbb{R})$ is not connected.

Let us look at the product space $(X \times Y, d)$ of two metric spaces (X, d_X) and (Y, d_Y). We want to know: When is the product space connected? Indeed, as it may be intuitively clear, whenever X and Y are connected, we would want the product space to be connected as well. The next result tells us that this is exactly when the product space is connected.

Theorem 6.28. *Let (X, d_X) and (Y, d_Y) be metric spaces and let $(X \times Y, d)$ be the product space. Then, the $X \times Y$ is connected if and only if the individual coordinate spaces X and Y are connected.*

Proof. First, let us consider that the product space is connected. Consider the projection map Π_X as in Definition 6.5. Since projection map is continuous, we have from Theorem 6.26, $\Pi_X(X \times Y) = X$ is connected. Similarly, Y is also connected.

Conversely, let X and Y be connected metric spaces. Then, we want to prove that $X \times Y$ is also connected. To see this, we will prove that for any two elements (x, y) and (x', y') in $X \times Y$, there is a connected set containing both. Consider the two subsets $\{x\} \times Y$ and $X \times \{y'\}$. Consider the maps $f : Y \to \{x\} \times Y$ defined as $f(z) = (x, z)$ and $g : X \to X \times \{y'\}$ defined as $g(z) = (z, y')$. Clearly, f and g are continuous and bijective so that $\{x\} \times Y$ and $X \times \{y\}$ are connected (since X and Y are connected).

Now, $(x, y), (x, y') \in \{x\} \times Y$ and $(x, y'), (x', y') \in X \times \{y'\}$. Therefore, $(\{x\} \times Y) \cap (X \times \{y'\}) \neq \emptyset$ and by Corollary 6.25.1, the set $(\{x\} \times Y) \cup (X \times \{y'\})$ is connected. This is precisely the set which contains both the points (x, y) and (x', y'). Hence, the product space $X \times Y$ is connected. □

Exercise 6.33. Let $f : (X, d_X) \to (Y, d_Y)$ be a continuous map, where X is connected. Prove that the graph, $G(f)$, is connected in the product metric. Here,

$$G(f) = \{(x, f(x)) \in X \times Y | x \in X\}.$$

Before moving on to further consequence of continuous functions on connected sets, let us look at a cliché example (found in almost every book on metric spaces and topology) which utilizes the fact that continuous images of connected sets are connected.

Example 6.18 (Topologist's sine curve). Consider the set $A = \left\{ (x, \sin \frac{1}{x}) \in \mathbb{R}^2 | 0 < x \le \frac{1}{\pi} \right\}$. We want to see if A is connected. Consider the function $f : (0, \frac{1}{\pi}] \to \mathbb{R}^2$ as

$$f(x) = \left(x, \sin \frac{1}{x} \right).$$

Let (x_n) be a sequence in $(0, \frac{1}{\pi}]$ such that $x_n \to x \in (0, \frac{1}{\pi}]$. Then, we have

$$f(x_n) = \left(x_n, \sin \frac{1}{x_n} \right) \to \left(x, \sin \frac{1}{x} \right) = f(x).$$

Hence, f is continuous and it is easy to see that $f\left((0, \frac{1}{\pi}] \right) = A$ so that A is connected.

However, we can move another step ahead! We know that closure of a connected set is connected. To see what the closure of A is, first let us observe $(0, 0)$. Notice that if we consider a sequence $\left(\frac{1}{n\pi} \right)_{n \in \mathbb{N}}$, then we have $\left((\frac{1}{n\pi}, \sin n\pi) \right)$, a sequence in A, which converges to $(0, 0)$. Therefore, $(0, 0) \in \bar{A}$.

What did we do here? We know that as $x \to 0$, the frequency of $\sin \frac{1}{x}$ crossing the same y-coordinate increases. Can we do this for any other point on the Y-axis? Indeed, if $(0, y) \in \mathbb{R}^2$ with $|y| \le 1$, we can always translate the whole plane so that this point becomes the origin and do the analysis as above. Therefore, every point in the set $\{(0, y) \in \mathbb{R}^2 | |y| \le 1\}$ is an accumulation point of A.

It is clear that any point of the plane other than these cannot be accumulation points since they are "far away" from the sine curve itself. Therefore, $\bar{A} = A \cup \{(0, y) \in \mathbb{R}^2 | |y| \le 1\}$ is also connected. The set A is called a *Topologist's sine curve* and can be seen in Figure 6.10.

Remark. It is astonishing at first sight that the set $\bar{A}$ as mentioned in Example 6.18 is connected. This is because we can see that the sine curve never actually touches the Y-axis, yet the union of the curve and a part of Y-axis is in a single piece. Indeed, this is an example where the disjoint union of two

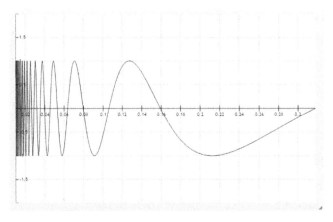

FIGURE 6.10: The Topologist's sine curve.

connected sets can be connected. The small observation of not touching the Y-axis will be dealt with later in the text, when the reader is introduced to a concept called *path connectedness*.

Now, we look at a consequence of continuous real-valued functions on connected sets in $\mathbb{R}$. The following theorem, called the intermediate value theorem, is well known in analysis and the reader might have encountered its proof earlier. However, we wish to show the role of connectedness.

Theorem 6.29 (Intermediate value theorem). *Let $f : [a, b] \to \mathbb{R}$ be a continuous function. Let y be a point between $f(a)$ and $f(b)$, i.e., let $f(a) \leq y \leq f(b)$ or $f(b) \leq y \leq f(a)$ hold. Then, $\exists x \in [a, b]$ such that $f(x) = y$.*

Proof. We know that $[a, b]$ is connected in $\mathbb{R}$ and hence $f([a, b])$ is also connected. As a consequence, $f([a, b])$ must be an interval. From the definition of interval, $y \in f([a, b])$ so that $\exists x \in [a, b]$ with $f(x) = y$. □

We now supply the reader with a consequence of intermediate value theorem.

Theorem 6.30. *Any polynomial with real coefficients and of odd degree has a real root.*

Proof. Let $p(x) = a_n x^n + a_{n-1} x^{n-1} + \cdots a_0$ be a polynomial where $a_i \in \mathbb{R}$ for $i = 0, 1, 2, \cdots, n$, $a_n \neq 0$ and n is odd. We may assume that $a_n = 1$, without loss of generality. Because if not, we can divide by a_n and get $p(x) = a_n q(x)$, where $q(x)$ is a polynomial of odd degree with leading coefficient 1 and the roots of p and q coincide. Hence, we shall consider the polynomial

$$p(x) = x^n + a_{n-1} x^{n-1} + \cdots + a_0.$$

For $x \neq 0$, we can rewrite this as $p(x) = x^n f(x)$, where

$$f(x) = 1 + \frac{a_{n-1}}{x} + \frac{a_{n-2}}{x^2} + \cdots + \frac{a_0}{x^n}.$$

For $|x| > 1$, we have

$$|f(x) - 1| \leq \frac{|a_{n-1}|}{|x|} + \frac{|a_{n-2}|}{|x|^2} + \cdots + \frac{|a_0|}{|x|^n}$$

$$\leq \frac{1}{|x|} \sum_{i=0}^{n-1} |a_i|.$$

Let $A = \sum_{i=0}^{n-1} |a_i|$. For $|x| > \max\{1, 2A\}$, we have

$$|f(x) - 1| < \frac{1}{2}$$

so that for such an x, we have $f(x) > 0$. So, we choose an $x_0 > \max\{1, 2A\}$. Then, $f(x_0) > 0$ and $f(-x_0) > 0$. However, $x_0^n > 0$ and $(-x_0)^n < 0$ so that $p(x_0) > 0$ and $p(-x_0) < 0$. Therefore, by intermediate value theorem, p has a root in $[-x_0, x_0]$. $\qquad\square$

Exercise 6.34. As a consequence of the intermediate value property, prove that for any $\alpha \in [0, \infty)$ and $n \in \mathbb{N}$, there is an $x \in [0, \infty)$ such that $x^n = \alpha$. This proves the existence of nth roots of non-negative real numbers. Can we apply this to negative real numbers?

Exercise 6.35. Let $f, g : [0, 1] \to \mathbb{R}$ be continuous functions. Assume that $f(x) \in [0, 1]$ for all $x \in [0, 1]$ and $g(0) = 0$, $g(1) = 1$. Show that $\exists x \in [0, 1]$ such that $f(x) = g(x)$.

Exercise 6.36. Give an example which explains that if X is not connected, then there is a function $f : X \to \mathbb{R}$ which does not have the intermediate value property.

Exercise 6.37. Let $f : [-1, 1] \to [-1, 1]$ be continuous. Show that $\exists x \in [-1, 1]$ such that $f(x) = x$.

Let us now look at an example which explains the importance of the above mentioned results.

Example 6.19. Suppose we want to construct a continuous function $f : \mathbb{R} \to \mathbb{R}$ such that f takes irrational values on rational points and rational values on irrational points. Is it possible? Let us assume that it is! Suppose we are able to construct such a function. What will be its consequences?

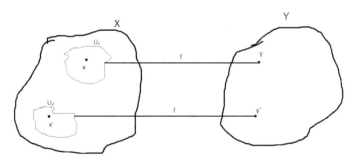

FIGURE 6.11: A locally continuous function.

First, observe that since $\mathbb{R}$ is connected, $f(\mathbb{R})$ will be an interval. Therefore, either $f(\mathbb{R})$ is a singleton or it is uncountable. Also, since f takes at least one rational value and one irrational value, it is not constant and $f(\mathbb{R})$ is not a singleton set. Therefore, $f(\mathbb{R})$ is uncountable.

Now, f can take each value from $\mathbb{Q}$ to a unique value in $f(\mathbb{Q})$. Therefore, $f(\mathbb{Q})$ is countable. Also, f takes irrational points to rationals so that $f(\mathbb{R} \setminus \mathbb{Q}) \subseteq \mathbb{Q}$ and therefore, $f(\mathbb{R} \setminus \mathbb{Q})$ is also countable.

Finally, we know that $\mathbb{R} = \mathbb{Q} \cup (\mathbb{R} \setminus \mathbb{Q})$ so that $f(\mathbb{R}) = f(\mathbb{Q} \cup (\mathbb{R} \setminus \mathbb{Q})) = f(\mathbb{Q}) \cup f(\mathbb{R} \setminus \mathbb{Q})$ is countable, which is a contradiction! Therefore, there is no such continuous function.

Definition 6.11 (Locally constant function). *A function $f : X \to Y$ is locally constant if for every $x \in X$, there is an open set $U_x \subseteq X$ such that $x \in U_x$ and f is constant on U_x (Figure 6.11).*

Exercise 6.38. Is the function $f : \mathbb{R} \to \mathbb{R}$, given by

$$f(x) = \begin{cases} -1, & \text{if } x < 0 \\ 0, & \text{if } x = 0 \\ 1, & \text{if } x > 0 \end{cases}$$

locally constant?

We now give a result for locally constant functions on connected spaces.

Theorem 6.31. *Let X be a connected metric space. Then, any locally constant function $f : X \to Y$ is constant on X.*

Proof. Let $x_0 \in X$ and consider the set

$$S = \{x \in X | f(x) = f(x_0)\}.$$

Notice that $S \neq \emptyset$ since $x_0 \in S$. Let $x \in S$. Then, $\exists U_x \subseteq X$, open such that $x \in U_x$ and f is constant on U_x. Hence, $\forall y \in U_x$, $f(y) = f(x) = f(x_0)$ so that $U_x \subseteq S$. Therefore, S is open.

On the other hand, let $x \in X \setminus S$. Again, $\exists U_x \subseteq X$, open such that $x \in U_x$ and f is constant on U_x. Therefore, $\forall y \in U_x$, we have $f(y) = f(x) \neq f(x_0)$ so that $U_x \subseteq X \setminus S$. Hence, $X \setminus S$ is open and S is closed.

This proves that S is both open and closed. Since X is connected and $S \neq \emptyset$, we must have $S = X$ so that f is constant. □

Exercise 6.39. Suppose that every locally constant function $f : X \to Y$ is constant for metric spaces X, Y with at least two distinct points. Is X connected?

Exercise 6.40. Let $f : X \to \mathbb{R}$ be a non-constant continuous function on a connected metric space. Show that $f(X)$ is uncountable and so is X.

Before we end this section, which enhances our understanding of connectedness through continuous function, we would like to introduce the reader to a stronger concept of connectedness, what is called *path connectedness*.

6.5.1 Path connectedness

In Chapter 5, we have seen what is meant by connectedness. Intuitively, it means to have the whole space in a single piece. Now, we wish to have another property in our metric space. We want to travel from one point to another. A first time reader may argue that this is simple! Just draw a straight line between two points and then travel along this line. However, it must be understood that in an arbitrary setting, this is not always possible. This is because to make "lines", we would require the two operations, called addition and scalar multiplication, which need not be at our hands.

Thus, our curiosity demands us to first understand how can we travel from point to point in a metric space. Suppose that we could do so. Then, we would be tracing a *path* between two points. Therefore, we first understand what is meant by a path.

Suppose x and y are two points of a set X. Then, the path must start at one of these points, say x for the time being, and should end at the other. Physically speaking, let the "time" taken to do so be 1 unit. Then, between the time 0 and 1, we should be somewhere along the path. Also, there must not be any breaks in the path, i.e., the path must be continuous. Based on this intuition, we now define a path in arbitrary metric space.

Definition 6.12 (Path). *Let (X, d) be a metric space and $x, y \in X$. A path from x to y is a continuous function $f : [0, 1] \to X$ such that $f(0) = x$ and $f(1) = y$.*

The following exercise gives the reader basic but important properties about paths.

Exercise 6.41. Prove that a path is uniformly continuous, connected, and compact.

Exercise 6.42. Prove that if in a metric space (X, d), there is a path from x to y, then there is also a path from y to x.

Based on the definition of path, we now define a path connected metric space.

Definition 6.13 (Path connectedness). *A metric space (X, d) is path connected if for any two points x and y of X, there is a path between them.*

An amazing consequence of this definition is that in a path connected metric space, we can travel from one point to the other.

Let us look at a few examples of path connected spaces.

Example 6.20. Consider the metric space $(\mathbb{R}, |\cdot|)$ and $x, y \in \mathbb{R}$. Define $f : [0, 1] \to \mathbb{R}$ as

$$f(z) = (y - x) z + x.$$

Clearly, f is continuous and $f(0) = x$, $f(1) = y$ so that f is a path. Hence, $\mathbb{R}$ is path connected.

Example 6.21. Consider the set $\mathbb{R}^2$, with any p-metric. We will check if this is path connected. Let $(x_1, y_1), (x_2, y_2) \in \mathbb{R}^2$. Then, we know we can always have a line segment from (x_1, y_1) to $(0, 0)$ and then from $(0, 0)$ to (x_2, y_2). The union of these two segments will give us our path. These two line segments can be given by the functions $f : [0, 1] \to \mathbb{R}^2$ and $g : [0, 1] \to \mathbb{R}^2$ defined as

$$f(z) = ((1 - z) x_1, (1 - z) y_1),$$

and

$$g(z) = (z x_2, z y_2).$$

We leave it to the reader to verify that there are continuous and indeed are paths. Thus, the function $h : [0, 1] \to \mathbb{R}^2$ defined as

$$h(z) = \begin{cases} f(2z), & 0 \le z \le \dfrac{1}{2} \\ g(2z - 1), & \dfrac{1}{2} \le z \le 1 \end{cases}$$

is well-defined since $f(1) = g(0)$. Also, h is continuous (verify!). Thus, h is the required path between (x_1, y_1) and (x_2, y_2), which proves that $\mathbb{R}^2$ is path connected.

Exercise 6.43. Is $\mathbb{R}^n$ path connected?

Let us now look at what happens to $\mathbb{R}$ when one point is removed.

Example 6.22. Consider $\mathbb{R}^*$. We know that it is disconnected. Therefore, we intuitively know that we cannot "cross" the 0-barrier, i.e., if we take a negative number and a positive number it would not be possible for us to make a path between them. Indeed if $f : [0,1] \to \mathbb{R}$ is a path between two points $x < 0 < y$, then by the intermediate value theorem, there must be a point $z \in [0,1]$ such that $f(z) = 0$. Since we do not have 0 to be mapped to, such a path is not possible in $\mathbb{R}^*$. Hence, $\mathbb{R}^*$ is not path connected.

Now, we can also ask what happens when a point from $\mathbb{R}^2$ is removed. Does it still stay path connected. Let us answer this question in the following example.

Example 6.23. Consider the set $\mathbb{R}^2$ with any p-metric. Define $\mathbb{R}^{2*} = \mathbb{R}^2 \setminus \{(0,0)\}$. We want to check if this set is path connected. In this example, we shall explain how to get paths between two points and leave it to the reader to write it rigorously. First, given any two points (x_1, y_1) and (x_2, y_2), either $(0,0)$ will lie on the line segment joining the two points, or it won't. If it does not lie on the line segment, this is the required path. If it does lie on the line segment (but the points are not on the axes), we will avoid origin by either traveling along the x-direction or the y-direction first and then the other direction to reach the required point. If we choose to travel along the x-direction first, then there will be a path between (x_1, y_1) and (x_2, y_1) and then another path between (x_2, y_1) and (x_2, y_2). The union of these two paths will give the required path. Now, suppose that the two points lie on the axes. We shall take the case where they lie on the X-axis. A similar idea can give a path between them when they lie on the Y-axis. If the points are $(x_1, 0)$ and $(x_2, 0)$ such that origin lies on the line segment joining them, then one of them is on the positive side and the other is on the negative side. Suppose that $(x_1, 0)$ is on the positive side. Then, we can have a straight line path between $(x_1, 0)$ and $(0, 1)$ and another straight line path between $(0, 1)$ and $(x_2, 0)$. The union of these two line segments will be our required path to complete the proof that $\mathbb{R}^{2*}$ is path connected. The complete idea is shown in Figure 6.12.

Exercise 6.44. Complete the proof that $\mathbb{R}^{2*}$ is path connected by following the steps of Example 6.23.

Remark. Since we can have translations in $\mathbb{R}$ and $\mathbb{R}^2$, the results would remain the same for removal of any point. There is nothing special about 0 and $(0,0)$. We chose to remove these points because it makes further analysis easier. For the removal of any other point, we can always translate the whole space to make the removed point as origin.

Exercise 6.45. Is the discrete space path connected?

In Example 6.21, we had an idea that since we know paths from each point to $(0,0)$, the whole space must be path connected. We now generalize this idea to arbitrary metric spaces.

Theorem 6.32. *Let (X, d) be a metric space and $x_0 \in X$ be fixed. Then, X is path connected if and only if there is a path between x and x_0 for each $x \in X$.*

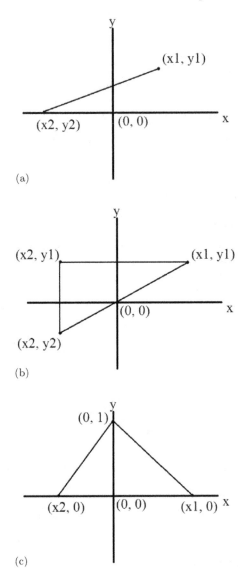

FIGURE 6.12: Geometrical representation of idea behind Example 6.23. (a) Path between (x_1, y_1) and (x_2, y_2) when $(0,0)$ does not lie on the line segment joining the two points. (b) Path between (x_1, y_1) and (x_2, y_2) when $(0,0)$ lies on the line segment joining the two points but the two points do not lie on either of the axes. (c) Path between (x_1, y_1) and (x_2, y_2) when $(0,0)$ lies on the line segment joining the two points and the two points lie on the X-axis.

Proof. First, if X is path connected, then it is clear that there is a path between x and x_0 for each $x \in X$.

Conversely, if there is a path between x and x_0 for each $x \in X$, we want to show that X is path connected. Let x and y be two points in X and let $f : [0,1] \to X$ and $g : [0,1] \to X$ be the paths between x and x_0 and x_0 and y. Define the function $h : [0,1] \to X$ as

$$
h(z) = \begin{cases} f(2z), & 0 \leq z \leq \dfrac{1}{2}. \\ g(2z-1), & \dfrac{1}{2} \leq z \leq 1. \end{cases}
$$

Then, h is well-defined, continuous, and $h(0) = x$, $h(1) = y$. Therefore, X is path connected. $\square$

In Example 6.22, we made a comment that since we know $\mathbb{R}^*$ is disconnected, we expect it to be path disconnected. Generalizing this idea, if we know that a metric space is disconnected, then it has its components far apart so that we cannot always travel from one point to another. This is equivalent to saying that if we can always travel from one point to another, then the whole space must be in a single piece, i.e., it must be connected. The next theorem tells us that this is indeed true.

Theorem 6.33. *A path connected space is connected.*

Proof. Let (X,d) be a path connected metric space. Fix a point $x \in X$. Since it is path connected, for every point $y \in X$, there is a path, say $f_y : [0,1] \to X$ joining x and y. Also, $x, y \in f_y([0,1])$. Therefore, $X = \bigcup_{y \in X} f_y([0,1])$. We also note that since each path f_y starts at x, $\bigcap_{y \in X} f_y[0,1] \neq \emptyset$. Hence, by Corollary 6.25.1, we can conclude that X is connected. $\square$

As usual, we now ask the converse question: Is every connected set path connected? If so, then we have not being studying anything different! The next example shows that this is not the case.

Example 6.24. Consider the closure of topologist's sine curve as in Example 6.18. We have seen that this is a connected set. Now, we will show that it is not path connected.

Let, if possible, it be path connected. Consider the two points $(0,0)$ and $(1,0)$ in $\bar{A}$. By our assumption, there should be a path $f : [0,1] \to \bar{A}$ such that f is continuous and $f(0) = (0,0)$ and $f(1) = (1,0)$. We shall consider the following notation: $B = \{(0,y) \in \mathbb{R}^2 | |y| \leq 1\}$. Then, B is a closed subset of $\bar{A}$ (why?) and $0 \in f^{-1}(B)$. Since $f^{-1}(B) \subseteq [0,1]$, it is bounded and hence has a least upper bound, say l and in fact, $l \in f^{-1}(B)$.

Since $(1,0) \notin B$, it is clear that $0 < l < 1$. Now, we shall denote $f(x) = (f_1(x), f_2(x))$, where $f_1(x) = x$ and $f_2(x) = \sin\left(\frac{1}{x}\right)$. We know that f is continuous if and only if f_1 and f_2 are both continuous. Let $\delta > 0$ such that

$l + \delta \leq 1$. Then, $f_1 (l + \delta) > 0$ so that by the Archimedean property, there is some $n \in \mathbb{N}$ with the property that $f_1 (l) < \frac{2}{(4n+1)\pi} < f_1 (l + \delta)$. Applying the intermediate value property, there is some z with $l < z < l + \delta$ such that $f_1 (z) = \frac{2}{(4n+1)\pi}$. Hence, $f_2 (z) = 1$ so that $|f_2 (l) - f_2 (z)| \geq 1$, although $|l - z| < \delta$. Hence, f_2 is not continuous at l, which is a contradiction to the fact that f was a path! Therefore, $\bar{A}$ cannot be path connected.

The next theorem gives the fate of path connected spaces under continuous functions.

Theorem 6.34. *Continuous image of path connected sets is path connected.*

Proof. Let $f : (X, d_X) \rightarrow (Y, d_Y)$ be a continuous function, where X is path connected. Let $y_1, y_2 \in f (X)$. Then, $\exists x_1, x_2 \in X$ such that $f (x_1) = y_1$ and $f (x_2) = y_2$. Since X is path connected, there is a path $g : [0, 1] \rightarrow X$ that joins x_1 and x_2. Consider the function $h : [0, 1] \rightarrow Y$ defined as $h \equiv f \circ g$. Then, h is continuous. Also, $h (0) = y_1$ and $h (1) = y_2$. Hence, h is a path and $f (Y)$ is path connected. $\square$

Exercise 6.46. Prove that the unit circle and parabola are path connected. Is the complete hyperbola path connected?

6.6   Equicontinuity and Arzela-Ascoli's Theorem

In this section, we shall study families of functions for continuity and then state an important theorem due to Arzela and Ascoli. We begin with a definition.

Definition 6.14 (Equicontinuity). *Let X, Y be metric spaces and $x_0 \in X$. A family $\mathscr{A}$ of functions from X to Y is said to be equicontinuous at x_0 if $\forall \epsilon > 0$, $\exists \delta > 0$ such that $\forall f \in \mathscr{A}$ and $\forall x \in X$ with $d_X (x, x_0) < \delta$, we have $d_Y (f (x), f (x_0)) < \epsilon$.*

A stronger version of equicontinuity is uniform equicontinuity, which can be defined as follows.

Definition 6.15 (Uniform equicontinuity). *Let X, Y be metric spaces. A family $\mathscr{A}$ of functions from X to Y is said to be uniformly equicontinuous if $\forall \epsilon > 0$, $\exists \delta > 0$ such that $\forall f \in \mathscr{A}$ and $\forall x, y \in X$ with $d_X (x, y) < \delta$, we have $d_Y (f (x), f (y)) < \epsilon$.*

Remark. Note that equicontinuity does not only mean that every member is continuous. Also, each member is continuous to the "same extent" around a point $x_0 \in X$. This is because we are first given an $\epsilon > 0$ for which we have to find a $\delta > 0$ such that the conditions of continuity work for this common δ on each $f \in \mathscr{A}$. Same is true for uniform equicontinuity. Indeed, every member of (uniformly) equicontinuous family is (uniformly) continuous.

Exercise 6.47. Let $f : X \to Y$ be a function, where X, Y are metric spaces. Give equivalent conditions for $\{f\}$ to be equicontinuous and uniformly equicontinuous.

Let us now look at an example which shows a family of equicontinuous functions.

Example 6.25. Let X be a compact metric space and $F : X \times X \to Y$ be continuous, when $X \times X$ is equipped with the product metric. For each $x \in X$, define $f_x : X \to Y$ as $f_x(u) = F(u, x)$ and consider the family of functions $\mathscr{A} = \{f_x | x \in X\}$. Indeed, each element of $\mathscr{A}$ is continuous (verify!). Fix a point $u_0 \in X$. We want to estimate $d_Y(f_x(u), f_x(u_0))$ for an arbitrary x. Since X is compact, F is uniformly continuous. Therefore, for a given $\epsilon > 0$, choose δ same as that for the uniform continuity of F, and we prove that $\mathscr{A}$ is equicontinuous.

Exercise 6.48. Let X, Y be metric spaces where X is compact. Let $\mathscr{A}$ be a family of equicontinuous functions from X to Y. Prove that $\mathscr{A}$ is uniformly equicontinuous.

We now state and prove the Arzela-Ascoli's theorem.

Theorem 6.35 (Arzela-Ascoli). *Let X be a compact metric space and let $\mathscr{C}(X, \mathbb{K})$, the set of all continuous functions from X to $\mathbb{K}$, be equipped with the metric d_∞. Here $\mathbb{K}$ is either $\mathbb{R}$ or $\mathbb{C}$. Then, a set $\mathscr{B} \subseteq \mathscr{C}(X, \mathbb{K})$ is compact if and only if $\mathscr{B}$ is bounded, closed, and equicontinuous.*

Proof. First, let us assume that $\mathscr{B}$ is compact. Then, we know that it is closed and bounded. In fact, it is totally bounded. All we need to prove is that $\mathscr{B}$ is equicontinuous. So, let $\epsilon > 0$ be given. By the compactness, there are finitely many functions $f_i \in \mathscr{B}$ for $1 \le i \le n$ (for some $n \in \mathbb{N}$) such that $\mathscr{B} \subseteq B(f_i, \frac{\epsilon}{3})$. This means $\forall f \in \mathscr{B}, \exists f_i \in \mathscr{B}$ such that $d_\infty(f, f_i) < \frac{\epsilon}{3}$. Hence, $\forall x \in X$, we have $d_\mathbb{K}(f(x), f_i(x)) \le d_\infty(f, f_i) < \frac{\epsilon}{3}$. Let $\delta_i > 0$ such that $\forall x, y \in X$ with $d_X(x, y) < \delta_i$, we have $d_\mathbb{K}(f_i(x), f_i(y)) < \frac{\epsilon}{3}$. This is possible because X is compact and hence each element of $\mathscr{B}$ is uniformly continuous. Now, define $\delta = \min_{1 \le i \le n} \{\delta_i\}$. Then, for any points $x, y \in X$ with $d_X(x, y) < \delta$ and $f \in \mathscr{B}$, we have

$$d_\mathbb{K}(f(x), f(y)) \le d_\mathbb{K}(f(x), f_i(x)) + d_\mathbb{K}(f_i(x), f_i(y)) + d_\mathbb{K}(f_i(y), f(y))$$

$$< \frac{\epsilon}{3} + \frac{\epsilon}{3} + \frac{\epsilon}{3}$$

$$= \epsilon.$$

Hence, $\mathscr{B}$ is uniformly equicontinuous and, in particular equicontinuous.

Now, conversely, assume that $\mathscr{B}$ is closed, bounded, and equicontinuous. Since X is compact, $\mathscr{B}$ is uniformly equicontinuous. Also, $(\mathscr{C}(X, \mathbb{K}), d_\infty)$ is complete so that $\mathscr{B}$ is complete. Therefore, to prove that $\mathscr{B}$ is compact, it will be sufficient to prove that it is totally bounded.

Let $\epsilon > 0$ be given. Let $M > 0$ be such that $|f(x)| \leq M$ for all $x \in X$ and $f \in \mathscr{B}$. Using the uniform equicontinuity of $\mathscr{B}$, $\exists \delta > 0$ such that $\forall f \in \mathscr{B}$ and $\forall x, y \in X$ with $d_X(x, y) < \delta$, we have $d_\mathbb{K}(f(x), f(y)) < \frac{\epsilon}{4}$.

Since X is compact, there are finitely many $x_i \in X$ for $1 \leq i \leq n$, for some $n \in \mathbb{N}$ such that $X = \bigcup\limits_{i=1}^{n} B_X(x_i, \delta)$. Also, the closed ball $\overline{B_\mathbb{K}(0, M)}$ is compact in $\mathbb{K}$ and hence is totally bounded. Hence, there are finitely many $y_i \in \mathbb{K}$ for $1 \leq i \leq m$, for some $m \in \mathbb{N}$ such that $\overline{B_\mathbb{K}(0, M)} \subseteq \bigcup\limits_{i=1}^{m} B_\mathbb{K}(y_i, \frac{\epsilon}{4})$.

Let A be the set of all functions from the set $\{x_i | 1 \leq i \leq n\}$ to $\{y_i | 1 \leq i \leq m\}$. Then, $\forall \alpha \in A$, define a set

$$U_\alpha = \left\{ f \in \mathscr{B} \,\middle|\, d_\mathbb{K}(f(x_i), \alpha(x_i)) < \frac{\epsilon}{4}, \text{ for } 1 \leq i \leq n \right\}.$$

Now, let $x \in X$ and $f, g \in U_\alpha$. Then, $x \in B_X(x_i, \delta)$ for some x_i. Therefore, we have

$$d_\mathbb{K}(f(x), f(x_i)) < \frac{\epsilon}{4},$$

$$d_\mathbb{K}(g(x), g(x_i)) < \frac{\epsilon}{4}.$$

Also,

$$d_\mathbb{K}(f(x_i), \alpha(x_i)) < \frac{\epsilon}{4},$$

$$d_\mathbb{K}(g(x_i), \alpha(x_i)) < \frac{\epsilon}{4}.$$

Hence

$$d_\mathbb{K}(f(x_i), g(x_i)) \leq d_\mathbb{K}(f(x_i), \alpha(x_i)) + d_\mathbb{K}(g(x_i), \alpha(x_i)) < \frac{\epsilon}{2}.$$

Finally, we can conclude that

$$d_\mathbb{K}(f(x), g(x)) \leq d_\mathbb{K}(f(x), f(x_i)) + d_\mathbb{K}(f(x_i), g(x_i)) + d_\mathbb{K}(g(x_i), g(x))$$

$$< \frac{\epsilon}{4} + \frac{\epsilon}{2} + \frac{\epsilon}{4} = \epsilon.$$

Hence, the diameter of U_α is at most ϵ. Finally, we shall show that $\mathscr{B} = \bigcup\limits_{\alpha \in A} U_\alpha$. For the same, for a given $f \in \mathscr{B}$ and any i, with $1 \leq i \leq n$, choose a j with $1 \leq j \leq m$ such that $f(x_i) \in B_\mathbb{K}(y_j, \frac{\epsilon}{4})$. Then, consider the function α defined so that $\alpha(x_i) = y_j$. Clearly, $d_\mathbb{K}(f(x_i), \alpha(x_i)) < \frac{\epsilon}{4}$ for every x_i and

$1 \leq i \leq n$. Hence, $f \in U_\alpha$ for some α, or what is the same, $\mathscr{B} = \bigcup_{\alpha \in A} U_\alpha$. Therefore, $\{U_\alpha | \alpha \in A\}$ is an ϵ-net for $\mathscr{B}$. Since A is finite (why?), $\mathscr{B}$ is totally bounded and hence compact. This completes the proof. □

6.7 Open and Closed Maps

Earlier, in characterization of continuity, we talked about inverse images of open and closed subsets. In this section, we shall see the images of open and closed subsets and then comment upon some properties of functions.

Definition 6.16 (Open and closed maps). *A function $f : X \to Y$, where X and Y are metric spaces, is open (closed) if the image of every open (closed) set in X is open (closed) in Y.*

Remark. It is natural for a first time reader to think that talking about open and closed maps is the same as talking about continuity of the map. Also, one may get confused that every open map is also closed and vice-versa. The following examples will help to clear this confusion.

Example 6.26. Consider a function $f : X \to Y$ defined as $f(x) = y_0$, for some $y_0 \in Y$. This is a constant function and $f(X) = \{y_0\}$ is closed in Y. Therefore, f takes closed sets in X to a closed set in Y but does not take open sets in X to open sets in Y. Hence, this is an example of a continuous map which is closed but not open.

Example 6.27. Consider the projection map $\Pi_1 : \mathbb{R}^2 \to \mathbb{R}$ defined as $\Pi_1(x, y) = x$. Let $U \subseteq \mathbb{R}^2$ be any open set and let $x \in \Pi_1(U)$. Let $y \in \mathbb{R}$ such that $(x, y) \in U$. Then, $\exists r > 0$ such that $B((x, y), r) \subseteq U$. Hence, $\Pi_1(B((x, y), r)) = (x - r, x + r) \subseteq \Pi_1(U)$ and therefore, $\Pi_1(U)$ is open. This means that Π_1 is an open map.

Now, consider the closed set $F = \{(x, y) \in \mathbb{R}^2 | x > 0 \text{ and } xy = 1\}$. Then, $\Pi_1(F) = (0, \infty)$ is not closed. Therefore, projection maps are examples of continuous maps which are open but not closed.

Exercise 6.49. Give an example of a continuous map which is

1. Both open and closed.

2. Neither open nor closed.

Exercise 6.50. Give examples, if possible, of discontinuous maps which are

1. Open but not closed.

2. Closed but not open.

3. Both open and closed.

4. Neither open nor closed.

Exercise 6.51. Prove that a bijection $f : X \to Y$ is open if and only if it is closed.

Exercise 6.52. Prove the following:

1. If $f : X \to Y$ is a continuous surjection, then a map $g : Y \to Z$ is open if $g \circ f$ is open.

2. If $g : Y \to Z$ is a continuous injection, then a map $f : X \to Y$ is open if $g \circ f$ is open.

3. Do these results remain true for closed maps?

6.8   Homeomorphism

Throughout this chapter, we have given a lot of importance to continuous functions. Earlier, we had given importance to bijections. Indeed, these are important in their own way and provide a lot of useful results. Before ending this chapter, we would now like to combine these two important properties that a function may exhibit and ask ourselves: What can we say about continuous bijections? Well, when we say the term "bijection", the inverse function immediately strikes our mind. Therefore, it is only natural to add continuity to the inverse as well. Based on this motivation, we now define the term *homeomorphism* rigorously.

Definition 6.17 (Homeomorphism). *A continuous bijection $f : X \to Y$ is said to be a homeomorphism if its inverse $f^{-1} : Y \to X$ is also continuous.*

Two metric spaces X and Y are homeomorphic if there is a homeomorphism between them.

Let us look at the simplest homeomorphism (other than the identity function) one can have. Define $f : (0,1] \to [1,\infty)$ as $f(x) = \frac{1}{x}$. We know that f is continuous, bijective, and in fact, it is self inverse (involution). Hence, f is a homeomorphism of $(0,1]$ and $[1,\infty)$. We observe a few things here. Notice that $(0,1]$ is bounded but $[1,\infty)$ is not. Also, $[1,\infty)$ is complete but $(0,1]$ is not. Therefore, we can conclude that homeomorphisms do not preserve boundedness and completeness.

It is natural to ask: What do homeomorphisms preserve? Earlier, we have seen that continuous images of compact sets are compact and those of (path) connected sets are (path) connected. Therefore, if X is homeomorphic to Y, then X and Y are together compact and (path) connected. These properties, which are preserved by homeomorphisms, are called *topological properties*.

Remark. It might have struck to the mind of every reader: What is the importance of homeomorphisms? They do not even preserve all properties! However, we advise the reader to look at homeomorphisms as *continuous deformations*. Of course, this will be much more clear when the reader takes a course in topology and sees some applications from algebraic topology or (differential) geometry. But, here we try to give a flavor of the subject. Usually, metric spaces (or in general, topological spaces) can be considered to be made of rubber/clay, in which open sets are special ingredients. Now, we can mold this rubber into whatever shape we want provided we do not disturb these special ingredients. Homeomorphisms help us achieve this! If the reader pursues general topology and/or algebraic topology in the future, they will be astonished to find that we can transform a bottle into a sphere without causing any cuts/tears (i.e. in a continuous fashion). Similarly, as most topologists give this example, we can transform a coffee mug into a doughnut in a continuous fashion. Although all this is theoretical, it can be achieved through homeomorphisms. Moreover, once we prove that two spaces are homeomorphic, we need to study only one of them. In a way, homeomorphisms create equivalence classes (see next exercise). Because of such fantastic abilities of topological (and hence metric) spaces, topology is also called *"geometry of rubber sheets"*.

Exercise 6.53. Prove that homeomorphism is an equivalence relation.

Exercise 6.54. Prove that a homeomorphism is both open and closed.

Theorem 6.36. *Two metrics d_1, d_2 on X are equivalent if and only if the identity map $id : (X, d_1) \to (X, d_2)$ is a homeomorphism.*

Proof. We already know that id is a bijection. Let d_1, d_2 be equivalent. We need to prove that id is continuous and its inverse is also continuous. Notice that $id^{-1} = id$. Now, for any open set $U \in (X, d_2)$, we have $id^{-1}(U) = U$ is open in (X, d_1). Therefore, id is continuous and by the same argument, it is a homeomorphism.

Conversely, let id be a homeomorphism. Then, it is both open and closed. Let U be an open set in (X, d_1). Then, $id(U) = U$ is open in (X, d_2). Also, $id^{-1} = id$ and therefore, every open set in (X, d_2) is also open in (X, d_1). Hence, d_1 and d_2 are equivalent. $\square$

Let us now look at some homeomorphic spaces.

Example 6.28. We know that any two (open) closed intervals are bijective and in fact in a continuous fashion. Therefore, they are homeomorphic. Also, there is a continuous bijection between $\mathbb{R}$ and $(-1, 1)$, given by $\tan\left(\frac{\pi}{2}x\right)$. It is easy to check that this is a homeomorphism.

Exercise 6.55. Is $(0, 1)$ homeomorphic to $[0, 1]$? What can you say about any open and a closed (bounded) interval? What can you say about $\mathbb{R}$ and a closed interval?

Example 6.29 (Stereographic projection). Consider $\mathbb{R}^{n+1}$ and the surface of the unit sphere $\mathbb{S}^n = \left\{ (x_1, x_2, \cdots, x_n, x_{n+1}) \in \mathbb{R}^{n+1} \,\middle|\, \sum_{i=1}^{n+1} x_i^2 = 1 \right\}$ in $\mathbb{R}^{n+1}$. We shall consider the case when $n = 2$, i.e., the sphere $\mathbb{S}^2 = \{ (x, y, z) \in \mathbb{R}^3 \,|\, x^2 + y^2 + z^2 = 1 \}$ in $\mathbb{R}^3$. We shall call the point $\mathbf{p} = (0, 0, 1)$ the "*North Pole*" for this sphere. Similarly, $\mathbf{p}' = (0, 0, -1)$ can be called the "*South Pole*".

Let $(x, y, z) \in \mathbb{S}^2$. Then, a line passing through $(0, 0, 1)$ and $(x, y, z) \neq (0, 0, 1)$ also crosses the XY plane at some point $(x', y', 0)$. Also, to each point (x, y, z), there is a unique line through this point and the North pole and therefore, a unique point on the XY plane which lies on this line. Therefore, our intuition tells us that the XY plane and $\mathbb{S}^2 \setminus \{\mathbf{p}\}$ must be bijective. Since the XY plane can be identified as $\mathbb{R}^2$, we can intuitively conclude that $\mathbb{S}^2 \setminus \{\mathbf{p}\}$ and $\mathbb{R}^2$ are bijective. This is shown in Figure 6.13. Therefore, it is only natural to ask whether this bijection is a homeomorphism. We shall answer this question soon. But first, let us construct this bijection explicitly.

Let us see what is the equation of line through the North pole and any point $(x, y, z) \neq \mathbf{p} \in \mathbb{S}^2$. Now, we shall use vector algebra to construct lines. Any vector on the line is parallel to the vector $(x, y, z - 1)$ and the line passes through $(0, 0, 1)$. Hence, the equation of the line in vector form will be given by

$$(x', y', z') = (\lambda x, \lambda y, \lambda z - \lambda + 1)$$

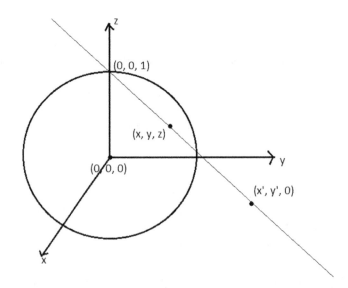

FIGURE 6.13: Construction of stereographic projection (strategy).

where $\lambda \in \mathbb{R}$ is a parameter. At the point where this line meets the XY plane, we have

$$\lambda (z - 1) + 1 = 0,$$

so that

$$\lambda = \frac{1}{1 - z}.$$

Therefore, we define a function $f : \mathbb{S}^2 \setminus \{\mathbf{p}\} \to \mathbb{R}^2$ as

$$f (x, y, z) = \left(\frac{x}{1 - z}, \frac{y}{1 - z} \right).$$

Similarly, if we consider the line through $\mathbf{p}$ and any point $(x, y, 0)$, then it intersects the sphere at a unique point. For this, we can define a function $g : \mathbb{R}^2 \to \mathbb{S}^2 \setminus \{\mathbf{p}\}$ as

$$g (x, y) = \left(\frac{2x}{x^2 + y^2 + 1}, \frac{2y}{x^2 + y^2 + 1}, \frac{x^2 + y^2 - 1}{x^2 + y^2 + 1} \right).$$

It is just a matter of computation to check that $f \circ g = id_{\mathbb{R}^2}$ and $g \circ f = id_{\mathbb{S}^2}$. Hence, both f and g are bijections and $g = f^{-1}$. Also, each "component" of f and g is a composition of continuous functions well-defined everywhere so that f and $g = f^{-1}$ are both continuous. Hence, $\mathbb{S}^2 \setminus \{\mathbf{p}\}$ and $\mathbb{R}^2$ are homeomorphic.

Exercise 6.56. Generalize stereographic projection for any $\mathbb{S}^n$ and $\mathbb{R}^n$. **Hint:** You will require help from linear algebra.

Example 6.30. Consider the sets $\overline{B_2 (0, 1)}$ and $\overline{B_\infty (0, 1)}$, the closed unit balls in $(\mathbb{R}^n, d_2)$ and $(\mathbb{R}^n, d_\infty)$, respectively. We study the case for $n = 2$, since it is easy to visualize. It is clear that $\overline{B_2 (0, 1)} \subseteq \overline{B_\infty (0, 1)}$. This is because if $x^2 + y^2 \leq 1$, then $|x| \leq 1$ and $|y| \leq 1$.

Now, consider $(0, 0) \neq (x, y) \in \overline{B_\infty (0, 1)}$ and the ray $\{t (x, y) \,|\, t \geq 0\}$. This ray intersects the boundary of $\overline{B_\infty (0, 1)}$ exactly once. To see this, suppose $(x_0, y_0), (x_1, y_1)$ be the points on the ray which are also points on $\partial B_\infty (0, 1)$. Then, we have $x_0 = t_0 x$, $y_0 = t_0 y$ and $x_1 = t_1 x$ and $y_0 = t_1 y$ for some $t_0, t_1 \geq 0$. Then, being on the boundary we will have $\max \{t_0 x, t_0 y\} = \max \{t_1 x, t_1 y\}$ and since $(x, y) \neq (0, 0)$, this is possible if and only if $t_0 = t_1$. Hence, the ray intersects the boundary at only one point. In fact, this point is $\frac{1}{\max\{|x|, |y|\}} (x, y)$.

Although the construction of these two unit closed balls are in different metrics, we shall now work only in one metric, namely d_2. First, observe that $\max \{|x|, |y|\} \leq \sqrt{x^2 + y^2}$ so that the function $g : \overline{B_\infty (0, 1)} \setminus \{(0, 0)\} \to \partial B_\infty (0, 1)$ defined by $g (x) = \frac{1}{\max\{|x|, |y|\}} (x, y)$ is continuous. Now, the line segment $[(0, 0), g (x)]$ needs to be "fit" inside $\overline{B_2 (0, 1)}$. We know how to do it. To bring the end point $g (x)$ on the boundary of $\overline{B_2 (0, 1)}$, we need to

scale it by its length. Hence, a natural choice will be to define a function
$f : \overline{B_\infty}(0, 1) \to \overline{B_2}(0, 1)$ as

$$f(x, y) = \begin{cases} \dfrac{\max\{|x|, |y|\}}{\sqrt{x^2 + y^2}}(x, y), & \text{when } (x, y) \neq (0, 0). \\[4mm] (0, 0), & \text{when } (x, y) = (0, 0). \end{cases}$$

Notice that the method of construction tells us that f is a bijection. Also,
$f^{-1} : \overline{B_2}(0, 1) \to \overline{B_\infty}(0, 1)$ is given by

$$f^{-1}(x, y) = \begin{cases} \dfrac{\sqrt{x^2 + y^2}}{\max\{|x|, |y|\}}(x, y), & \text{when } (x, y) \neq (0, 0). \\[4mm] (0, 0), & \text{when } (x, y) = (0, 0). \end{cases}$$

Clearly, both f and f^{-1} are continuous when $(x, y) \neq (0, 0)$. When $(x, y) = (0, 0)$, we have

$$d_2(f(x, y), f(0, 0)) \leq \sqrt{x^2 + y^2},$$

since $\frac{\sqrt{x^2+y^2}}{\max\{x,y\}} \geq 1$. Hence, continuity of f follows. Similarly, f^{-1} is also continuous. Therefore, the two sets $\overline{B_2}(0, 1)$ and $\overline{B_\infty}(0, 1)$ are homeomorphic in $(\mathbb{R}^2, d_2)$.

Remark. The example above is quite non-trivial. Therefore, we advise the reader to take a minute, go through the proof again, if required, and assimilate it. Also, draw diagrams (similar to those we have shown for stereographic projection) for better understanding of the construction of homeomorphisms. We have purposefully left out the diagram for this example so that the reader gets in the habit of making the diagram by themselves.

We now give an important result about continuous functions and their graphs.

Theorem 6.37. *Let $f : X \to Y$ be a continuous function, then the graph of f, defined as $G(f) = \{(x, f(x)) \in X \times Y | x \in X\}$ as a subset of $X \times Y$, is homeomorphic to X.*

Proof. There is a natural bijection $\phi : X \to G$ given by $\phi(x) = (x, f(x))$. Also, its inverse is the projection Π_1, which is continuous. We only need to prove that ϕ is continuous. We have already proved that Π_1 is an open map and since $\phi = \Pi_1^{-1}$, we have the continuity of ϕ. Therefore, G is homeomorphic to X. □

As an immediate consequence of the theorem, we have the following example.

Example 6.31. The parabola $\{(x, x^2) \in \mathbb{R}^2 | x \in \mathbb{R}\}$ is homeomorphic to $\mathbb{R}$ since x^2 is a continuous function.

Exercise 6.57. Prove that a circle and an ellipse are homeomorphic.

Exercise 6.58. Prove that separability of a metric space is a topological property.

We end this section with an important result.

Theorem 6.38. $\mathbb{R}$ *and* $\mathbb{R}^2$ *are not homeomorphic.*

Proof. Assume that there is a homeomorphism $f : \mathbb{R} \to \mathbb{R}^2$. Let $f(0) = (a, b)$. Then, $f : \mathbb{R} \setminus \{0\} \to \mathbb{R}^2 \setminus \{(a, b)\}$ is also a homeomorphism. However, this is not possible since $\mathbb{R} \setminus \{0\}$ is disconnected but $\mathbb{R}^2 \setminus \{(a, b)\}$ is connected. Hence, the two sets cannot be homeomorphic. □

Exercise 6.59. Using topological properties, prove that the circle $x^2 + y^2 = 1$, the parabola $y = x^2$ and the hyperbola $x^2 - y^2 = 1$ are mutually non-homeomorphic.

Problem Set

1. Define $\mathbb{S}^1 \subseteq \mathbb{C}$ as $\mathbb{S}^1 = \{z \in \mathbb{C} | |z| = 1\}$. Can we have a continuous function $f : \mathbb{C}^* \to \mathbb{S}^1$, where $\mathbb{C}^* = \mathbb{C} \setminus \{0\}$ and both the sets are equipped with the subspace topology of $\mathbb{C}$ having usual metric?

2. Investigate if the following functions are continuous:

   (a) The conjugation map $f : \mathbb{C} \to \mathbb{C}$ defined as $f(z) = \bar{z}$, where $\mathbb{C}$ is equipped with the usual metric.

   (b) In a metric space X, fix a point $x \in X$. Define $f : X \to \mathbb{R}$ as $f(y) = d(x, y)$.

3. Consider $\mathscr{C}^{(1)}[0, 1]$, the set of all differentiable functions, as a subspace of $\mathscr{C}[0, 1]$ equipped with d_∞. Is the derivative $D : \mathscr{C}^{(1)}[0, 1] \to \mathscr{C}[0, 1]$, defined as $D(f) \equiv f'$ continuous?

4. Let $f : \mathbb{R}^n \to \mathbb{R}^m$ be continuous. Show that f takes bounded sets to bounded sets.

5. Let $f : X \to (0, \infty)$ be a continuous function on a compact space X. Show that there is an $\epsilon > 0$ such that for every $x \in X$ we have $f(x) \geq \epsilon$.

6. Show that a function $f : X \to Y$, where Y is compact, is continuous if and only if the graph of f is a closed subset of $X \times Y$ under the product metric.

7. Show that the unit sphere $\mathbb{S}^n = \left\{ (x_1, x_2, \cdots, x_n) \in \mathbb{R}^n | \sum_{i=1}^{n} x_i^2 = 1 \right\}$ is path connected.

8. Show that an annulus $\{ (x, y) \in \mathbb{R}^2 | a < d_2 ((x, y), (0, 0)) < b \}$, where $0 < a < b$, is path connected.

9. Prove that for a compact space X, there are points $x, y \in X$ such that $d(x, y) = \operatorname{diam}(X)$.

10. Let S be a compact subset of a metric space X. If $y \in S^c$, prove that there is a point $a \in S$ such that $d(a, y) \le d(x, y)$ for all $x \in S$. Give an example to show that the conclusion may fail if "compact" is replaced by "closed".

11. Give an example of metric spaces X and Y and a continuous function f from X onto Y such that Y is compact but X is not compact.

12. Let X be a compact metric space and $f : X \to X$ be a continuous function. Show that there is a non-empty set $A \subseteq X$ such that $f(A) = A$.

Biographical Notes

Johann Peter Gustav Lejeune Dirichlet (13 February, 1805 to 5 May, 1859) was a German mathematician who made deep contributions to number theory (including creating the field of analytic number theory), and to the theory of Fourier series and other topics in mathematical analysis; he is credited with being one of the first mathematicians to give the modern formal definition of a function. It is in his honor that the Dirichlet function was named.

Heinrich Franz Friedrich Tietze (August 31, 1880 to February 17, 1964) was an Austrian mathematician, famous for the Tietze extension theorem on functions from topological spaces to the real numbers. He also developed the Tietze transformations for group presentations, and was the first to pose the group isomorphism problem. Tietze's graph is also named after him; it describes the boundaries of a subdivision of the Möbius strip into six mutually-adjacent regions, found by Tietze as part of an extension of the four color theorem to non-orientable surfaces.

Rudolf Otto Sigismund Lipschitz (14 May, 1832 7 to October, 1903) was a German mathematician who made contributions to mathematical analysis (where he gave his name to the Lipschitz continuity condition) and differential geometry, as well as number theory, algebras with involution, and classical mechanics.

Ulisse Dini (14 November, 1845 to 28 October, 1918) was an Italian mathematician and politician, born in Pisa. He is known for his contribution to real analysis, partly collected in his book *"Fondamenti per la teorica delle funzioni di variabili reali"*. Dini worked in the field of mathematical analysis during a time when it was begun to be based on rigorous foundations. In addition to his books, he wrote about sixty papers. He proved the Dini criterion for the convergence of Fourier series and investigated the potential theory and differential geometry of surfaces, based on work by Eugenio Beltrami. His work on the theory of real functions was also important in the development of the concept of the measure on a set. The implicit function theorem is known in Italy as the Dini's theorem.

Cesare Arzela (6 March, 1847 to 15 March, 1912) was an Italian mathematician who taught at the University of Bologna and is recognized for his contributions in the theory of functions, particularly for his characterization of sequences of continuous functions, generalizing the one given earlier by Giulio Ascoli in the Arzel-Ascoli theorem.

Giulio Ascoli (20 January, 1843 to 12 July, 1896) was a Jewish-Italian mathematician. He made contributions to the theory of functions of a real variable and to Fourier series. For example, Ascoli introduced equicontinuity in 1884, a topic regarded as one of the fundamental concepts in the theory of real functions. In 1889, Italian mathematician Cesare Arzel generalized Ascoli's theorem into the Arzelà-Ascoli theorem, a practical sequential compactness criterion of functions.

Chapter 7

Banach Fixed Point Theorem and Its Applications

Historically the idea of successive approximation in solving differential equations appeared in some work of E. Picard. But, it was Banach who rightly observed the power of metric structure and its properties, making it a suitable tool and introducing his famous theorem, which is now known as Banach contraction principle that has a wide range of applications in pure and applied mathematics.

In this chapter, we discuss mainly Banach contraction theorem and illustrate its applicability in the theory of differential equations, integral equations, algebraic equations, as well as in proving implicit function theorem.

7.1 Banach Contraction Theorem

We shall first define a fixed point and contraction mapping, which will be supplemented by some examples and other preliminaries.

Definition 7.1 (Fixed point). *Let X be a non-empty set, and $f : X \to X$ be a mapping. The point $x \in X$ is called a fixed point of f if x mapped to itself, i.e., $f(x) = x$.*

Example 7.1.

1. Constant mapping from a metric space to itself has a fixed point. (Can you tell what is the fixed point?)

2. Identity mapping has a fixed point. In fact, in this case every point is a fixed point.

3. A translation has no fixed point.

Definition 7.2 (Contraction). *Let (X, d) be a metric space. A mapping f of X to itself is said to be contraction (or contraction mapping) if there exists a fixed constant $0 \leq \alpha < 1$, such that for all $x, y \in X$,*

$$d(f(x), f(y)) \leq \alpha d(x, y). \tag{7.1}$$

A contraction mapping is also known as Banach contraction. If Equation (7.1) has a strict inequality and it holds for $\alpha = 1$ and $x \neq y$, then f is called contractive (or strict contractive). If Equation (7.1) holds for $\alpha = 1$, then f is called nonexpansive, and if Equation (7.1) holds for a fixed $\alpha > 0$, then f is called Lipschitz continuous, as we have seen earlier. Clearly, for any mapping f, the following implications hold:

Contraction $\Rightarrow$ Contractive $\Rightarrow$ Nonexpansive $\Rightarrow$ Lipschitz continuous.

Exercise 7.1.

1. Show that if $x = (x_n) \in \ell^2$ then $f(x) = \left(\frac{x_n}{2}\right)$ is a contraction mapping on ℓ^2.

2. Let $f : [a, b] \to [a, b]$ be a differentialable function. Assume that there exist a real number $k < 1$ such that $|f'(x)| \leq k$, for all $x \in [a, b]$. Show that f is a contraction.

3. Is the mapping $f : [0, \infty) \to [0, \infty)$ defined by $f(x) = \frac{1}{1+x^2}$, $\forall x \in [0, \infty)$ a contraction?

Any contraction mapping is clearly continuous, and in fact, uniformly continuous (why?).

Next, we consider an example which shows that a contraction mapping may not have a fixed point.

Example 7.2. Let $X = \left(0, \frac{1}{4}\right)$ be a metric space with the usual metric on X. Let $f : X \to X$ be given by $f(x) = x^2$. Then

$$|f(x) - f(y)| = |x^2 - y^2| = (x+y)|x-y| < \frac{1}{2}|x-y|, \text{ for all } x, y \in X.$$

So, f is a contraction mapping. If x_0 is a fixed point of f then we must have $f(x_0) = x_0$, i.e., $x_0^2 = x_0$. So, $x_0 = 0$ or 1. Neither 0 nor 1 lies in $X = \left(0, \frac{1}{4}\right)$. So f has no fixed point in X.

In the above example we have shown that a contraction mapping may not have a fixed point. It may be noted that the space X is not complete here, and we will see later that it is not a coincidence.

Definition 7.3. *Consider $f : X \to X$, where X is a metric space. Let $x_0 \in X$. We determine successively the elements of a sequence starting from x_0 as follows:*

$$x_1 = f(x_0), \quad x_2 = f(x_1), \quad x_3 = f(x_2), \quad \ldots, \quad x_{n+1} = f(x_n), \quad \ldots$$

This procedure of constructing a sequence of elements from an element is called iteration method.

Clearly,

$$x_1 = f(x_0),$$
$$x_2 = f(x_1) = f(f(x_0)) = f^2(x_0),$$

and in general,

$$x_n = f^n(x_0). \tag{7.2}$$

The following theorem is known as *Banach's fixed point theorem*. Sometimes it is also called *Banach contraction principle* (BCP).

Theorem 7.1 (Banach). *Let (X, d) be a complete metric space and $f : X \to X$ be a contraction. Then, f has a unique fixed point in X.*

Proof. Let $x_0 \in X$ be an arbitrary element. Starting from x_0 we form the iteration as Equation (7.2), i.e., $x_n = f^n(x_0) = f(x_{n-1})$ for all $n \in \mathbb{N}$. Now, we verify that (x_n) is a Cauchy sequence. We have

$$d(x_1, x_2) = d(f(x_0), f(x_1)) \le \alpha d(x_0, x_1) = \alpha d(x_0, f(x_0)),$$
$$d(x_2, x_3) = d(f(x_1), f(x_2)) \le \alpha d(x_1, x_2) \le \alpha^2 d(x_0, f(x_0)),$$
$$d(x_3, x_4) = d(f(x_2), f(x_3)) \le \alpha d(x_2, x_3) \le \alpha^3 d(x_0, f(x_0)).$$

In general, for any positive integer n, we can prove by principle of mathematical induction that

$$d(x_n, x_{n+1}) \le \alpha^n d(x_0, f(x_0)).$$

Also, for any positive integer p,

$$d(x_n, x_{n+p}) \le d(x_n, x_{n+1}) + d(x_{n+1}, x_{n+2}) + \cdots + d(x_{n+p-1}, x_{n+p})$$
$$\le \alpha^n d(x_0, f(x_0)) + \alpha^{n+1} d(x_0, f(x_0)) + \cdots$$
$$+ \alpha^{n+p-1} d(x_0, f(x_0))$$
$$= (\alpha^n + \alpha^{n+1} + \cdots + \alpha^{n+p-1}) d(x_0, f(x_0))$$

$$\leq \frac{\alpha^n - \alpha^{n+p}}{1 - \alpha} d\left(x_0, f\left(x_0\right)\right) \tag{7.3}$$

$$\leq \frac{\alpha^n}{1 - \alpha} d\left(x_0, f\left(x_0\right)\right). \qquad (\text{since } 0 < \alpha < 1) \tag{7.4}$$

Because $\alpha < 1$, the relation Equation (7.4) shows that $d(x_n, x_{n+p}) \to 0$ as $n \to \infty$. Therefore (x_n) is a Cauchy sequence. Since X is complete, the sequence (x_n) is convergent. Suppose

$$\lim_{n \to \infty} x_n = x^* \in X. \tag{7.5}$$

We show that x^* is a fixed point of the mapping f, i.e., $f\left(x^*\right) = x^*$. We have

$$d\left(x^*, f\left(x^*\right)\right) \leq d(x^*, x_n) + d\left(x_n, f\left(x^*\right)\right)$$

$$= d(x^*, x_n) + d\left(f\left(x_{n-1}\right), f\left(x^*\right)\right)$$

$$\leq d(x^*, x_n) + \alpha d(x_{n-1}, x^*).$$

In view of Equation (7.5), we get $d\left(x^*, f\left(x^*\right)\right) = 0$. So that $f\left(x^*\right) = x^*$. Therefore, x^* is a fixed point of f.

We now verify the uniqueness of fixed point of f. Suppose $y^* \in X$ be such that $f\left(y^*\right) = y^* \neq x^*$. Then

$$d(x^*, y^*) = d\left(f\left(x^*\right), f\left(y^*\right)\right) \leq \alpha d(x^*, y^*) < d\left(x^*, y^*\right). \tag{7.6}$$

This is a contradiction and hence, f has a unique fixed point in X. $\qquad\square$

Note. We consider Equation (7.3).

$$d(x_n, x_{n+p}) \leq \frac{\alpha^n - \alpha^{n+p}}{1 - \alpha} d\left(x_0, f\left(x_0\right)\right).$$

Fix $n \in \mathbb{N}$, then as $p \to \infty$, the right hand side tends to $\frac{\alpha^n}{1-\alpha} d\left(x_0, f\left(x_0\right)\right)$ (since $\alpha < 1$) and the left hand side tends to $d(x_n, x^*)$ because $x_{n+p} \to x^*$. So

$$d(x_n, x^*) \leq \frac{\alpha^n}{1 - \alpha} d\left(x_0, f\left(x_0\right)\right) \text{ for all } n \in \mathbb{N}. \tag{7.7}$$

The relation Equation (7.7) gives an estimation for the error of the nth approximation.

Remark. It is important that the contractive constant α should be strictly less than 1 in the proof of BCP. It gives us control over the rate of convergence of $(x_n) = (f^n\left(x_0\right))$ to the fixed point since $\alpha^n \to 0$ as $n \to \infty$. If we assume f a contractive mapping instead of a contraction, then we lose that control over the convergence, and then the fixed point of f need not exist.

Example 7.3. Consider the usual metric space $(\mathbb{R}, |\cdot|)$. Define $f : \mathbb{R} \to \mathbb{R}$ as:

$$f(x) = x + \frac{\pi}{2} - \tan^{-1} x.$$

Then for $x, y \in \mathbb{R}$, with $x < y$, we have

$$f(y) - f(x) = (y - x) - (\tan^{-1} y - \tan^{-1} x)$$
$$= y - x - \frac{y - x}{1 + z^2}$$

where $x < z < y$, by Lagrange's mean value theorem. Therefore

$$f(y) - f(x) = (y - x)\left(1 - \frac{1}{1 + z^2}\right).$$

So,

$$|f(x) - f(y)| = |x - y|\left|1 - \frac{1}{1 + z^2}\right| < |-y|.$$

Therefore, the mapping f is contractive. But f has no fixed point. For, if there exists $p \in \mathbb{R}$ such that $f(p) = p$, then

$$p + \frac{\pi}{2} - \tan^{-1} p = p$$

$$\text{i.e. } \tan^{-1} p = \frac{\pi}{2},$$

which is not true because $\tan^{-1} x < \frac{\pi}{2}$ for every $x \in \mathbb{R}$.

Note. The mapping f in Example 7.3 is contractive but is not a contraction. Because, if it were a contraction, $\mathbb{R}$ being a complete metric space, f would have a fixed point.

We have seen that a contractive mapping need not have a fixed point. But the following theorem due to Edelstein shows that if the space is considerably restricted (in fact compact) then a unique fixed point of a contractive mapping is ensured.

Theorem 7.2 (Edelstein). *Let (X, d) be a compact metric space and $f : X \to X$ be such that*

$$d(f(x), f(y)) < d(x, y), \forall x, y \in X, \ x \neq y. \tag{7.8}$$

Suppose that there exists a point $x_0 \in X$ such that the sequence of iterates $(f^n(x_0))$ has a subsequence converging to $x \in X$. Then x is the unique fixed point of f.

Similarly, a nonexpansive mapping on a complete metric space need not have a fixed point. For instance, consider the translation operator by a nonzero vector in a $\mathbb{R}^n$ (or, any other vector space equipped with a metric), which is clearly a nonexpansive, fixed point free mapping. On the other hand, a fixed point of a nonexpansive map need not be unique. For instance, consider an identity operator, which is nonexpansive and rich with fixed points. Thus, the fixed point theory of nonexpansive mappings is fundamentally different from that of the contraction mapping, and thus we shall not discuss it in this chapter.

7.2 Applications of Banach Contraction Principle

7.2.1 Root finding problem

Let $T\colon I \to I$ be a map, where I is any closed interval of real numbers. We consider the problem of finding a solution of the equation $T(x) = x$. It is clear that the solution of this equation is a fixed point of T.

Theorem 7.3. *Let $T\colon I \to I$, where $I \subseteq \mathbb{R}$ is a closed interval, be a differentiable function and there exists $\lambda > 0$ such that*

$$|T'(x)| \le \lambda < 1$$

for all $x \in I$, where T' stands for the first derivative of T with respect to x, then the equation $T(x) = x$ has a unique solution in I and for any $x_0 \in I$ the sequence (x_n) defined by $x_n = T(x_{n-1})$, $n \in \mathbb{N}$ converges to the solution of the equation.

Proof. We consider that I is equipped with the usual metric, $|\cdot|$. Then $(I, |\cdot|)$ is a complete metric space. By Lagrange's mean-value theorem we have $\forall x, y \in I$, with $x < y$, $\exists w \in (x, y)$ such that

$$T(x) - T(y) = (x - y)T'(w).$$

Therefore,

$$|T(x) - T(y)| = |(x - y)T'(w)|$$
$$= |x - y|\,|T'(w)|$$
$$\le \lambda\,|x - y|.$$

Since $\lambda < 1$, T is a contraction mapping on I and by BCP, T has a unique fixed point in I, and the sequence (x_n) converges to this fixed point. $\qquad\square$

Let us look at an application of Theorem 7.3. Suppose we have to find a root of the equation $f(x) = 0$. Then, we rewrite this equation in the form $T(x) = x$, in such a way that $|T'(x)| \le \lambda < 1$ for some $\lambda > 0$ and for all $x \in I$. Then, by an application of Theorem 7.3 we can find the solution of $T(x) = x$ which is a root of $f(x) = 0$. The following example will help the reader understand this situation thoroughly.

Example 7.4. Consider the equation $f(x) = \cos^2 x - 3x = 0$. We rewrite this equation in the following form:

$$T(x) = \frac{\cos^2(x)}{3} = x$$

It is clear that there is a root of equation $f(x) = 0$ in the interval $I = [0, 1]$ (why?). Also, we can find $\lambda > 0$ such that $|T'(x)| \leq \lambda < 1$ for all $x \in I$. In particular, $\lambda = \frac{1}{3}$. Therefore, by Theorem 7.3, there is a unique solution of the equation $T(x) = x$ which is the root of f. Now one can find the root as a limit of sequence (x_n), $x_n = \frac{\cos^2(x_{n-1})}{3}$, for $n \in \mathbb{N}$, with arbitrary $x_0 \in I$. We encourage the reader to actually do the iterations and plot a few of them to see the convergence.

Remark. The readers who might have taken a course in numerical methods or numerical analysis would know that this is a common technique used for finding the numerical roots of a function. It is often termed as "*fixed point iteration method*".

7.2.2   Solution of system of linear algebraic equations

We consider the following system of linear algebraic equations:

$$\left.\begin{array}{rcl} a_{11}x_1 + a_{12}x_2 + \cdots + a_{1n}x_n &=& b_1, \\ a_{21}x_1 + a_{22}x_2 + \cdots + a_{2n}x_n &=& b_2, \\ &\vdots& \\ a_{n1}x_1 + a_{n2}x_2 + \cdots + a_{nn}x_n &=& b_n. \end{array}\right\} \tag{7.9}$$

A rearrangement of the above system is:

$$x_1 = x_1 - [a_{11}x_1 + a_{12}x_2 + \cdots + a_{1n}x_n] + b_1,$$

$$x_2 = x_2 - [a_{21}x_1 + a_{22}x_2 + \cdots + a_{2n}x_n] + b_2,$$

$$\vdots$$

$$x_n = x_n - [a_{n1}x_1 + a_{n2}x_2 + \cdots + a_{nn}x_n] + b_n.$$

In matrix form, the above system can be written as

$$\mathbf{x} = \mathbf{x} - A\mathbf{x} + \mathbf{b} \tag{7.10}$$

where $A = [a_{ij}]_{n \times n}$ is the coefficient matrix, $\mathbf{x} = [x_1, x_2, \ldots, x_n]^T$, $\mathbf{b} = [b_1, b_2, \ldots, b_n]^T$. Denote $\alpha_{ij} = -a_{ij} + \delta_{ij}$, $1 \leq i, j \leq n$, where δ_{ij} is the *Kronecker delta* defined by

$$\delta_{ij} = \begin{cases} 1, & \text{if } i = j. \\ 0, & \text{otherwise.} \end{cases} \qquad (7.11)$$

Then the system Equation (7.10) can be written as

$$\mathbf{x} = A\mathbf{x} + \mathbf{b} \qquad (7.12)$$

where $A = [\alpha_{ij}]_{n \times n}$. We first convert the above problem into a fixed point problem. Define a mapping $T \colon \mathbb{R}^n \to \mathbb{R}^n$ by:

$$T(\mathbf{x}) = A\mathbf{x} + \mathbf{b}.$$

It is obvious that $\mathbf{x} \in \mathbb{R}^n$ is a solution of Equation (7.9) if and only if it is a fixed point of T.

Theorem 7.4. *If $\sum\limits_{j=1}^{n} |\alpha_{ij}| \leq \lambda < 1$, $1 \leq i \leq n$ for some positive number λ, then the system Equation (7.9) has a unique solution. Furthermore, for any $\mathbf{y}_0 \in \mathbb{R}^n$ the sequence $(\mathbf{y}_m)$ defined by $\mathbf{y}_m = A\mathbf{y}_{m-1} + \mathbf{b}$, $m \in \mathbb{N}$ converges to the solution of the system.*

Proof. Consider the space $\mathbb{R}^n$ equipped with the metric d_∞. We know that $(\mathbb{R}^n, d_\infty)$ is a complete metric space. Therefore, we only need to show that the mapping T is a contraction on X. Indeed, $\forall x, y \in X$ we have

$$d_\infty (T(x), T(y)) = d_\infty (A\mathbf{x} + \mathbf{b}, A\mathbf{y} + \mathbf{b})$$

$$= \max_{1 \leq i \leq n} \left\{ \left| \sum_{j=1}^{n} \alpha_{ij}(x_j - y_j) \right| \right\}$$

$$\leq \max_{1 \leq i \leq n} \left\{ \sum_{j=1}^{n} |\alpha_{ij}| \, |x_j - y_j| \right\}$$

$$\leq \max_{1 \leq i \leq n} \left\{ \sum_{j=1}^{n} |\alpha_{ij}| \right\} \max_{1 \leq p \leq n} \{|x_p - y_p|\}$$

$$= \max_{1 \leq i \leq n} \left\{ \sum_{j=1}^{n} |\alpha_{ij}| \right\} d_\infty(x, y)$$

$$\leq \lambda d_\infty(x, y).$$

This shows that T is a contraction, and the existence and uniqueness of the solution of system Equation (7.9) follows from BCP. □

Example 7.5. Consider the following system of equations:

$$
\left.
\begin{aligned}
-20x_1 + 4x_2 + 3x_3 &= -32, \\
12x_1 - 45x_2 + 20x_3 &= 300, \\
3x_1 - 4x_2 - 16x_3 &= -36.
\end{aligned}
\right\} \tag{7.13}
$$

The above system can be written as:

$$
\left.
\begin{aligned}
x_1 &= -\frac{1}{4}x_1 + \frac{1}{4}x_2 + \frac{3}{16}x_3 + 2, \\
x_2 &= \frac{1}{5}x_1 + \frac{1}{4}x_2 + \frac{1}{3}x_3 - 5, \\
x_3 &= \frac{1}{4}x_1 - \frac{1}{3}x_2 - \frac{1}{3}x_3 + 3.
\end{aligned}
\right\} \tag{7.14}
$$

Here, $\mathcal{A} = \begin{bmatrix} -\frac{1}{4} & \frac{1}{4} & \frac{3}{16} \\ \frac{1}{5} & \frac{1}{4} & \frac{1}{3} \\ \frac{1}{4} & -\frac{1}{3} & -\frac{1}{3} \end{bmatrix}$, $\mathbf{x} = [x_1, x_2, x_3]^T$, $\mathbf{b} = [2, -5, 3]$, and

$\sum_{j=1}^{3} |\alpha_{1j}| = \frac{11}{16} < 1$, $\sum_{j=1}^{3} |\alpha_{2j}| = \frac{47}{60} < 1$, $\sum_{j=1}^{3} |\alpha_{3j}| = \frac{11}{12} < 1$. Therefore,

$\sum_{j=1}^{3} |\alpha_{ij}| \le \frac{11}{12} < 1$ for $1 \le i \le 3$ and by Theorem 7.4, System Equation (7.13) has a unique solution which is the limit of the sequence $(\mathbf{y}_n)$ defined by $\mathbf{y}_n = \mathcal{A}\mathbf{y}_{n-1} + \mathbf{b}$, $n \in \mathbb{N}$ and $\mathbf{y}_0 \in \mathbb{R}^3$ is arbitrary.

Corollary 7.4.1. *Let $\alpha_{ij} = -a_{ij} + \delta_{ij}$, $1 \le i, j \le n$, $a_{ij} \in \mathbb{R}$, where δ_{ij} is the Kronecker delta. If $\sum_{j=1}^{n} |\alpha_{ij}| \le \lambda < 1$, $1 \le i \le n$ for some positive number λ, then the following homogeneous system of the linear algebraic equations has no nontrivial solution:*

$$
\begin{aligned}
a_{11}x_1 + a_{12}x_2 + \cdots + a_{1n}x_n &= 0, \\
a_{21}x_1 + a_{22}x_2 + \cdots + a_{2n}x_n &= 0, \\
&\vdots \\
a_{n1}x_1 + a_{n2}x_2 + \cdots + a_{nn}x_n &= 0.
\end{aligned}
$$

Proof. Assuming $\mathbf{b} = 0$ in Theorem 7.4, and using the fact that the given system always has trivial solution, the proof follows. □

Note. The reader who has gone through numerical methods, might recall that the iterates discussed above are similar to those in Gauss-Seidel method.

7.2.3 Picard existence theorem for differential equations

Many problems in a course is elementary differential equation involve the solution of equation of the form

$$\frac{dy}{dx} = f(x, y) \tag{7.15}$$

with an initial condition,

$$y(x_0) = y_0. \tag{7.16}$$

Here, f is, indeed, some real valued function defined on all or a part of $\mathbb{R}^2$.

Finding a solution means to construct a function ϕ with domain containing some interval $[x_0 - \delta, x_0 + \delta]$ such that $\phi(x_0) = y_0$ and

$$\phi'(x) = f(x, \phi(x)), \text{ for } |x - x_0| \leq \delta. \tag{7.17}$$

Indeed, this is equivalent (via integration) to the equation

$$\phi(x) = y_0 + \int_{x_0}^{x} f(t, \phi(t)) \, dt, \text{ for } |x - x_0| \leq \delta. \tag{7.18}$$

Thus the question of the existence of a solution to the problem posed by Equations (7.15) and (7.16) is equivalent to the existence of a function ϕ satisfying Equation (7.18) for some δ.

We now prove a theorem, due to Picard, which gives condition on f sufficient to ensure both the existence and the uniqueness of a function ϕ satisfying Equation (7.18).

Theorem 7.5. *If f is continuous on some rectangle*

$$D = \{(x, y) : |x - x_0| \leq a, \ |y - y_0| \leq b\},$$

for some $a, b > 0$, whose interior contains (x_0, y_0) and if there exist $k > 0$ such that $\forall (x, y_1), (x, y_2) \in D$, we have

$$|f(x, y_1) - f(x, y_2)| \leq k|y_1 - y_2|, \tag{7.19}$$

then there exists $\delta > 0$ and a unique function $\phi : [x_0 - \delta, x_0 + \delta] \to \mathbb{R}$, such that Equation (7.18) holds.

Proof. Since f is continuous on the compact set D, the image of D under f is bounded in $\mathbb{R}$. Therefore, $\exists M > 0$ such that $\forall (x, y) \in D$, we have

$$|f(x, y)| \leq M.$$

Choose $\delta > 0$ such that

$$k\delta < 1 \tag{7.20}$$

where k is as in (7.19) and the rectangle D' defined as

$$D' = \left\{ (x, y) \in \mathbb{R}^2 \mid |x - x_0| \leq \delta, |y - y_0| \leq M\delta \right\}, \tag{7.21}$$

is contained in D.

Let $\mathscr{C}^*$ be the subset of $\mathscr{C}[x_0 - \delta, x_0 + \delta]$ consisting of all functions ϕ which are continuous on $[x_0 - \delta, x_0 + \delta]$ and such that

$$|\phi(x) - y_0| \leq M\delta, \text{ for } |x - x_0| \leq \delta.$$

This implies, $\mathscr{C}^*$ is complete metric subspace of $(\mathscr{C}[x_0 - \delta, x_0 + \delta], d_\infty)$. Note that by (7.21) we have $(t, \phi(t)) \in D$ if $|t - x_0| \leq \delta$ and $\phi \in \mathscr{C}^*$.

We now define a function T on $\mathscr{C}^*$ as follows:

For $\phi \in \mathscr{C}^*$ define $T(\phi) = \psi$ as:

$$\psi(x) = y_0 + \int_{x_0}^{x} f(t, \phi(t)) \, dt \text{ for } |x - x_0| \leq \delta. \tag{7.22}$$

Then ϕ satisfies (7.18) if and only if $T(\phi) = \phi$.

We shall show that T is contraction on the complete metric space $\mathscr{C}^*$. First we show that T maps $\mathscr{C}^*$ into $\mathscr{C}^*$. Indeed, if $\phi \in \mathscr{C}^*$ and $\psi = T(\phi)$ it is easy to show that ψ is continuous on $[x_0 - \delta, x_0 + \delta]$. Moreover, if $|x - x_0| \leq \delta$, then

$$|\psi(x) - y_0| \leq \left| \int_{x_0}^{x} f(t, \phi(t)) \, dt \right| \leq M|x - x_0| \leq M\delta.$$

Hence, $|\psi(x) - y_0| \leq M\delta$ if $|x - x_0| \leq \delta$, and so $\psi \in \mathscr{C}^*$. Hence $T : \mathscr{C}^* \to \mathscr{C}^*$.

To show that T is contraction suppose $\phi_1, \phi_2 \in \mathscr{C}^*$ and let $\psi_1 = T(\phi_1)$, $\psi_2 = T(\phi_2)$. Then, from (7.22), if $|x - x_0| \leq \delta$ we have

$$\psi_1(x) - \psi_2(x) = \int_{x_0}^{x} (f(t, \phi_1(t)) - f(t, \phi_2(t))) \, dt,$$

and we get,

$$\left|\psi_1(x) - \psi_2(x)\right| \leq \int_{x_0}^{x} |f(t, \phi_1(t)) - f(t, \phi_2(t))| \, dt.$$

Using (7.19) we then obtain

$$\left|\psi_1(x) - \psi_2(x)\right| \leq k \int_{x_0}^{x} |\phi_1(t) - \phi_2(t)| \, dt$$

$$\leq k\delta d_\infty(\phi_1, \phi_2).$$

Thus,

$$d_\infty(\psi_1, \psi_2) \leq k\delta d(\phi_1, \phi_2),$$

or, what is the same,

$$d_\infty(T(\phi_1), T(\phi_2)) \leq k\delta d_\infty(\phi_1, \phi_2).$$

In view of Equation (7.20) this proves that T is a contraction on $\mathscr{C}^*$. Hence, there is precisely one $\phi \in \mathscr{C}^*$ such that $T(\phi) = \phi$. But the definition Equation (7.22) of T shows that $T(\phi) = \phi$ means that Equation (7.18) holds. This completes the proof.   □

Remark. The Lipschitz condition (Equation 7.19) is necessary in order to prove the uniqueness of the solution in Theorem 7.5.

Let us look at an example to understand the necessity of Lipschitz condition.

Example 7.6. Consider the initial value problem

$$\frac{dy}{dx} = 3y^{\frac{2}{3}}. \tag{7.23}$$

$$y(0) = 0. \tag{7.24}$$

Clearly, $\phi_1 \equiv 0$ is a solution to the problem, since $\phi_1^{\frac{2}{3}} \equiv 0 \equiv \phi_1'$. Also, $\phi_2(x) = x^3$ is a solution to the same problem. Why did this happen? Our guess (from the proof of Theorem 7.5) is that $3y^{\frac{2}{3}}$ is not Lipschitz continuous, as said in the remark. Indeed, this is because the difference quotient

$$\left|\frac{f(0, y) - f(0, 0)}{(y - 0)}\right| = \frac{3y^{\frac{2}{3}}}{y} = \frac{3}{y^{\frac{1}{3}}},$$

is unbounded in every neighborhood at $(0, 0)$.

Remark. In Theorem 7.5, if we drop the Lipschitz condition, and assume only that $f(x, y)$ is continuous on $\mathbb{R}^2$ or some part of $\mathbb{R}^2$, then it is still possible to prove that the initial value problem (7.15) and (7.16) has a solution. This result is known as Peano's theorem. However, its proof depends on more sophisticated arguments than those we have used above.

Exercise 7.2.

1. Show that ϕ is a solution to

$$\frac{dy}{dx} = x + y, \quad y(0) = 0$$

iff $\phi(x) = \int_0^x (t + \phi(t))dt$.

2. Show that $f(x, y) = y^{\frac{1}{2}}$ does not satisfy Lipschitz condition on the rectangle $|x| \leq 1$ and $0 \leq y \leq 1$.

7.2.4 Solutions of integral equations

Suppose $K_1, K_2 \colon [a, b] \times [a, b] \times \mathbb{R} \to \mathbb{R}$ and $g \colon [a, b] \to \mathbb{R}$ are given functions. We consider the following integral equation:

$$u(t) = \beta A(t, u(t)) + \gamma B(t, u(t)) + g(t), \ t \in [a, b], \qquad (7.25)$$

where

$$A(t, u(t)) = \int_a^t K_1(t, s, u(s))ds,$$

$$B(t, u(t)) = \int_a^t K_2(t, s, u(s))ds,$$

and $\beta, \gamma \in \mathbb{R}$. We assume that $|\beta| + |\gamma| > 0$, otherwise the above equation reduced into a trivial form. To study the existence and uniqueness of solution of integral Equation (7.25), we first reduce it into a fixed point problem. Let $\mathscr{C}(I, \mathbb{R})$ be the space of all real-valued continuous functions on I, where $I = [a, b]$ with the Bielecki metric.[1] This metric d on $\mathscr{C}(I, \mathbb{R})$ is given by:

$$d(u, v) = \sup_{t \in I} \left\{ |u(t) - v(t)|e^{-\theta t} \right\}, \text{ for all } u, v \in \mathscr{C}(I, \mathbb{R}),$$

for some fixed $\theta > 0$.

We now define a function $T \colon \mathscr{C}(I, \mathbb{R}) \to \mathscr{C}(I, \mathbb{R})$ by:

$$(T(u))(t) = \beta A(t, u(t)) + \gamma B(t, u(t)) + g(t), \ t \in I. \qquad (7.26)$$

---

[1]This metric is induced by a norm, called the Bielecki norm, given by $\|u\|_\theta :=$ $\sup_{t \in I} \left\{ |u(t)|e^{-\theta t} \right\}$, $u \in C(I, \mathbb{R})$, where $\theta > 0$.

A function u^* is called a solution of Equation (7.25), if $u^* \in \mathscr{C}(I, \mathbb{R})$ and it satisfies Equation (7.25). It is clear that the problem of finding the solution of Equation (7.25) is equivalent to the problem of finding the fixed point of T. We now ensure the existence and uniqueness of fixed point of function T by applying some constraints on the functions K_1 and K_2.

Theorem 7.6. *Suppose the functions $K_1, K_2 \colon I \times I \times \mathbb{R} \to \mathbb{R}$ and $g \colon I \to \mathbb{R}$ are continuous and the following condition is satisfied: $\forall t, s \in I, \forall u(s), v(s) \in \mathbb{R}$, $\exists L_1, L_2 \colon I \times I \times \mathbb{R} \times \mathbb{R} \to \mathbb{R}^+$, continuous, such that:*

$$|K_i(t, s, u(s)) - K_i(t, s, v(s))| \leq L_i(t, s, u(s), v(s)) |u(s) - v(s)|, \ i = 1, 2.$$
(7.27)

If there exists $\alpha > 0$ such that $L_i(t, s, u(s), v(s)) \leq \alpha$, $i = 1, 2$, $\forall s, t \in I$, then the integral Equation (7.25) has a unique solution in $C(I, \mathbb{R})$.

Proof. It is well-known that $(\mathscr{C}(I, \mathbb{R}), d)$ is a complete metric space. Define $T \colon \mathscr{C}(I, \mathbb{R}) \to \mathscr{C}(I, \mathbb{R})$ by Equation (7.26). We shall show that the function T is a contraction on $\mathscr{C}(I, \mathbb{R})$. If $u, v \in \mathscr{C}(I, \mathbb{R})$, then by the definition of T we have

$$d(T(u), T(v)) = \sup_{t \in I} \left\{ |\beta A(t, u(t)) + \gamma B(t, u(t)) - \beta A(t, v(t)) - \gamma B(t, v(t))| e^{-\theta t} \right\}$$

$$= \sup_{t \in I} \left\{ \left| \beta \int_a^t [K_1(t, s, u(s)) - K_1(t, s, v(s))] \, ds \right. \right.$$

$$\left. \left. + \gamma \int_a^t [K_2(t, s, u(s)) - K_2(t, s, v(s))] \, ds \right| e^{-\theta t} \right\}$$

$$\leq \sup_{t \in I} \left\{ \left(|\beta| \int_a^t |K_1(t, s, u(s)) - K_1(t, s, v(s))| \, ds \right. \right.$$

$$\left. \left. + |\gamma| \int_a^t |K_2(t, s, u(s)) - K_2(t, s, v(s))| \, ds \right) e^{-\theta t} \right\}$$

$$\leq \sup_{t \in I} \left\{ \left(|\beta| \int_a^t L_1(t, s, u(s), v(s)) \mid u(s) - v(s) \mid ds \right. \right.$$

$$\left. \left. + |\gamma| \int_a^t L_2(t, s, u(s), v(s)) \mid u(s) - v(s) \mid ds \right) e^{-\theta t} \right\}$$

$$\leq d(u, v) \sup_{t \in I} \left\{ \int_a^t (|\beta| L_1(t, s, u(s), v(s)) + \right.$$

$$|\gamma| L_2(t, s, u(s), v(s))) e^{\theta(s-t)} \, ds \Big\}$$

$$\leq d(u, v) \alpha(|\beta| + |\gamma|) \sup_{t \in I} \left\{ \int_a^t e^{\theta(s-t)} \, ds \right\}$$

$$\leq d(u, v) \frac{\alpha(|\beta| + |\gamma|)}{\theta} \left[1 - e^{-\theta(b-a)} \right].$$

For $\theta = \alpha(|\beta| + |\gamma|)$, we get T as a contraction. By BCP, there exists a unique fixed point u^* of T in $\mathscr{C}(I, \mathbb{R})$, which is the unique solution of integral Equation (7.33) in $\mathscr{C}(I, \mathbb{R})$. $\qquad\square$

Example 7.7. Let $I = [0, \pi/2]$ and we have to find a function $u \colon I \to \mathbb{R}$ such that:

$$u(t) = \sin(t) + \int_0^t \ln(1 + s + |u(s)|)ds, \ t \in I. \tag{7.28}$$

Then, we consider the space $\mathscr{C}(I, \mathbb{R})$ with the suitable metric Bielecki metric, d. Define a function $T \colon \mathscr{C}(I, \mathbb{R}) \to \mathscr{C}(I, \mathbb{R})$ by:

$$(T(u))(t) = \sin(t) + \int_0^t \ln(1 + s + |u(s)|)ds.$$

By Lagrange mean value theorem we have

$$|\ln(1 + s + |u(s)|) - \ln(1 + s + |v(s)|)| \le |u(s) - v(s)|, \forall s \in I, \ \text{and} \ \forall u, v \in \mathbb{R}.$$

Therefore, all the conditions of Theorem 7.6 are satisfied with

$$L_1(t, s, u(s), v(s)) = L_2(t, s, u(s), v(s)) = \alpha = \beta = 1,$$

$\gamma \in \mathbb{R}$, $g(t) = \sin(t)$, $K_1(t, s, u(s)) = \ln(1 + s + |u(s)|)$, $K_2(t, s, u(s)) = 0$ for all $s, t \in I$, $u \in \mathbb{R}$. Thus, there exists a unique solution of integral Equation (7.28) which is the limit of sequence (u_n) defined by $u_n = T(u_{n-1})$, $n \in \mathbb{N}$ and $u_0 \in \mathscr{C}(I, \mathbb{R})$ is arbitrary.

Next, we consider the following integral equation:

$$u(t) = u(a) + \int_a^t K(s, u)ds, \tag{7.29}$$

where $K \colon [a, b] \times \mathbb{R} \to \mathbb{R}$ is a continuous function.

Corollary 7.6.1. *Suppose, the following condition is satisfied:* $\forall t, s \in I = [a, b]$, $\forall u(s), v(s) \in \mathbb{R}$, $\exists L \colon I \times \mathbb{R} \times \mathbb{R} \to \mathbb{R}^+$, *continuous, such that:*

$$|K(s, u(s)) - K(s, v(s))| \le L(s, u(s), v(s)) |u(s) - v(s)|.$$

If there exists $\alpha > 0$ such that $L(s, u(s), v(s)) \le \alpha$ for all $s \in I$, then the integral Equation (7.29) has a unique solution in $\mathscr{C}(I, \mathbb{R})$.

Remark. In the above corollary, one can take any $t_0 \in [a, b]$ instead of a as lower limit of the integral in Equation (7.29), the conclusion remains the same.

We now consider the following Fredholm type integral equation:

$$u(t) = \beta A(t, u(t)) + \gamma B(t, u(t)) + g(t), \ t \in [a, b], \tag{7.30}$$

where

$$A(t, u(t)) = \int_a^b K_1(t, s, u(s))ds, \ B(u(t), t) = \int_a^b K_2(t, s, u(s))ds, \beta, \gamma \in \mathbb{R},$$

and $K_1, K_2 \colon [a, b] \times [a, b] \times \mathbb{R} \to \mathbb{R}$ and $g \colon [a, b] \to \mathbb{R}$ are given functions. Again, we assume that $|\beta| + |\gamma| > 0$, and d as the Bielecki metric on $\mathscr{C}(I, \mathbb{R})$. Define $T \colon \mathscr{C}(I, \mathbb{R}) \to \mathscr{C}(I, \mathbb{R})$ by:

$$(T(u))(t) = \beta A(t, u(t)) + \gamma B(t, u(t)) + g(t), \ t \in I, \tag{7.31}$$

where $I = [a, b]$. A function u^* is called a solution of Equation (7.30), if $u^* \in \mathscr{C}(I, \mathbb{R})$ and it satisfies Equation (7.30). Then, the problem of finding the solution of Equation (7.30) is reduced to the problem of finding the fixed point of T, i.e., u^* is a solution of Equation (7.30) if and only if it is a fixed point of T.

Theorem 7.7. *Suppose the functions $K_1, K_2 \colon I \times I \times \mathbb{R} \to \mathbb{R}$ and $g \colon I \to \mathbb{R}$ are continuous and the following condition is satisfied:* $\forall t, s \in I, \forall u(s), v(s) \in \mathbb{R}$, $\exists L_1, L_2 \colon I \times I \times \mathbb{R} \times \mathbb{R} \to \mathbb{R}^+$, *continuous, such that:*

$$|K_i(t, s, u(s)) - K_i(t, s, v(s))| \le L_i(t, s, u(s), v(s)) |u(s) - v(s)|, \ i = 1, 2. \tag{7.32}$$

If $L_i(t, s, u(s), v(s)) \le e^{(t-b)\theta_1}, \ i = 1, 2, \ \forall s, t \in I,$ where $\theta_1 = |\beta| + |\gamma|$, then the integral Equation (7.30) has a unique solution in $\mathscr{C}(I, \mathbb{R})$.

Proof. Following the same lines as in the proof of Theorem 7.6, we obtain:

$$d(T(u), T(v)) \le \sup_{t \in I} \left\{ \left(|\beta| \int_a^b L_1(t, s, u(s), v(s)) |u(s) - v(s)| \, ds \right. \right.$$

$$\left. \left. + |\gamma| \int_a^b L_2(t, s, u(s), v(s)) |u(s) - v(s)| \, ds \right) e^{-\theta t} \right\}$$

$$\le d(u, v) \sup_{t \in I} \left\{ \int_a^t (|\beta| L_1(t, s, u(s), v(s)) \, ds + \right.$$

$$|\gamma| L_2(t, s, u(s), v(s))) e^{\theta(s-t)} \, ds \right\}$$

$$\le d(u, v)(|\beta| + |\gamma|) \sup_{t \in I} \left\{ \int_a^b e^{(t-b)\theta_1} e^{\theta(s-t)} \, ds \right\}.$$

For $\theta = \theta_1$, above equation yields:

$$d\left(T\left(u\right), T\left(v\right)\right) \leq \left[1 - e^{-\theta(b-a)}\right] d(u, v) = \lambda d(u, v)$$

where $\lambda = \left[1 - e^{-\theta(b-a)}\right] < 1$. Thus, T is a contraction on $\mathscr{C}(I, \mathbb{R})$. The existence of unique fixed point, i.e., the unique solution of Equation (7.30) follows from BCP. □

7.2.5 Solutions of initial value and boundary value problems

We consider the Cauchy initial value problem. Suppose $K \colon [a, b] \times \mathbb{R} \to \mathbb{R}$ is a continuous function and we want to find a differentiable function u on $[a, b]$ such that:

$$\frac{du}{dt} = K(t, u); \ u(t_0) = u_0. \tag{7.33}$$

Such a problem is called the *Cauchy initial value problem*.

Theorem 7.8. *Suppose, the following condition is satisfied:* $\forall t, s \in I = [a, b]$, *and* $\forall u\left(s\right), v\left(s\right) \in \mathbb{R}$, $\exists L \colon I \times \mathbb{R} \times \mathbb{R} \to \mathbb{R}^{+}$, *continuous, such that:*

$$|K(s, u(s)) - K(s, v(s))| \leq L(s, u(s), v(s)) \, |u(s) - v(s)| \, .$$

If there exists $\alpha > 0$ *such that* $L(s, u(s), v(s)) \leq \alpha, \forall s \in I$, *then Equation (7.33) has a unique solution.*

Proof. To prove the existence and uniqueness of initial value problem Equation (7.33), we first convert it into an integral equation. Integrating Equation (7.33) from t_0 to $t \in [a, b]$:

$$u(t) = u_0 + \int_{t_0}^{t} K(s, u) ds. \tag{7.34}$$

Therefore, to find a solution of problem Equation (7.33) is equivalent to find a solution of the integral Equation (7.34). Let $X = \mathscr{C}[a, b]$. Then (X, d) is a complete metric space, where d is a suitable Bielecki metric on $\mathscr{C}[a, b]$. Now, the existence and uniqueness of solution follows from Corollary 7.6.1. □

Example 7.8. Consider the following initial value problem:

$$\frac{du}{dt} = u, \ u(0) = 1, \ t \in [0, T], \ T > 0.$$

For this problem, we have $K(t, u) = u$, and so,

$$|K(s, u) - K(s, v)| = |u - v|.$$

Therefore, all the conditions of Theorem 7.8 are satisfied with $L(s, u(s), v(s)) = \alpha = 1$, and so, the initial value problem has a unique solution. The solution of initial value problem is the limit of sequence (z_n), $z_n = T(z_{n-1})$, $n \in \mathbb{N}$ and $z_0 \in \mathscr{C}[0, T]$ is arbitrary with

$$(T(u))(t) = u(0) + \int_0^t u(s)ds.$$

Suppose we start with $z_0 \equiv u(0) = 1$, then we obtain the following terms of the sequence:

$$z_1(t) = (T(z_0))(t) = 1 + \int_0^t 1 \cdot ds = 1 + t,$$

$$z_2(t) = (T(z_1))(t) = 1 + \int_0^t z_1(s) \, ds = 1 + t + \frac{t^2}{2},$$

$$\vdots$$

$$z_n(t) = (T(z_{n-1}))(t) = 1 + \int_0^t z_{n-1}(s) \, ds = \sum_{i=0}^n \frac{t^i}{i!}.$$

The limit of this sequence, i.e., $\lim_{n \to \infty} z_n(t) = e^t$ for all $t \in [0, T]$, is the unique solution of the initial value problem.

Example 7.9. Consider the following initial value problem:

$$\frac{du}{dt} = t + \tan^{-1}(u), \ u(0) = 0, \ t \in [0, \pi/2].$$

In this problem, we have $K(t, u) = t + \tan^{-1}(u)$, and so, by Lagrange mean value theorem we have

$$|K(s, u) - K(s, v)| = |\tan^{-1}(u) - \tan^{-1}(v)| \le |u - v|.$$

Therefore, all the conditions of Theorem 7.8 are satisfied with $L(s, u(s), v(s)) = \alpha = 1$, and so, the initial value problem has a unique solution. The solution of the problem is the limit of sequence (z_n), $z_n = T(z_{n-1})$, $n \in \mathbb{N}$ and $z_0 \in \mathscr{C}[0, T]$ is arbitrary with

$$(T(u))(t) = \int_0^t [s + \tan^{-1}(u(s))] \, ds.$$

Note that, unlike the previous example, the calculation of the terms of sequence is not an easy task. Yet, the BCP ensures the existence and the uniqueness of the solution of the problem.

7.2.6   Implicit function theorem

The classical proof of the implicit function theorem uses the theory from calculus (the Taylor's expansion, the mean-value theorem, and estimates on derivatives). Here, we use contraction mapping principle to get a quick, soft, and easy (although abstract form) proof of this theorem. Here, we consider this theorem in two-dimensional Euclidean space $\mathbb{R}^2$.

Theorem 7.9. *Let f be a continuous real valued function of two variables x and y with a continuous partial derivative f_y in a square $D \subseteq \mathbb{R}^2$ with a center (x_0, y_0). Assume that*

$$f(x_0, y_0) = 0 = f_y(x_0, y_0). \tag{7.35}$$

Then there exists unique function ϕ continuous in a neighborhood of x_0, such that

$$\phi(x_0) = y_0 \ and \ \phi(x) = \phi(x, \phi(x)) + y_0. \tag{7.36}$$

Proof. Since, $f_y(x_0, y_0) = 0$ and f_y is continuous on D, which is bounded, with center (x_0, y_0), we infer that $f_y(x, y)$ is "small" for all (x, y) "near" (x_0, y_0).

We choose $\delta_1 > 0$ and $\delta_2 > 0$ so that whenever $|x - x_0| < \delta_1$ and $|y - y_0| < \delta_2$, then

$$|f_y(x, y)| \leq \frac{1}{2}. \tag{7.37}$$

Since, $f(x, y_0)$ is continuous at $x = x_0$ and $f(x_0, y_0) = 0$, we can assume, by choosing "smaller" δ_1, that $|f(x, y_0) - y_0| \leq \frac{\delta_2}{2}$ for all x with $|x - x_0| < \delta_1$.

Let X be set of continuous functions ϕ on $[x_0 - \delta_1, x_0 + \delta_1]$ such that $\phi(x_0) = y_0$ and $|\phi(x) - y_0| \leq \delta_2$ if $|x - x_0| \leq \delta_1$.

We let $d(\phi_1, \phi_2) = \sup_{|x - x_0| \leq \delta_1} \{|\phi_1(x) - \phi_2(x)|\}$. Then we know that (X, d) is a metric space.

We shall now introduce an operator T on X. For $\phi \in X$, we define $(T(\phi))(x) = f(x, \phi(x)) + y_0$, for $|x - x_0| \leq \delta_1$. Then we have,

$$|\left(T\left(\phi\right)\right)\left(x\right) - y_0| = |f(x, \phi(x))|$$

$$\leq |f(x, \phi(x)) - f(x, y_0)| + |\phi(x, y_0)|$$

$$\leq |\phi(x) - y_0||f_y(x, y')| + \frac{\delta_2}{2}, \qquad (\phi(x) < y' < y_0)$$

$$\leq \frac{\delta_2}{2} + \frac{\delta_2}{2} = \delta_2.$$

Thus T maps X to itself.

Now, for $\phi_1(x) < \eta < \phi_2(x)$, we have

$$|\left(T\left(\phi_1\right)\right)\left(x\right) - \left(T\left(\phi_2\right)\right)\left(x\right)| = \sup_{x \in [x_0 - \delta, x_0 + \delta]} \{|f(x, \phi_1(x)) - f(x, \phi_2(x))|\}$$

$$= \sup_{x \in [x_0 - \delta, x_0 + \delta]} \{|\phi_1(x) - \phi_2(x)||f_y(x, \eta)|\}$$

$$\leq \frac{1}{2}|\phi_1(x) - \phi_2(x)|.$$

Therefore, T is contraction on X and hence by Banach contraction theorem, there exist unique $\phi \in X$ such that $T(\phi) = \phi$, i.e., $\phi(x_0) = y_0$ and $\phi(x) = f(x, \phi(x)) + y_0$ for $x \in [x_0 - \delta_1, x_0 + \delta_1]$. This completes the proof. $\square$

Problem Set

For the problems involving $\mathbb{R}$, use the usual metric.

1. Check whether or not the following maps are contractions on the corresponding metric spaces. If yes, find the associated fixed point.

   (a) $X = \mathbb{R}^+$, $f(x) = \sqrt{x + 1}$.
   (b) $X = \mathbb{R}$, $f(x) = \sqrt{x^2 + 1}$.

2. Show that $f(x) = \ln(x + 1)$ has a unique fixed point in $[0, 2]$.

3. Show that $f(x) = x^3$ is a contraction on $\left(0, \frac{1}{2}\right)$ but it has no fixed point in this interval. Does this contradict Banach contraction principle?

4. Show that $f : \left[1, \frac{5}{2}\right] \to \left[1, \frac{5}{2}\right]$ defined by

$$f(x) = 1 + x - \frac{x^3}{10}$$

is a contraction mapping on $\left[1, \frac{5}{2}\right]$. What is the fixed point of f?

5. Let $\mathscr{B}[0,1]$ denote the set of all bounded real-valued functions defined on $[0,1]$. Define $T : \mathscr{B}[0,1] \to \mathscr{B}[0,1]$ as

$$(T(u))(x) = \frac{u(x) + x^2}{2}.$$

Is T a contraction mapping? If so, find its unique fixed point. Here $\mathscr{B}[0,1]$ is equipped with the metric d_∞.

6. Let $T : X \to X$ be a contraction mapping where X is a discrete metric space. Show that T is constant.

7. Let $X = \{x \in \mathbb{Q} | x \geq 1\}$ and let $f : X \to X$ be defined by $f(x) = \frac{x}{2} + \frac{1}{x}$. Show that f is a contraction mapping and that f has no fixed point in X.

8. Let $f : [a,b] \to [a,b]$ be differentiable over $[a,b]$. Show that f is a contraction mapping if and only if there is a number $K < 1$ such that $|f'(x)| \leq K$ for every $x \in (a,b)$.

9. Show that $T : (\mathscr{C}[0,1], d_\infty) \to (\mathscr{C}[0,1], d_\infty)$ defined as

$$(T(f))(x) = \int_0^x (x-t)f(t)\ dt$$

is a contraction mapping.

10. Consider the metric space $(\mathscr{C}[0,1], d_\infty)$. Define $f : \mathscr{C}[0,1] \to \mathscr{C}[0,1]$ as $(F(f))(x) = \int_0^x f(t)\ dt$. Show that for all $f, g \in \mathscr{C}[0,1]$ and for all $x \in [0,1]$ we have:

   (a) $(F(f))(x) - (F(g))(x) \leq x d_\infty(f,g)$.
   (b) $(F^2(f))(x) - (F^2(g))(x) \leq \frac{x^2}{2} d_\infty(f,g)$.

   Here F^2 denotes the composition $F \circ F$. Hence deduce that F^2 is a contraction mapping, however, F is not.

11. Use Banach Fixed Point Theorem to show that

$$x_1 = \frac{1}{3}x_1 - \frac{1}{4}x_2 + \frac{1}{4}x_3 - 1,$$

$$x_2 = -\frac{1}{2}x_1 + \frac{1}{5}x_2 + \frac{1}{4}x_3 + 2,$$

$$x_3 = \frac{1}{5}x_1 - \frac{1}{3}x_2 + \frac{1}{4}x_3 - 2$$

has a unique solution.

Biographical Notes

Prof Charles Émile Picard (24 July, 1856 to 11 December, 1941) was a French mathematician. Picard's mathematical papers, textbooks, and many popular writings exhibit an extraordinary range of interests, as well as an impressive mastery of the mathematics of his time. He made important contributions in the theory of differential equations, including work on Picard-Vessiot theory, Painlevé transcendents, and his introduction of a kind of symmetry group for a linear differential equation. He also introduced the Picard group in the theory of algebraic surfaces, which describes the classes of algebraic curves on the surface modulo linear equivalence. Apart from these works, he has also made significant contributions in complex variables, function theory, and algebraic topology.

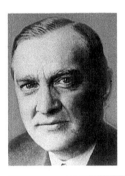

Stefan Banach (30 March, 1892 to 31 August, 1945) was a Polish mathematician who is generally considered one of the world's most important and influential twentieth-century mathematicians. He was the founder of modern functional analysis. Some of the notable mathematical concepts that bear Banach's name include Banach spaces, Banach algebras, Banach measures, the Banach-Tarski paradox, the Hahn-Banach theorem, the Banach-Steinhaus theorem, the Banach-Mazur game, the Banach-Alaoglu theorem, and the Banach fixed-point theorem.

Brook Taylor (18 August, 1685 to 29 December, 1731) was an English mathematician who is best known for Taylor's theorem and the Taylor series. Taylor was elected a fellow of the Royal Society early in 1712, and in the same year sat on the committee for adjudicating the claims of Sir Isaac Newton and Gottfried Leibniz, and acted as secretary to the society from 13 January, 1714 to 21 October, 1718. From 1715, his studies took a philosophical and religious bent.

Appendix A

Inner Products, Norms, and Metrics

Before concluding the book, we would like to introduce the reader to two important concepts other than metric: norms and inner products. Indeed, we had promised in this book (Chapter 2) that we will not talk about norms while dealing with metric spaces. However, as one moves ahead in the study of a particular subject, it is very difficult to isolate metrics. Many applications of metric spaces, especially those on the fixed point theory, arise on sets which have certain special properties of addition and scaling. We call such sets *vector spaces*, and the elements of these sets as *vectors*. Once we have the notion of vectors, we can think of measuring their lengths and angles between them, as we used to do in the plane $\mathbb{R}^2$ and space $\mathbb{R}^3$.

We begin with recalling that vectors in plane and space had two operations defined on them: *vector addition* and *scalar multiplication*. Based on the properties of these operations, we shall define formally what is called a vector space. But first, to define "scalar multiplication", we would need to define what are *scalars*. For the same, we give the concept of *field*.

Definition A.1 (Field). *A set $\mathbb{F}$ equipped with two operations $+ : \mathbb{F} \times \mathbb{F} \to \mathbb{F}$ and $\cdot : \mathbb{F} \times \mathbb{F} \to \mathbb{F}$ is called a field if:*

1. *For each $x, y, z \in \mathbb{F}$, we have $x + (y + z) = (x + y) + z$.   [Associativity of addition]*

2. *There is an element (denoted as) $0 \in \mathbb{F}$ such that $\forall x \in \mathbb{F}$, we have $0 + x = x + 0 = x$. This element is called the additive identity.*

3. *For each $x \in \mathbb{F}$, there is some $y \in \mathbb{F}$ such that $x + y = y + x = 0$. The element y is called the additive inverse of x. The element y is denoted as $-x$.*

4. *For each $x, y \in \mathbb{F}$, we have $x + y = y + x$.   [Commutativity of addition]*

5. *For each $x, y, z \in \mathbb{F}$, we have $x \cdot (y \cdot z) = (x \cdot y) \cdot z$.   [Associativity of multiplication]*

6. *There is an element (denoted as) $1 \in \mathbb{F}$ such that for each $x \in \mathbb{F}$, we have $1 \cdot x = x \cdot 1 = x$. This element is called multiplicative identity.*

7. *For each $0 \neq x \in \mathbb{F}$, there is some $y \in \mathbb{F}$ such that $x \cdot y = y \cdot x = 1$. The element y is called multiplicative inverse of x.*

8. *For each $x, y \in \mathbb{F}$, we have $x \cdot y = y \cdot x$.*   *[Commutativity of multiplication]*

9. *For each $x, y, z \in \mathbb{F}$, we have $x \cdot (y + z) = x \cdot y + x \cdot z$ and $(x + y) \cdot z = x \cdot z + y \cdot z$.*   *[Distributivity of multiplication over addition]*

Here, the elements of $\mathbb{F}$ are called *scalars*. By experience, we know that $\mathbb{R}$, $\mathbb{C}$, and $\mathbb{Q}$ are fields. Once we have scalars, we can now define scalar multiplication and hence a vector space.

Definition A.2 (Vector space). *A set V equipped with two operations $+ : V \times V \to V$ and $\cdot : \mathbb{F} \times V \to V$ is called a vector space if:*

1. *For any $x, y, z \in V$ we have $(x + y) + z = x + (y + z)$.*   *[Associativity]*

2. *For any $x, y \in V$, we have $x + y = y + x$.*   *[Commutativity]*

3. *There is an (denoted as) element $\mathbf{0} \in V$ such that for every $x \in V$ we have $x + \mathbf{0} = \mathbf{0} + x = x$. This element is called the additive identity or the zero vector.*

4. *For each $x \in V$, there is an element $y \in V$ such that $x + y = y + x = \mathbf{0}$. The element y is called the additive inverse of x. The element y is denoted as $-x$.*

5. *For each $\alpha \in \mathbb{F}$ and $x, y \in V$, we have $\alpha \cdot (x + y) = \alpha \cdot x + \alpha \cdot y$.*

6. *For each $\alpha, \beta \in \mathbb{F}$ and $x \in V$, we have $(\alpha + \beta) \cdot x = \alpha \cdot x + \beta \cdot x$.*

7. *For each $\alpha, \beta \in \mathbb{F}$ and $x \in V$, we have $\alpha \cdot (\beta \cdot x) = (\alpha \beta) \cdot x = \beta \cdot (\alpha \cdot x)$.*

8. *For each $x \in V$, we have $1 \cdot x = x$, where $1 \in \mathbb{F}$ is the multiplicative identity of $\mathbb{F}$.*

When the field $\mathbb{F}$ is associated with the vector space V, we say that V is defined *over* $\mathbb{F}$. Most of the times, we deal with either vector spaces over $\mathbb{R}$ or $\mathbb{C}$, i.e., the field associated with vector spaces is either the set of real or complex numbers. Immediately, one can observe that every field is a vector space over itself. In this book, we have dealt with three major vector spaces: $\mathbb{R}^n$, ℓ^p, and $\mathscr{C}[a, b]$. In fact, in further study of topics related to metric spaces, these vector spaces always come into picture at some point. Therefore, the reader is advised to get themselves familiar with these spaces.

Now, from the experience in $\mathbb{R}^2$ and $\mathbb{R}^3$, we know that we can measure *lengths* of vectors. Based on the properties that these lengths follow, we shall now define what is called a *norm*.

Definition A.3 (Norm). *A function* $\| \cdot \| : V \to [0, \infty)$, *where V is a vector space is called a norm if:*

1. *For each $x \in V$, $\|x\| = 0$ if and only if $x = \mathbf{0}$.*

2. *For each $x \in V$ and $\alpha \in \mathbb{F}$, we have $\|\alpha \cdot x\| = |\alpha|\, \|x\|$.*   *[Homogeneity]*

3. *For each $x, y \in V$, we have $\|x + y\| \le \|x\| + \|y\|$.*   *[Triangle inequality]*

A vector space equipped with a norm is called a *normed linear space* (NLS). The study of NLS is important, especially in fixed point theory and answering questions about existence of solutions to differential and integral equations.

Notice that since in vector spaces, we can have lines and line segments, we can measure the distance between two vectors using the line segment joining them. Therefore, given a norm $\| \cdot \|$ on a vector space V, we immediately have a metric $d : V \times V \to [0, \infty)$ as

$$d(x, y) = \|x - y\|.$$

The reader should verify that this indeed forms a metric. We say that d is *induced* by the norm $\|\cdot\|$. Notice that the p-metrics defined for $\mathbb{R}^n$, ℓ^p, and $\mathscr{C}[a, b]$ are all norm induced. We give the p-norm in $\mathbb{R}^n$ and the reader is supposed to construct the norms for the other two spaces. For $\mathbf{x} = (x_1, x_2, \cdots, x_n) \in \mathbb{R}^n$, we define

$$\|\mathbf{x}\|_p = \left(\sum_{i=1}^{n} |x_i|^p \right)^{\frac{1}{p}}. \tag{A.1}$$

We can also have

$$\|\mathbf{x}\|_\infty = \max_{1 \le i \le n} \{|x_i|\}. \tag{A.2}$$

Thus, once we know how to measure lengths of vectors in a vector space, we can talk not only about the algebraic properties of the space, but also about the topological properties such as compactness, connectedness, completeness, etc. We would like to mention here that a complete normed linear space is called a *Banach space*. Therefore, $\mathbb{R}^n$, ℓ^p, ℓ^∞, and $(\mathscr{C}[a, b], d_\infty)$, studied in this book, are Banach spaces.

The norm induced metrics follow certain *"good"* properties. We mention those here without a proof, since it is trivial.

Theorem A.1. *If a metric d is induced by a norm $\| \cdot \|$, then*

1. *For each $x, y \in V$ and a fixed $a \in V$, we have $d(x + a, y + a) = d(x, y)$.*
   [Translational invariance]

2. *For each $x, y \in V$ and $\alpha \in \mathbb{F}$, we have $d(\alpha \cdot x, \alpha \cdot y) = |\alpha|\, d(x, y)$.*
   [Homogeneity]

These are in fact, necessary conditions for a metric to be norm induced. Indeed, the reader can verify that the discrete metric does not enjoy homogeneity and hence is not a norm induced metric.

Now, we are not only limited to finding lengths of vectors. By experience in the plane and space, we know that given two vectors, we can find the projections and hence the angle between them. The projections were a result of the *dot product* defined in $\mathbb{R}^2$ and $\mathbb{R}^3$. Again, keeping in mind its properties, we define what is called an *inner product* on an arbitrary vector space V.

Definition A.4 (Inner product). *A function $\langle \cdot, \cdot \rangle : V \times V \to \mathbb{R}$ is called an inner product if:*

1. *For each $x \in V$, we have $\langle x, x \rangle \geq 0$.*

2. *For each $x, y, z \in V$ we have $\langle x + y, z \rangle = \langle x, z \rangle + \langle y, z \rangle$.*   *[Additivity]*

3. *For each $x, y \in V$, we have $\langle x, y \rangle = \langle y, x \rangle$.*   *[Symmetry]*

4. *For each $x, y \in V$ and $\alpha \in \mathbb{R}$, we have $\lambda \alpha \cdot x, y \rangle = \alpha \langle x, y \rangle$.* *[Homogeneity]*

Notice that the definition stated above is valid for vector spaces over $\mathbb{R}$. If a vector space is over $\mathbb{C}$, then $\mathbb{R}$ is replaced by $\mathbb{C}$ everywhere, and the symmetry is changed to conjugate symmetry, i.e.,

$$\langle x, y \rangle = \overline{\langle y, x \rangle}.$$

A vector space equipped with an inner product is called an *inner product space* (IPS). Again, the study of IPS is interesting and a vast subject in itself. Therefore, we do not go to depth of it in this book. However, we would like to mention that once we have an inner product, we can measure lengths of vectors (by taking projections of vectors on themselves), i.e., norms and eventually get topology due to the norm induced metric. If $\langle \cdot, \cdot \rangle$ is the given inner product, then a norm can be defined as

$$\|x\| = \sqrt{\langle x, x \rangle}.$$

The reader is advised to verify that this is indeed a norm. Just like complete NLS are called Banach spaces, complete IPS are called *Hilbert spaces*.

Before concluding, we would like to mention that since we have topological properties once we have inner products, vector spaces are of major interest to many mathematicians of varied fields. These spaces play roles in algebra, topology, differential equations (both ordinary and partial), and integral equations.

Bibliography

[1] Ansari, Q. H., *Metric Spaces Including Fixed Point Theory and Set-Valued Maps*, Alpha Science International, 2009.

[2] Apostol, T. M., *Mathematical Analysis, 2nd ed.*, Narosa Publications, 1985.

[3] Baccala, B., *The Hölder and Minkowski Inequalities*, 2008.

[4] Geobel, K., Kirk, W. A., *Topics in Metric Fixed Point Theory*, Cambridge University Press, 1990.

[5] Giles, J. R., *Introduction to the Analysis of Normed Linear Spaces*, Cambridge University Press, 2000.

[6] Gopal, D., Kumam, P., Abbas, M., *Background and Recent Developments in Metric Fixed Point Theory*, CRC Press, 2017.

[7] Jain, P. K., Ahmed, K., *Metric Spaces, 2nd ed.*, Narosa, 2004.

[8] Kaplansky, I., *Set Theory and Metric Spaces*, Allyn and Bacon, 1972.

[9] Khamsi, M. A., Kirk, W. A., *An Introduction to Metric Spaces and Fixed Point Theory*, Wiley-Interscience, 2001.

[10] Kreyszig, E., *Introductory Functional Analysis with Applications*, Wiley, 2007.

[11] Kumaresan, S., *Topology of Metric Spaces, 2nd ed.*, Narosa Publications, 2011.

[12] Limaye, B. V., *Functional Analysis*, New Age International Private Limited, 2014.

[13] Munkres, J., *Topology, 2nd ed.*, Pearson Publications, 2015.

[14] Rosen, K. H., *Discrete Mathematics and Its Applications, 7th ed.*, McGraw-Hill, 2012.

[15] Shirali, S., Vasudeva, H. L., *Metric Spaces*, Springer, 2006.

[16] Simmons, G. F., *Introduction to Topology and Modern Analysis*, McGraw-Hill Education, 1963.

[17] *The Baire Category Theorem and Its Applications* (Lecture Handout), 2013.

Index

Printed in the USA
CPSIA information can be obtained
at www.ICGtesting.com
LVHW021732041124
795688LV00039B/1185